CFD Modelling for Wastewater Treatment Processes

IWA Working Group on Computational Fluid Dynamics

Scientific and Technical Report Series No. 30

CFD Modelling for Wastewater Treatment Processes

IWA Working Group on Computational Fluid Dynamics

Edited by

Julien Laurent, Randal Samstag, Jim Wicks and Ingmar Nopens

IWA PUBLISHING

Published by **IWA Publishing**
Unit 104–105, Export Building
1 Clove Crescent
London E14 2BA, UK
Telephone: +44 (0)20 7654 5500
Fax: +44 (0)20 7654 5555
Email: publications@iwap.co.uk
Web: www.iwapublishing.com

First published 2022

British Library Cataloguing in Publication Data
A CIP catalogue record for this book is available from the British Library

ISBN: 9781780409023 (Paperback)
ISBN: 9781780409030 (eBook)
ISBN: 9781780409047 (ePub)

Contents

About the Chapter Authors

JULIEN LAURENT

ICube, UMR 7357, ENGEES, CNRS, Université de Strasbourg
2 rue Boussingault,
F-67000 Strasbourg, France
Email: julien.laurent@engees.unistra.fr
ORCID: 0000-0002-9966-3386

Julien Laurent is a full professor at Ecole Nationale du Génie de l'Eau et de l'Environnement de Strasbourg (ENGEES). He performs his research at ICube joint research unit within the Fluid Mechanics research team. Since 2014, he has chaired the IWA Working Group on CFD for wastewater.

INGMAR NOPENS

BIOMATH, Department of Data Analysis and Mathematical Modelling
Ghent University
Coupure Links 653
9000 Gent, Belgium
Email: ingmar.nopens@ugent.be
ORCID: 0000-0001-6670-3700

Co-founder and COO at AM-Team
Dendermondsesteenweg 48/1, 9000 Gent
Ingmar.Nopens@AM-Team.com

Ingmar Nopens is a full professor at Ghent University leading the BIOMATH research group dealing with model-based analysis and optimization of (bio)-processes. He is a co-founder of AM-TEAM, a Ghent University spin-off and scale-up active in the field of advanced modelling for, among others, the water market and applying CFD for improved model-based design and operation. He is the former chair of the IWA Specialist Group on Modelling and Integrated Assessment (MIA). He co-founded the IWA Working Group on CFD for wastewater and chaired it at its initiation and during the first years of its existence.

RANDAL SAMSTAG

Randal Samstag was an independent consultant for wastewater treatment analysis and modelling. He served as project manager or process engineer on over 60 planning projects for wastewater treatment plants. He served as project manager, design, or process engineer on over 40 projects for the design of wastewater treatment plants. He was project manager and partial author of a report evaluating appropriate technology for sewage pollution control for the wider Caribbean region for the United Nations Environment Programme. He completed evaluations and designs for nitrogen removal for over a dozen wastewater treatment plants including recent evaluations for the three largest treatment plants discharging to Puget Sound and the first two nitrogen-removing facilities in Washington State. He led field test evaluation and/or computational fluid dynamics modelling of aeration, mixing, and sedimentation at 25 sites. He developed integrated computer software for treatment plant design and cost estimation. He was active in development of numerical models for CFD and treatment processes. He is the principal author of over 50 technical reports and studies and over 60 book chapters, peer-reviewed journal articles, papers for national and international conferences, and other presentations.

In his over 35-year career in municipal and consulting engineering, he worked on projects throughout the United States, in Latin America and the Caribbean, the Middle East, and the Republic of Singapore. He completed a series of water supply and community development projects supported by NGOs in Latin America and Asia as leader or team member.

He was the leader and a co-founder of the IWA Working Group on Computational Fluid Dynamics.

JIM WICKS

The Fluid Group, The Magdalen Centre,
Robert Robinson Avenue, The Oxford Science Park,
Oxford OX4 4GA, UK
Email: jim.wicks@thefluidgroup.com
ORCID: 0000-0003-4539-7884

Jim Wicks is a physicist, a materials scientist specializing in the study of liquids at the atomic scale, and the Founder and Research Director of The Fluid Group, Oxford, providing CFD for the water, chemical and environmental engineering industry. Jim co-founded the IWA Working Group on CFD for wastewater and is Vice-Chair of the management team.

DAMIEN BATSTONE

Australian Centre for Water and Environmental Biotechnology
University of Queensland
Brisbane, Australia
Email: d.batstone@uq.edu.au
ORCID: 0000-0002-4294-3942

Professor Damien Batstone leads the CFD and process modelling group at the Australian Centre for Water and Environmental Biotechnology (ACWEB), The University of Queensland. He was a founding member of the IWA Working Group on CFD.

CHRISTOPHER DEGROOT

Department of Mechanical and Materials Engineering, Western University
London, ON, Canada
Email: cdegroo5@uwo.ca

Christopher DeGroot is an assistant professor at Western University (London, Canada) and CEO of Maple Key Labs, Inc. His academic research work focuses on CFD model development, mainly using OpenFOAM, for wastewater unit processes. Maple Key Labs is a CFD consulting company that works mainly in the water/wastewater sectors.

DAVID FERNANDES DEL POZO

BIOMATH, Department of Data Analysis and Mathematical Modelling
Ghent University
Coupure Links 653
9000 Gent, Belgium
Email: david.fernandesdelpozo@ugent.be/fernandesdelpozodavid@gmail.com
ORCID: 0000-0002-8652-6825

Dr. David Fernandes del Pozo is a postdoctoral researcher at the BIOMATH group at Ghent University. He is very passionate about modelling and has performed research into both single and multiphase CFD modelling with an emphasis on model calibration and validation.

ALONSO G. GRIBORIO

Director of Clarification Technology
Hazen and Sawyer
4000 Hollywood Blvd, 750N, North Tower
Hollywood, FL 33021 USA
Email: agriborio@hazenandsawyer.com

Alonso Griborio is Hazen and Sawyer's Director of Clarification Technology. He specializes in the design and optimization of water reclamation facilities. Dr. Griborio is a co-developer of the clarifier CFD model '2Dc' and is a co-founder of the IWA Working Group on CFD.

RAINIER HREIZ

Laboratoire Réactions et Génie des Procédés, Université de Lorraine, CNRS, LRGP, F-54000 Nancy, France
Email: rainier.hreiz@univ-lorraine.fr

Rainier Hreiz is associate professor at the Chemical Engineering National High School (ENSIC, Université de Lorraine, France) and performs his research at the Reactions and Chemical Engineering Laboratory (LRGP). His research focuses on CFD and transport phenomena in the context of chemical engineering.

ANNA M. KARPINSKA PORTELA

Senior Modelling Engineer – Operational Planning & Improvement
Southern Water
Southern House Falmer
Lewes Road, Falmer, Brighton
BN1 9PY, East Sussex, UK
Email: anna.karpinska-portela@southernwater.co.uk
ORCID: 0000-0002-9615-3858

Anna Karpinska Portela is a Senior Modelling Engineer within the Technical Improvement Team at Southern Water. She is a technical lead in OPEX saving initiatives, capital schemes, and R&D projects linked to the Activated Sludge Process, operational efficiency framework, and Net Zero Carbon strategies. Her work includes building both holistic plant-wide models, and unit process-focused CFD models, targeting improved treatment performance, real-time control, and optimization. Since 2012, she has been a member of the management team of the IWA Working Group on CFD for wastewater.

OLIVIER POTIER

Laboratoire Réactions et Génie des Procédés, Université de Lorraine, CNRS, LRGP, F-54000 Nancy, France
Email: olivier.potier@univ-lorraine.fr
ORCID: 0000-0003-3143-4015

Olivier Potier has been involved in modelling and simulation of reactors for the last two decades: systemic, compartmental modelling and CFD simulation of WWTPs. He founded a German-French dual master's degree in engineering (double-diplôme d'ingénieur) at the Université de Lorraine and has been directing this course for 20 years. He is President of the Scientific Council of ENGEES (National Graduate School of Environment and Water Engineering in Strasbourg); Head of International Relationships at the Graduate Engineering School (ENSGSI) at the Université de Lorraine; Co-President of the Eastern Section of the French Chemical Engineering Society (SFGP), and currently Co-Chair of the international TRACER series of conferences. He founded the Working Group on 'Innovation and Innovative Design' for the French Chemical Engineering Society in 2010; contributed to the earliest articles that implemented the biological kinetic model ASM1 in the CFD simulation of wastewater treatment plant (WWTP) reactors and published the first paper on the implementation of C-K theory (an innovative design theory) in chemical engineering. He still works extensively in all these areas.

USMAN REHMAN

AM-TEAM
Frieda Saeysstraat 1B
9052 Gent (Zwijnaarde), Belgium
Email: Usman.Rehman@AM-TEAM.com

Usman Rehman is co-founder and CTO at AM-TEAM, leading the team of modellers for the water and wastewater sector. AM-TEAM is active in the field of advanced modelling for, among others, the water market and applying CFD for improved model-based design and operation. He has been a member of the IWA Working Group on CFD for wastewater since 2014.

STEPHEN SAUNDERS

IBIS Group Inc.
Postal: 69 Harbour View Lane, Chamcook, NB E5B 3G8 Canada
Email: steve@ibisgroupcfd.com

Steve Saunders is a consulting engineer specializing in CFD modelling of water and environmental applications at IBIS Group Inc.

TEWODROS MELESS TESHOME

Department of Hydraulic Engineering
Hawassa University, Institute of Technology

Hawassa, Ethiopia
Email: teddymett@hu.edu.et

Tewodoros Teshome obtained his bachelor's degree in hydraulic engineering from Arbaminch Water Technology Institute, Ethiopia in 2003. He worked as a Project Engineer in government office for two years and joined Hawassa University. After three years of teaching, he pursued his master's degree in UNESCO-IHE, Institute for Water Studies, the Netherlands, in the field of Municipal Water and Infrastructure specializing in Integrated Urban Engineering. He then began lecturing in Hawassa University, Civil Engineering Department and engaged as a Sanitary Engineer in a private consulting firm. In 2020, he received his PhD in fluid mechanics from the University of Strasbourg, France, and is currently working as assistant professor of hydraulic engineering in Hawassa University, Institute of Technology.

MARIA ELENA VALLE-MEDINA

ICube, UMR 7357, ENGEES, CNRS, Université de Strasbourg
2 rue Boussingault
F-67000 Strasbourg, France
Email: mariaelena.valle-medina@engees.unistra.fr

Maria Elena Valle-Medina holds a bachelor's degree in chemical engineering with environmental specialty from 'Instituto Tecnológico de Aguascalientes' (Mexico) as well as a master's degree in water, soils and waste treatment and management from 'AgroParisTech' (France). She received her PhD in 2019 at the University of Strasbourg (France). Her research focuses on the improvement of CFD models for secondary settling tanks. She developed a specific workflow using ultrasonic measurements both in batch settling columns and full-scale clarifiers to gather experimental data for model calibration and validation. She is now a research engineer at Icube joint research unit where she continues to conduct CFD projects on clarifier modelling.

ED WICKLEIN

Carollo Engineers,
1200 Fifth Avenue, Suite 900,
Seattle, WA 98101, USA
Email: ewicklein@carollo.com

Edward (Ed) Wicklein is a principal technologist at Carollo Engineers with over two decades of experience in hydraulic design and analysis of municipal facilities using CFD modelling. He has been actively involved in the IWA Working Group on CFD Modelling since its inception. Through activities with IWA and other organizations, he is dedicated to expanding and promoting CFD modelling applications to improve municipal water treatment for communities everywhere.

MIN YANG

BIOMATH, Department of Data Analysis and Mathematical Modelling
Ghent University
Coupure Links 653
9000 Gent, Belgium
Email: Yang.Min@UGent.be

Dr. Min Yang is a senior CFD expert at AM-TEAM. He is also a part-time postdoctoral researcher at the BIOMATH group at UGent, where he received a Marie Sklodowska-Curie Individual Fellowship. Dr. Min Yang has extensive knowledge and rich practical experience in CFD, kinetics and the population balance model (PBM). He has dedicated himself to the development and application of advanced mechanistic and data-driven models and methodologies for the simulation and optimization of various biological, chemical, and environmental engineering systems.

REVIEWERS

JAVIER CLIMENT

Manager at HYDRENS (Hydrodynamic & Environmental Services)
Av. del Mar, 53, bajo. 12003, Castelló de la Plana (Spain)
+34 627 674 859
Email: javier.climent@hydrens.com

NELSON MARQUES

Director of Operations at FSD blueCAPE
Invited Lecturer at Instituto Superior de Engenharia de Lisboa
+351 963 440 984
Email: nelson.marques@fsdynamics.pt

Preface

Whether we like it or not, full-scale water resource recovery facilities (WRRFs) are often poor replicas of a bench-scale process. Most process models, painstakingly created over decades, assume simple 1D or 2D hydrodynamics, either plug flow or completely mixed, extended to simulate plants of the real world by stringing units together to create a tanks-in-series (TIS) solution.

The reality is more complex. Circular aeration tanks generate volumes of highly aerated water and significant sectors with depleted oxygen. Internal flows and various substrate concentrations recirculate in tanks along multiple directions. Complex pathways form naturally within granulated media. Anaerobic digesters, fundamental for energy/nutrient recovery, have complex internal flow patterns, and recent work suggests slow or 'inefficient' mixing could enhance methane production.

Computational fluid dynamics (CFD) is an accepted method for process analysis in a diverse range of industries from aeronautics to ocean engineering. Since Larsen's pioneering study presented the first CFD model for activated sludge sedimentation incorporating solids transport and settling (Larsen, 1977), it has been used for analysis and design of water and wastewater treatment plant process units.

A limitation is that there are currently too few CFD modelling experts within the environmental educational and academic sectors. This lack of educational support can lead to software misuse by inexperienced users. Misuse leads to erroneous outcomes, which can result in poor designs or processes that result in suboptimal hydraulic design, contributing to distrust of CFD models in the wastewater resource recovery field.

In addition, the relevance of precise details within wastewater treatment plants were historically not as significant when discharge limits were less stringent, energy input was less critical, and the facilities were sized for future conditions. The necessity to minimize carbon footprint and energy requirements as well as the paradigm shift towards resource recovery is definitely advancing CFD to the heart of a better unit processes understanding and optimization.

While there are several textbooks available on the general use of CFD modelling, currently there is no comprehensive textbook or scientific report available that addresses both CFD fundamentals and practical applications of CFD specifically for WRRFs. Based on this lack of available guidelines, in 2012, the International Water Association (IWA) chartered a working group on CFD modelling to address application of CFD to wastewater resources recovery engineering as well as training of CFD engineers.

In contrast, other industries, with a longer history of CFD model use and more stringent fluid flow requirements, have established clear guidelines for quality modelling. Key references include

Roache (1972), the 1998 American Institute of Aeronautics and Astronautics Guide to Verification and Validation of CFD Models (AIAA, 1998), and the ERCOFTAC Best Practice Guidelines (Casey & Wintergerste, 2000) for CFD modelling.

Nevertheless, in 2019, the American Society of Civil Engineers (ASCE) released the book 'Computational Fluid Dynamics: Applications in Water, Wastewater, and Stormwater Treatment' (Liu & Zhang, 2019). This book was probably the first one covering this subject area. As noted by Hajimirzaie and Ansar (2020), it is a valuable and timely contribution to the CFD literature. However, it constitutes an introduction and overview of computational fluid dynamics and how it can be used in the water and wastewater industry, and is more dedicated to engineers.

The present Scientific and Technical Report (STR) is more focused on WRRF unit process CFD simulations. We believe that it is complementary to ASCE's book as it provides more in-depth fundamentals and guidelines for several typical WRRF unit processes, as well as literature reviews and future research directions.

This STR is organized into 10 chapters. The first two illustrate the interest of CFD models for WRRF applications, as well as fundamentals and good modelling practice guidelines. Then, chapters 3 to 8 describe in detail the specificities of CFD models used for unit processes, ranging from headworks, biological reactors and sedimentation tanks to disinfection. Finally, the last two chapters open the discussion on two fundamental topics: experimental validation of CFD simulations, and the complementarity between CFD and Chemical Reaction Engineering approaches.

Water resource recovery modelling is not just about lab-scale processes. Now and in the future it is about improving our understanding of (and designing better) full-scale facilities!

REFERENCES

AIAA (1998). Guide for the verification and validation of computational fluid dynamics simulations. American Institute of Aeronautics and Astronautics. AIAA-G-077-1998, Reston, VA.

Casey M. and Wintergerste T. (2000). ERCOFTAC – Special interest group on "quality and trust in industrial CFD" – Best practice guidelines. *European Research Community on Flow, Turbulence and Combustion.*

Hajimirzaie S. M. and Ansar M. (2020). Review of Computational Fluid Dynamics: Applications in Water, Wastewater, and Stormwater Treatment edited by Xiaofeng Liu and Jie Zhang. *Journal of Hydraulic Engineering*, **146**(4), 07520001.

Larsen P. (1977). On the Hydraulics of Rectangular Settling Basins Experimental and Theoretical Studies. Department of Water Resources Engineering, Lund Institute of Technology, University of Lund, Lund, Sweden.

Liu X. and Zhang J. (2019). Computational Fluid Dynamics: Applications in Water, Wastewater, and Stormwater Treatment. American Society of Civil Engineers, Reston, VA.

Roache P. J. (1972). Computational fluid dynamics. Hermosa Publishers, Albuquerque.

Foreword

Computational fluid dynamics (CFD) is now an essential component of wastewater system design and optimization. The chapters of this report (CFD for Wastewater, Scientific and Technological Report (STR)) have been prepared by professionals who have pioneered the development of CFD applications in the wastewater field.

In 1997 the International Association on Water Quality (IAWQ) published STR No. 6 which had a chapter that presented the state of the art of secondary clarifier CFD modelling at that time. Since then, there have been major advances in the application of computational fluid dynamics software in a broad range of aspects of wastewater engineering. In addition to the review of the fundamentals of CFD modelling, this STR presents the latest of these advances in areas like good modelling practices, plant flow distribution, pumping stations, sedimentation, aeration, mixing, biological processes, disinfection, and digestors. Numerous case studies are given that will provide the reader with insights into how the models are set up and how the results can be usefully displayed. To aid researchers, the authors have provided research needs for each topic in this STR.

The chapters on CFD modelling fundamentals provide the reader with a good background on modelling principles from black box to glass box types. Governing equations including the Reynolds-averaged Navier-Stokes (RANS) and the relevant transport equations with various turbulence closure sub-models are presented for 2D and 3D models. Standard numerical methods for solving the governing equations that are common to many CFD packages are reviewed. Of special interest and importance is the section on good modelling practice. In addition, the chapter on model validation is an essential part of good modelling. These chapters on fundamentals should be essential reading for all beginners, experienced professional modellers and project managers in the field of CFD applications in the wastewater field.

The chapter on headworks, which includes examples of flow distribution and pump intakes, shows the practical application of CFD that can aid in making improvements to existing plants or in designing new headworks.

Four chapters deal with CFD in biological environments. Two of these chapters are devoted to coupling aerobic treatment processes with the associated hydrodynamics. The coverage includes tanks with suspended bacteria and algae. In addition to specific examples of biological-CFD coupling, these chapters have extensive summaries of research in this area. Similarly, a chapter has been devoted to the role of CFD in anaerobic digesters; the three-phase nature of flow in these tanks is discussed and

Editor(s): Julien Laurent, Randal Samstag, Jim Wicks and Ingmar Nopens

modelling simplifications are suggested. The chapter on disinfection gives a detailed application of CFD to UV disinfection.

The chapter on sedimentation extends the clarifier modelling in IAWQ STR 6 to include the application of 3D CFD applications. This chapter provides the approach needed to overcome the drawbacks of using 2D models for highly 3D flows such as the flows in the inlet zones of clarifiers. A hybrid approach using 2D and 3D CFD models is often desirable.

Based on the progress in the application of CFD over the past two decades, as highlighted by this STR, what can we expect in the near future? One of the problems faced by smaller and medium-sized engineering firms is the access to high-speed parallel computing and the expertise to make use of it. There is some promise for improvement here since many undergraduate and master's engineering programmes are now including an introduction to CFD. This STR will be a good reference for these courses. Cloud computing is providing more reasonably priced access to the computer power required to run the modern CFD packages. Cheaper central processing unit (CPU) costs will permit the running of fully unsteady 3D simulations that have execution times that are competitive with the current 2D models but will require less calibration.

The advancement in the capabilities of future CFD models will change the nature of field testing. The role of field studies will need to be more focused on determining both the conventional and more advanced parameters that CFD models will require. The advanced CFD models will require rheological and more detailed sludge settling and floc density inputs. Flocculation parameters for primary wastewater will become more important. Model calibration data may become less demanding but validation testing of such stress test data will continue to be important and may require additional observations such as water temperatures and atmospheric data. On the other hand, the CFD models for flow distribution are already highly reliable and, in the future, may require less field validation.

Finally, it is most appropriate that this STR should be dedicated to Randal Samstag whose contributions, persistence, and enthusiasm for CFD modelling over nearly four decades has inspired many wastewater modellers including myself.

J. Alex McCorquodale
Professor Emeritus
Department of Civil Engineering
University of New Orleans

Acknowledgements

This work is dedicated in memory of Randal Samstag. His passion for wastewater CFD modelling goes back decades, with early insights leading him to develop his own CFD software to evaluate and optimize secondary clarifiers. Randal was a wonderful communicator and mentor to early career researchers, leading delivery of workshops and training courses. He was largely responsible for bringing this group together through the IWA and WEF conferences, then directly inspiring and advocating for this publication. Randal saw the need to clearly document the current state of the art in wastewater CFD modelling, to develop and document good modelling practices and to provide a roadmap for the next modelling frontiers. He showed us it was possible to make this happen through pulling together key chapters early in the development of this work. His contribution, energy and friendship will be missed, but we hope the next generation will be appropriately inspired to continue his legacy.

doi: 10.2166/9781780409030_0001

Chapter 1

Why CFD?

Randal Samstag[1] and Jim Wicks[2]

[1]Civil and Sanitary Engineer, Bainbridge Island, WA, USA
[2]The Fluid Group, Oxford, UK
Email: jim.wicks@thefluidgroup.com

1.1 INTRODUCTION

> 'If we know precisely what is happening within the vessel, thus if we have a complete velocity distribution map for the fluid, then we are able to predict the behavior of a vessel as a reactor. Though fine in principle, the attendant complexities make it impractical to use this approach' (Levenspiel, 1972).

When Levenspiel wrote these words in his famous textbook, it was 'impractical' to base models of reactors on a 'complete' velocity distribution map for the fluid. Setting aside this goal, Levenspiel went on to present descriptions of models which estimated some of the effects of hydraulic geometry on a reactor, but without attempting to establish a full velocity profile.

From the outset, it is useful to characterize a reactor model based on the amount of knowledge of the underlying velocity field to which they aspire. Models determine outputs given inputs and disturbances. For a fluid model, inputs are generally boundary conditions (which may be time variant), and outputs are boundary velocities, boundary concentrations, or internal velocity fields, depending on the objective. Major types of models are illustrated in Figure 1.1.

The first type, the 'black box' model, assumes absolutely no knowledge about the internal hydraulics or geometry of a reactor. The black box model simply takes data from inputs and outputs and fits a statistical relation to the two. Examples of this type of model are the autoregressive-integrated-moving-average (ARIMA) model of Box *et al.* (2015) and residence time distribution (RTD) theory first suggested by Danckwerts (1953). These models can provide overall estimates of outputs, since there is usually some kind of input function. An ARIMA empirical model relates current outputs to current inputs and historical inputs and flows via weighted values. This is a linear representation of a non-linear function (i.e., velocity profiles), that can be quite variable. The ARIMA model tells us nothing about the effect of the reactor's geometry/design on its hydrodynamics and on reactions and has no predictive capability. For this we need a model of internal hydraulics. The RTD approach utilizes an empirical distribution function to represent dispersion of a property through a compartment. It provides a simple input/output picture of the fluid flow characteristics of a reactor, but it is zero-dimensional (the length dimension is reduced to a point function) and does not contain mechanistic information (though the function can be developed from a mechanistic basis) (Rodrigues, 2021). Mechanistic information is lost, and since the function is zero-dimensional, it contains insufficient information to uniquely

Editor(s): Julien Laurent, Randal Samstag, Jim Wicks and Ingmar Nopens

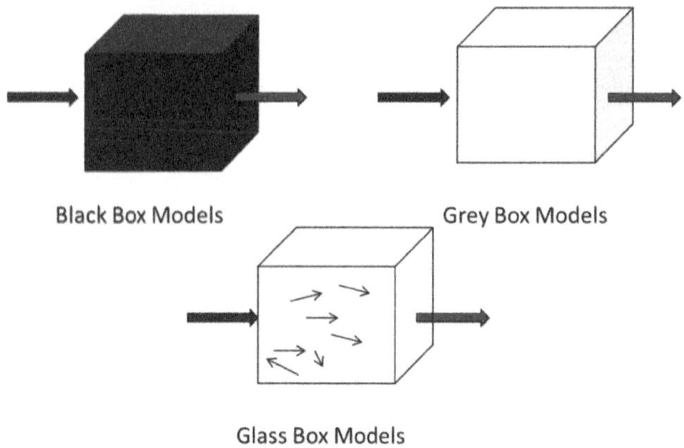

Figure 1.1 Types of reactor models.

specify a three-dimensional solution (although any 3D model i.e. mathematically consistent with a measured 1D solution is at least one of many possible correct answers).

'Grey box' models utilize knowledge about actual tank flow within the reactor to predict transport within the tank, but generally simplified to a single dimension or with a coarse dimensional solution. Examples of this type of model are the axial dispersion model and the tanks-in-series (TIS) model, both described in detail by Levenspiel (1972). The axial dispersion model permits advection in one dimension only, hence it is sometimes called the dispersed 'plug-flow' model. Turbulent spreading or dispersion has been modelled in both axial and radial directions in pipes and fluidized beds. When only dispersion in the axial or longitudinal direction is modelled, it is called the 'axial' dispersion model.

Levenspiel often advocated using a network of reactors with classical flow characteristics (the 'systemic' approach) each of which could be fitted together in series and parallel, including recirculation, to simulate a complex three-dimensional flow solution from first principles. The TIS model (which is well-established in the wastewater treatment industry) is the simplest form of this and assumes that transport in a reactor can be modelled by replacement of the real tank with an 'equivalent' number of completely mixed tanks operated in series. The IWA Activated Sludge Models (ASMs) (Henze *et al.*, 2006) and related biokinetic models are generally implemented in a number of reactor cells assumed to correspond to the general geometry of a full-scale reactor.

Computational fluid dynamics (CFD) changes this picture entirely, by generating an internally consistent, velocity field simulation (possibly at the mm scale) using the equations of fluid momentum (Navier-Stokes), viscosity, density, and turbulence. Today, CFD provides a practical way to model reactors, thus allowing prediction of their behaviour with different geometric shapes and at varying throughput flows, from first principles. These models inherently contain more fundamental information than either 'black box' or 'grey box' models because they contain elements of physics at a local level. The difficulty is that their complexity can make them unwieldy, and the ultimate challenge is to combine their strength in terms of mechanistic representation, with the speed required to find an appropriate solution. While noting that CFD contains simplifications needed to represent elements such as turbulence, it enables truly 'glass box' models, which allow us to track the course of reactants and products throughout a reactor. Since Levenspiel, these models have already been proven very useful in the field of wastewater treatment analysis, and advances in computing power and approaches are increasing the motivation to use them for a range of wastewater related problems.

This *Scientific and Technical Report* (STR) will explore how CFD can be used to improve our understanding of treatment processes and allow us to optimize the geometry of those processes for better operation and control. It will provide an up-to-date summary of the current techniques and models applicable to wastewater treatment unit processes. The following paragraphs introduce the types of analysis that can be accomplished using computational fluid dynamics.

1.2 IMPROVING HYDRAULIC DESIGN

Wastewater hydraulic design is critical to maintain a hydraulic profile throughout the treatment plant, often through passive flow splitters, and under variable flows. The classic approach is the static design of systems using relevant basic principles and geometric approximations, such as hydraulic diameter, designing the system in parts, generally using non-dimensional analysis (Chanson, 2004). However, this may not address time variant and static interactions between control volumes, particularly where scale varies widely. CFD has been relatively widely used for simulating hydraulic components in the water industry: flow splitters, open channel conveyances and pumping wet wells. The effectiveness of any treatment plant depends on an efficient and accurate flow split between process units. CFD provides a way to calculate more precise estimates of head losses, to predict the influence of channel configuration on split of flow and solids in distribution manifolds, and to predict potential vortex formation upstream of pump intakes.

Figure 1.2 presents an example outcome from a CFD analysis of a typical flow-splitting configuration in a wastewater treatment plant developed with OpenFOAM (The OpenFOAM Foundation, 2021) open source software using the volume of fluid method of CFD analysis. These techniques are discussed in full detail in subsequent chapters of this STR (Chapter 2).

1.3 OPTIMIZING TANK GEOMETRY

Sedimentation tanks in wastewater treatment plants come in a bewildering variety of shapes and configurations. Traditional design is based on one-dimensional settling velocity analysis (flux curve state point analysis), together with rules of thumb for geometry (Metcalf & Eddy *et al.*, 2014). We know from practical experience that the performance of these units can vary widely depending on the details of their geometric design. Different designs for inlets, outlets, tank depth, and tank baffling

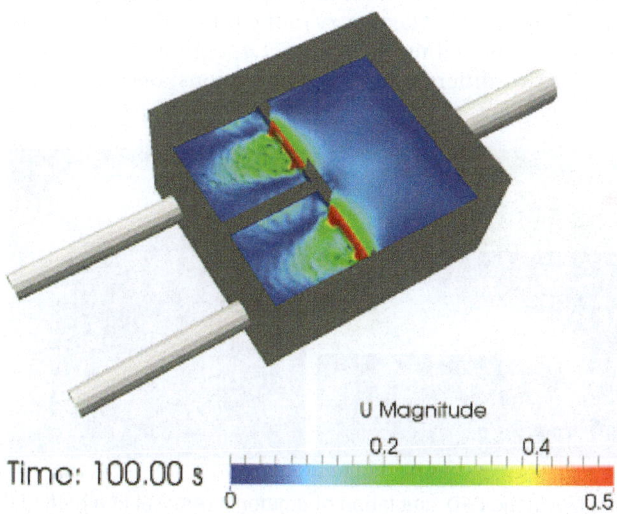

Figure 1.2 Wastewater flow split simulation. Colour scale represents velocity magnitude (Marques, 2015).

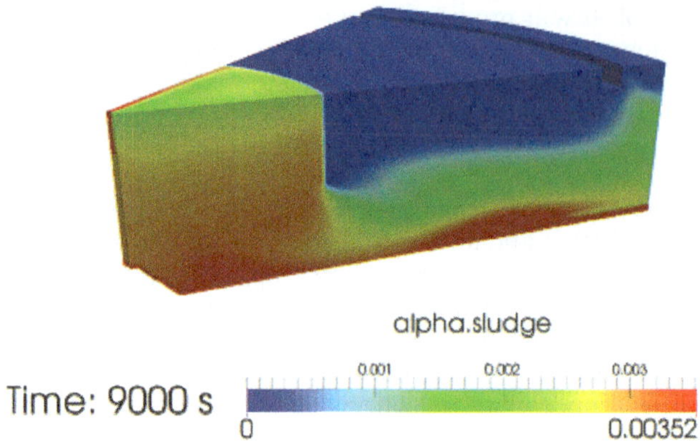

Figure 1.3 Secondary sedimentation simulation in OpenFOAM. Colour scale represents the dispersed phase (sludge) volume fraction (Marques, 2015).

can result in differences in capacity of these units by factors of two or more based on the same unit process area.

CFD has been used since at least the 1990s for modelling of sedimentation tanks. Today it is practical to optimize sedimentation tank geometry for a range of flows and solids loading so that capacity is maximized, and effluent quality maintained. Figure 1.3 presents a three-dimensional CFD model of a secondary sedimentation tank prepared with OpenFOAM open source software using a drift flux method for simulation of solids settling and transport. Simulations like these can be used to confirm that the geometry that an engineer is proposing for a given installation will produce good results.

1.4 IMPROVING MODEL PREDICTIONS

Estimates of the performance of biological treatment in activated sludge aeration tanks are routinely obtained using TIS models. But the exact geometry of the full-scale tank can have a significant impact on the results of the biokinetic modelling. In Figure 1.4, contour maps (in plan view) of ammonia concentration are shown for two different tank configurations: on the left a baffled tank with three

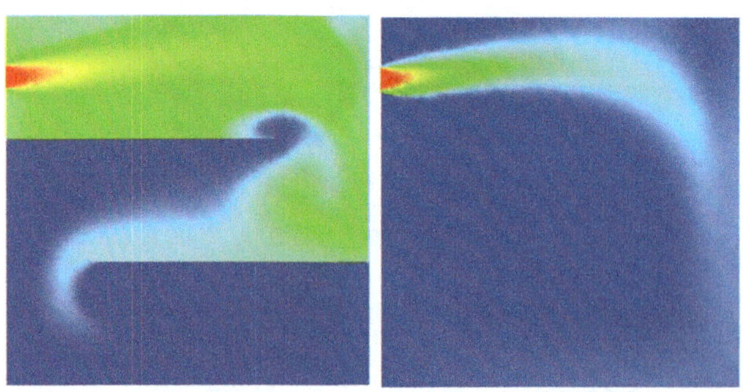

Figure 1.4 Comparison of a biokinetic CFD simulation of ammonia removal in a baffled (left) versus an un-baffled (right) tank. Colour scale represents ammonia concentration (J D Wicks, presented at 3rd IWA/WEF Wastewater Treatment Modelling seminar, 'WWTmod2012', Mont-Sainte-Anne, Quebec, Canada February 26–28, 2012).

reaction zones and on the right an un-baffled tank. These results were developed using a commercial (Fluent) CFD model using an ASM2d biokinetic model (Henze *et al.*, 2006) imposed over the tank hydraulic calculations using a user defined function. The results showed a tank effluent ammonia concentration of 0.70 mg/L for the baffled tank and 0.82 mg/L for the un-baffled tank. Are these differences significant considering the additional cost of the baffles? That depends on the cost of the baffles, but the example illustrates that by applying CFD, a more realistic estimate of the consequences of design choices can be made than is possible using simpler methods.

1.5 USE FOR CALIBRATION OF SIMPLER MODELS

CFD-based biokinetic models can also be used in a calibration scheme for simpler models. Figure 1.5 presents a possible approach towards this end (Laurent *et al.*, 2014). CFD requires significant

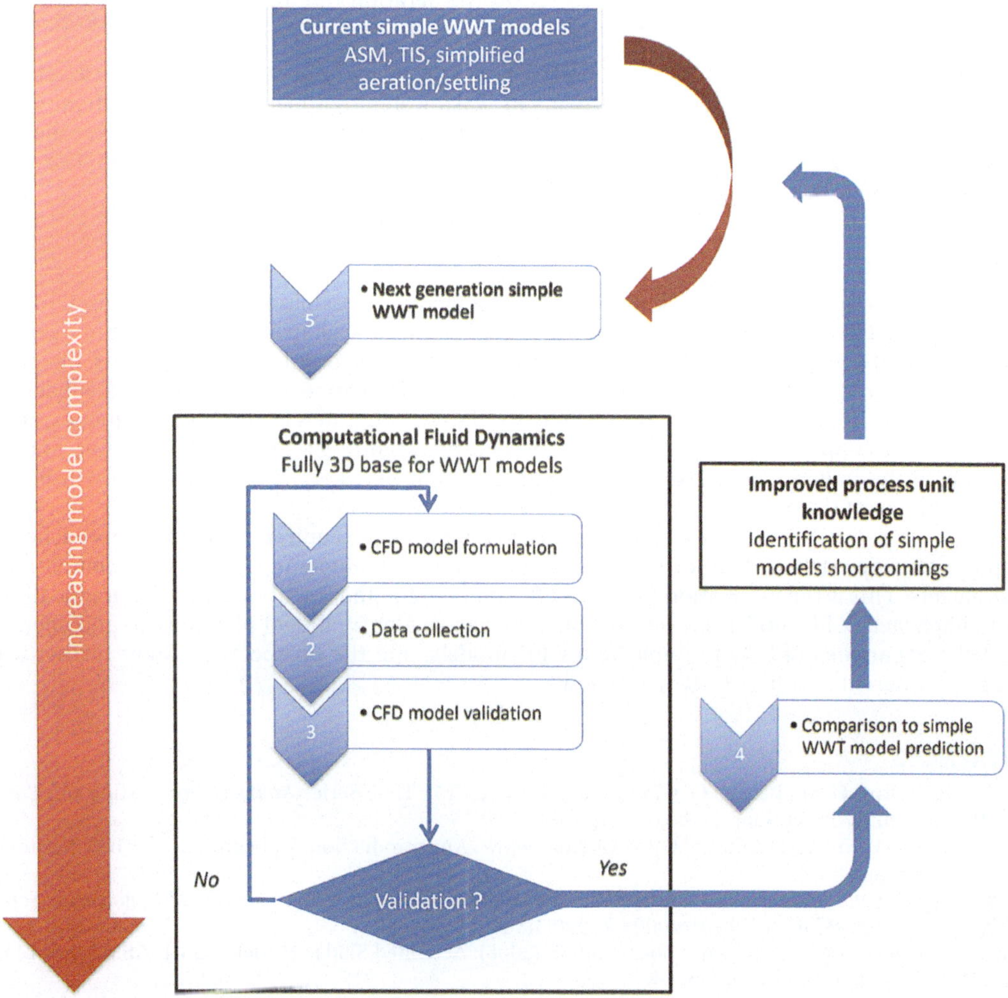

Figure 1.5 Implementation map for use of CFD as a calibration tool for simpler biokinetic models (Laurent *et al.*, 2014).

computational resources, especially for simulation of dynamically varying flows, and TIS and other simpler biokinetic models can accomplish approximate simulations in a fraction of the time required for a full-scale CFD simulation. The concept here is to use a calibrated CFD model as a source for comparison with results from simpler models. Availability of the CFD results can allow tuning parameters in the simpler models to be confirmed.

1.6 OPTIMIZE PROCESS CONTROL

In a study using CFD to study the effect of sensor location on the cost and effectiveness of dissolved oxygen control in an activated sludge aeration tank Rehman *et al.* (2015) showed that

> 'a dissolved oxygen controller with a fixed setpoint performs differently as a function of the location of the sensor. When placed in a poorly mixed location, the controller increases the aeration intensity to its maximum capacity leading to higher aeration costs. When placed just above the aerated zone, the controller decreases the aeration rate resulting in lower dissolved oxygen concentrations in the remainder of the system, compromising effluent quality.'

Application of CFD to these kinds of problems is still in its infancy, but it is an example of an exceptionally promising area for improvement in wastewater treatment operation, where a better understanding of the spatial distribution of flow velocity and concentration (variation over the length and width, or diameter and depth) can improve performance and energy efficiency and, last but not least, save money.

1.7 CONCLUSIONS

'Why CFD?' was the question posed in the title of this chapter. Possibly the answer is this: CFD is the most promising analytical tool yet devised for optimization of wastewater treatment tank geometry and the accurate prediction of the impact of physical and hydraulic configurations on process performance. It retains the greatest level of mechanistic information, while allowing feedback from relevant data. Although aspects of the model will need to be calibrated by relevant techniques, and the results used with appropriate caution, there is no question that this tool will be used more and more frequently in the industry over the next 30 years, and that our designs and operation will be made better by the use of them.

The greatest barrier is accessibility; both computational requirements and approaches. An inappropriate approach will either result in a poor predictive outcome, or excessive computational requirements. This is the motivation for this STR, and the remainder will focus on the practicalities of how to implement CFD models for improvement of wastewater treatment plant design and operation. Detailed comparisons of how to implement CFD models, and the prospects for their further use in analysis of different unit processes will follow.

REFERENCES

Box G. E. P., Jenkins G. M., Reinsel G. C. and Ljung G. M. (2015). Time Series Analysis: Forecasting and Control, 5th edn. John Wiley & Sons, Hoboken, NJ, USA.

Chanson H. (2004). The Hydraulics of Open Channel Flow: An Introduction, 2nd edn. Elsevier B.V. Amsterdam, The Nederlands.

Danckwerts P. V. (1953). Continuous flow systems: distribution of residence times. *Chemical Engineering Science*, **2**(1), 1–13, https://doi.org/10.1016/0009-2509(53)80001-1

Henze M., Gujer W., Mino T. and van Loosdrecht M. (2006). Activated Sludge Models ASM1, ASM2, ASM2d and ASM3. IWA Publishing, London, UK.

Laurent J., Samstag R. W., Ducoste J. M., Griborio A., Nopens I., Batstone D. J., Wicks J. D., Saunders S. and Potier O. (2014). A protocol for the use of computational fluid dynamics as a supportive tool for wastewater treatment plant modelling. *Water Science and Technology*, **70**(10), 1575–1584, https://doi.org/10.2166/wst.2014.425

Levenspiel O. (1972). Chemical Reaction Engineering, 1st edn. John Wiley & Sons, Hoboken, NJ, USA.

Marques N. (2015). OpenFOAM® in Wastewater Applications, in: Watermatex 2015 Workshop: Computational Fluid Dynamics – Hands-on CFD Modeling for Water Resource Recovery Facilities. 9th IWA Symposium on Systems Analysis and Integrated Assessment (Watermatex) from 14 to 17th June 2015. Surfers Paradise, Queensland, Australia. https://rsamstag.wordpress.com/downloads-2/iwa-cfd-working-group-documents/?frame-nonce=729478071b&preview=true&iframe=true&calypso_token=65d8e63c-675c-4480-9bc7-81586328dbd4

Metcalf & Eddy, Tchobanoglous G., Stensel H. D., Tsuchihashi R. and Burton F. (2014). Wastewater Engineering: Treatment and Resource Recovery, 5th edn. McGraw-Hill, New-York, NY, USA.

Rehman U., Vesvikar M., Maere T., Guo L., Vanrolleghem P. A. and Nopens I. (2015). Effect of sensor location on controller performance in a wastewater treatment plant. *Water Science and Technology*, **71**(5), 700–708, https://doi.org/10.2166/wst.2014.525

Rodrigues A. E. (2021). Residence time distribution (RTD) revisited. *Chemical Engineering Science*, **230**, 116188, https://doi.org/10.1016/j.ces.2020.116188

The OpenFOAM Foundation. (2021). OpenFOAM v9 User Guide. https://cfd.direct/openfoam/user-guide (accessed 12.20.21)

doi: 10.2166/9781780409030_0009

Chapter 2
Fundamentals

Randal Samstag[1], Edward Wicklein[2], Rainier Hreiz[3], Julien Laurent[4], Jim Wicks[5] and Damien Batstone[6]

[1]Civil and Sanitary Engineer, Bainbridge Island, WA, USA
[2]Carollo Engineers, Seattle, WA, USA; Email: ewicklein@carollo.com
[3]Laboratoire Réactions et Génie des Procédés, Université de Lorraine, CNRS, LRGP, F-54000, Nancy, France; Email: rainier.hreiz@univ-lorraine.fr
[4]ICube, ENGEES, CNRS, Université de Strasbourg, Strasbourg, France; Email: julien.laurent@engees.unistra.fr
[5]The Fluid Group, Oxford, UK; Email: jim.wicks@thefluidgroup.com
[6]Australian Centre for Water and Environmental Biotechnology, University of Queensland, Brisbane, Australia; Email: d.batstone@uq.edu.au

2.1 INTRODUCTION

Computational fluid dynamics (CFD) provides an estimate of the fluid flow field and constituent transport through a defined domain using numerical methods. In this chapter we present some fundamentals of CFD with emphasis on techniques and models that are most pertinent in the case of wastewater treatment processes. We will give an overview of the governing equations for CFD for wastewater treatment, provide an indication of some of the most popular numerical methods, and develop elements of best modelling practice for CFD of wastewater treatment units.

2.2 GOVERNING EQUATIONS

2.2.1 The transport equation

The most fundamental equation in CFD is an equation that defines transport of a property through a computational domain. Typical transported properties include momentum, mass (solids, solutes), or energy. Properties can be represented as extensive (momentum, mass, energy) or, generally in a more usable form, intensive (velocity, concentration, temperature) (Hangos & Cameron, 2001). Scalar concentrations can represent particulates, including biological solids, or solutes, including reactant concentrations like chemical oxygen demand, oxygen or nutrients such as ammonia and phosphorus. The basic intensive form of the transport equation, which expresses a multidimensional equation set in a compact form, is expressed as follows in a Cartesian coordinates system (Equation (2.1)):

$$\frac{\partial}{\partial t}(\rho\varphi) + \frac{\partial}{\partial r_j}(\rho U_j \varphi) = \frac{\partial}{\partial r_i}\left(\Gamma \frac{\partial \varphi}{\partial r_j}\right) + S \quad (2.1)$$

This first term of this partial differential equation represents the local rate of change of a transported property, φ, with respect to time. The second term corresponds to advective transport in space due to the fluid motion, and the third term represents transport in space by diffusive mechanisms. S is the source/sink term representing the generation/consumption of φ. For example, it represents the influence of external forces like gravity on transport, or change resulting from reaction or decay. The Greek letter ρ here represents density; t represents time; r_j represents the multidimensional spatial domain; U_j represents the multidimensional velocity vector; the Greek letter Γ represents a diffusion coefficient. In the more familiar Cartesian coordinates form, the advection and diffusion terms are written as follows:

Advection:

$$\frac{\partial}{\partial x_j}(\rho U_j \varphi) = \frac{\partial}{\partial x}(\rho u \varphi) + \frac{\partial}{\partial y}(\rho v \varphi) + \frac{\partial}{\partial z}(\rho w \varphi) \tag{2.2}$$

Diffusion:

$$\frac{\partial}{\partial x_j}\left(\Gamma \frac{\partial \varphi}{\partial x_j}\right) = \frac{\partial}{\partial x}\left(\Gamma \frac{\partial \varphi}{\partial x}\right) + \frac{\partial}{\partial y}\left(\Gamma \frac{\partial \varphi}{\partial y}\right) + \frac{\partial}{\partial z}\left(\Gamma \frac{\partial \varphi}{\partial z}\right) \tag{2.3}$$

where u, v and w represent the velocity vector components in the x, y and z directions, respectively (Figure 2.1).

2.2.2 The Navier-Stokes equations

The Navier-Stokes equations are the basic model for fluid motion. For incompressible flow, they combine a continuity equation (representing mass conservation) with a transport equation for momentum as follows (compact form) (Andersson *et al.*, 2011):

Continuity:

$$\frac{\partial (U_i)}{\partial r_j} = 0 \tag{2.4}$$

Momentum (assuming an incompressible Newtonian flow with constant density and viscosity):

$$\frac{\partial U_i}{\partial t} + U_j \frac{\partial U_i}{\partial r_j} = -\frac{1}{\rho}\frac{\partial P}{\partial r_i} + \upsilon\left(\frac{\partial^2 U_i}{\partial r_j^2}\right) + F_i \tag{2.5}$$

Here the Greek letter, υ, represents the kinematic viscosity (molecular viscosity, μ, divided by the fluid density, ρ) and F_i represents the influence of external forces on the domain, typically gravity (only acting in z direction, see Equation (2.9)). Note that in Equation (2.5), j should be summed over all dimensions, that is $j = 1$, 2 and 3, and I appears in all terms and for three dimensions (Andersson *et al.*, 2011). In three Cartesian spatial coordinates, x, y and z, and their corresponding velocities, u, v,

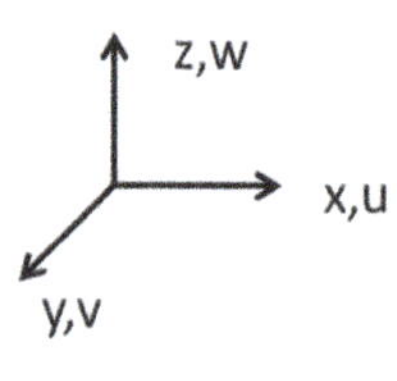

Figure 2.1 Three-dimensional coordinate system.

and w, and with gravity as the only body force, the Navier-Stokes equations may be written as follows (Equation (2.6) for continuity, Equation (2.7), Equation (2.8) and Equation (2.9) for momentum):

$$\frac{\partial \rho}{\partial t}+\left(\frac{\partial}{\partial x}(\rho u)+\frac{\partial}{\partial y}(\rho v)+\frac{\partial}{\partial z}(\rho w)\right)=0 \quad \text{(compressible flow)}$$
$$\frac{\partial}{\partial x}u+\frac{\partial}{\partial y}v+\frac{\partial}{\partial z}w=0 \quad \text{(incompressible flow)} \tag{2.6}$$

$$\frac{\partial u}{\partial t}+u\frac{\partial u}{\partial x}+v\frac{\partial u}{\partial y}+w\frac{\partial u}{\partial z}=-\frac{1}{\rho}\frac{\partial P}{\partial x}+v\left(\frac{\partial^2 u}{\partial x^2}+\frac{\partial^2 u}{\partial y^2}+\frac{\partial^2 u}{\partial z^2}\right) \tag{2.7}$$

$$\frac{\partial v}{\partial t}+u\frac{\partial v}{\partial x}+v\frac{\partial v}{\partial y}+w\frac{\partial v}{\partial z}=-\frac{1}{\rho}\frac{\partial P}{\partial y}+v\left(\frac{\partial^2 v}{\partial x^2}+\frac{\partial^2 v}{\partial y^2}+\frac{\partial^2 v}{\partial z^2}\right) \tag{2.8}$$

$$\frac{\partial w}{\partial t}+u\frac{\partial w}{\partial x}+v\frac{\partial w}{\partial y}+w\frac{\partial w}{\partial z}=-\frac{1}{\rho}\frac{\partial P}{\partial z}+v\left(\frac{\partial^2 w}{\partial x^2}+\frac{\partial^2 w}{\partial y^2}+\frac{\partial^2 w}{\partial z^2}\right)+\rho\frac{(\rho-\rho_w)}{\rho_w}g_z \tag{2.9}$$

2.2.3 Turbulence

In the above Navier-Stokes equations it is assumed that kinematic molecular viscosity ($\nu=\mu/\rho$) serves as the coefficient of momentum diffusion. Almost all flows of interest in wastewater treatment facilities are turbulent, which prevents the numerical resolution of the Navier-Stokes equations. Hinze (1959) defines turbulence as follows:

> 'Turbulent fluid motion is an irregular condition of flow in which the various quantities show a random variation with time and space coordinates, so that the statistically distinct average values can be discerned.'

It was Osborne Reynolds in the nineteenth century who suggested that turbulence could be appropriately modelled by breaking the variable properties in the Navier-Stokes equations into separate parts: time-averaged parts and fluctuating parts. For example, velocity can be considered as follows:

$$U_i=\overline{U_i}+u_i \tag{2.10}$$

where, U_i is the instantaneous velocity, $\overline{U_i}$ the time averaged part and u_i the fluctuating component. With this approach, the momentum equations become, along with the continuity equation, the Reynolds-averaged Navier-Stokes (RANS) equations. When all the variables are treated in this manner and inserted into separate equations for the time-averaged and fluctuating parts and then combined, the incompressible momentum equation becomes:

$$\frac{\partial}{\partial x_i}\bar{U}_i=0$$
$$\frac{\partial \bar{U}_i}{\partial t}+\bar{U}_j\frac{\partial \bar{U}_i}{\partial x_j}=-\frac{\partial \bar{P}}{\partial x_i}+\frac{\partial}{\partial x_j}\left(\mu\frac{\partial \bar{U}_i}{\partial x_j}-\rho\overline{u_i u_j}\right)+\bar{F}_i \tag{2.11}$$

For a complete description of the Reynolds averaging approach to modelling the effect of turbulent flows, see standard texts in fluid mechanics, for example, Hinze (1959) and Versteeg and Malalasekera (2007). It was Boussinesq who suggested that the influence of the fluctuating terms could be modelled by replacing all of the effects of turbulence by the product of the mean velocity gradient and a quantity called the 'turbulent' or 'eddy viscosity':

$$-\rho\overline{u_i u_j} = \mu_t\left(\frac{\partial\bar{U}_i}{\partial x_j} + \frac{\partial\bar{U}_j}{\partial x_i}\right) - \frac{2}{3}\rho k\delta_{ij} \tag{2.12}$$

where k is the turbulence kinetic energy (see Section 2.2.3.2) and δ_{ij} the Kronecker delta (equal to 1 if $i=j$ and 0 elsewhere). Hence the modelled turbulent momentum equations for incompressible flow become:

$$\frac{\partial\bar{U}_i}{\partial t} + \bar{U}_j\frac{\partial\bar{U}_i}{\partial x_j} = -\frac{1}{\rho}\frac{\partial\bar{P}}{\partial x_i} + \frac{\partial}{\partial x_j}\left[(\nu+\nu_t)\left(\frac{\partial\bar{U}_i}{\partial x_j} + \frac{\partial\bar{U}_j}{\partial x_i}\right)\right] - \frac{2}{3}\frac{\partial k}{\partial x_i} + F_i \tag{2.13}$$

The task becomes finding an appropriate expression for ν_t. Since this variable is not constant like molecular viscosity, but varies depending on the flow condition within the domain, it must itself be modelled. Launder and Spalding (1972) described several approaches to doing so. Two of the most popular approaches in early modelling of wastewater treatment unit processes have been Prandtl's mixing length hypothesis and two-equation models like the k-epsilon model which are still widely used today in CFD for wastewater applications.

2.2.3.1 Prandtl's mixing length hypothesis

Launder and Spalding (1972) nominated the Prandtl mixing length hypothesis (MLH) as the 'pre-eminent' one for modelling 'nearly-two-dimensional boundary-layer' turbulence. It was originally proposed by Prandtl. Launder and Spalding say of this, 'The hypothesis is that the turbulent viscosity is equal to the local product of the density, of the magnitude of the mean rate of strain, and of the square of a characteristic length scale of the turbulent motion.' Their formula for this expression is as follows:

$$\upsilon_t = l_m^2\left[\frac{\partial u}{\partial y}\right] \tag{2.14}$$

Here l_m is the mixing length, ∂u is the horizontal velocity increment, ∂y is the vertical increment, and for 'plane flows' $l_m/\delta = 0.07$, where δ is the layer width.

The MLH has been used to model turbulence in secondary sedimentation CFD, for example in the widely used 2DC model developed by Griborio (2004) and McCorquodale *et al.* (2004). This Prandtl mixing length model had been previously used in sedimentation models (Abdel-Gawad & McCorquodale, 1985, 1984a, 1984b; Larsen, 1977).

2.2.3.2 k-epsilon

Launder and Spalding (1972) provided extensive comparisons between the MLH and more complex one and two-equation models for turbulence in industrial flows. They concluded that while the MLH is useful for many purposes, 'there is almost no successful experience of predicting recirculating flows with its aid' and that in the MLH 'the turbulent viscosity vanishes whenever the mean velocity gradient is zero. But, along the centre-line of a plane channel, for example, experiments have shown that μ_t is high.' Also, the MLH 'takes no account of processes of convection and diffusion of turbulence.' For these reasons, these researchers went on to explore more complex models.

A two-equation model was developed at Imperial College in London, the k-epsilon model (Launder & Spalding, 1974). This model is now probably the most commonly used model for turbulent eddy viscosity, in use today in both custom CFD codes and commercial models. With increasing speed

Table 2.1 Suggested values for the constants in the k-epsilon model.

c_μ	$c_{1\varepsilon}$	$c_{2\varepsilon}$	σ_k	σ_ε
0.09	1.44	1.92	1.0	1.3

of modern computers, the computational overhead of this more complex model has become less concerning. As presented in Rodi (2000) the standard *k*-epsilon model is expressed as follows:

$$\nu_t = c_\mu \frac{k^2}{\varepsilon}, \Gamma = \frac{\nu_t}{\sigma_t} \tag{2.15}$$

$$\frac{\partial k}{\partial t} + U_i \frac{\partial k}{\partial x_i} = \frac{\partial}{\partial x_i}\left(\frac{\nu_t}{\sigma_k}\frac{\partial k}{\partial x_i}\right) + \underbrace{\nu_t\left(\frac{\partial U_i}{\partial x_j} + \frac{\partial U_j}{\partial x_i}\right)\frac{\partial U_i}{\partial x_j}}_{P} + \underbrace{\beta g_i \frac{\nu_t}{\sigma_t}\frac{\delta \phi}{\delta x_i}}_{G} - \varepsilon \tag{2.16}$$

$$\underbrace{\frac{\partial \varepsilon}{\partial t}}_{\substack{\text{rate of}\\ \text{change}}} + \underbrace{U_i \frac{\partial \varepsilon}{\partial x_i}}_{\text{advection}} = \underbrace{\frac{\partial}{\partial x_i}\left(\frac{\nu_t}{\sigma_\varepsilon}\frac{\partial \varepsilon}{\partial x_i}\right)}_{\text{diffusion}} + \underbrace{c_{1\varepsilon}\frac{\varepsilon}{k}(P+G)(1+c_{3\varepsilon}R_f) - c_{2\varepsilon}\frac{\varepsilon^2}{k}}_{\text{generation-destruction}} \tag{2.17}$$

Here the turbulent eddy viscosity, ν_t, is defined in relation to solution of two transport equations for, respectively, the turbulence kinetic energy, *k* and its dissipation rate, ε. The eddy viscosity divided by the turbulent Prandtl/Schmidt Number, σ_t, appears in both equations as a diffusion coefficient.

The transport equation for *k* contains a stress production term, *P*, and a buoyancy term, *G*, combining the gravitation term, the heat flux, $\delta\phi/\delta x_i$ and an expansion coefficient, β. Rodi (2000) suggests a value for β of 10 following (Munk & Anderson, 1948). The transport equation for dissipation, ε, contains the usual unsteady, advection (convection) and diffusion terms plus a 'generation-destruction' term combining the effects of the term groupings, *P* and *G*, with the flux Richardson number, R_f, and three empirical coefficients, $c_{1\varepsilon}$, $c_{2\varepsilon}$ and $c_{3\varepsilon}$. Rodi (2000) suggests values for empirical coefficients as shown in Table 2.1. Rodi (2000) cites the value given in Hossain (1980) for $c_{3\varepsilon}$ of 0.8 but also mentions that 'its value is somewhat controversial.'

Zhou and McCorquodale (1992a) say that they omitted the buoyancy correction terms in the k-epsilon equations for their work, since 'the buoyancy correction term involving the flux Richardson number in the ε equation is suitable only for shear layer flows.'

2.2.3.3 Other approaches to turbulence modelling

The MLH and k-epsilon models are certainly not the only attempts to capture the effects of turbulence on flow in reactors and channels likely to be found in wastewater treatment plants. Launder and Spalding (1972) describe one-equation transport models for energy and a variety of two-equation and multi-equation models. Rodi (2000) describes one-equation models, like Bradshaw's model, and alternative two-equation models like algebraic stress/flux models. A recent review by Karpinska and Bridgeman (2016) discusses the characteristics of several alternative turbulence models available in commercial CFD packages, several of which have been applied to activated sludge reactor flows using the Reynolds averaging approach: standard, renormalized group (RNG) and realizable k-epsilon models; standard and shear stress transport (SST) k-omega models; and the Reynolds stress model (RSM). They present a table with advantages and disadvantages of each of these approaches. They also discuss alternative approaches like large eddy simulation (LES) and direct numerical simulation (DNS). They conclude that LES and DNS are both very computationally expensive for activated sludge reactor CFD. Among the RANS models they recognize that the standard k-epsilon model is 'robust, economic in terms of computational effort and (has) satisfactory accuracy in diverse turbulent flow issues' but that it gives 'poor performance in some scenarios (strong streamline curvature, vortices, rotating flows, flow separation, adverse pressure gradients)' and 'poor prediction of the lateral expansion in 3D wall jets'.

2.2.4 Scalar transport

A very important topic in CFD applied to wastewater treatment tanks is consideration of treatment of transported properties of the wastewater such as suspended solids; biochemical oxygen demand; nutrients like phosphorus and nitrogen species; and biological organisms and chemical properties that affect chemical reaction and biological growth, importantly alkalinity and pH. In addition, aeration is important in many wastewater treatment processes and it is important to be able to adequately characterize transport of dissolved and gaseous oxygen in wastewater treatment tanks. These properties are crucial to our understanding of the effects of hydraulic conditions on reaction and transport. This transport can be handled in several ways: as a scalar quantity whose volume is considered to be insignificant to the behaviour of the fluid mixture or as a separate phase in the mixture transport. In this section we consider the case of scalar transport. This has been widely used for simulation of solids settling and mixing in sedimentation tanks and activated sludge aeration tanks, but also for modelling of biokinetic reactions in wastewater reactors. Another application of the scalar transport equation is the operation of virtual tracer tests that enable residence time distribution (RTD) of a given system to be derived (see Chapters 1 and 10). This is useful for developing more accurate simple models as proposed in Laurent *et al.* (2014) but also for validating CFD model results (see Chapter 9).

2.2.4.1 Solids transport

The first applications of CFD to wastewater treatment were applied to the problem of sedimentation. In early work in this field, solids settling and transport were modelled as a scalar transport. The fate of solids in reactors is important because solids behaviour is inherent to wastewater treatment goals and outcomes. An additional concern for solids transport, however, is that solids concentration gradients can have a significant effect on the overall tank velocity profile because of their effect on spatially- and time-variant fluid density.

In general, while fully coupled approaches can be used, the normal approach is scalar transport, where the solid or solute is a scalar state, solved using the mass form of the general conservation equation. Density is accounted for by the impact of the scalar on fluid density. The scalar property is represented by mass or volume fraction concentration (C) as an intensive property.

$$\frac{\partial}{\partial t}(C) = -U_j \frac{\partial C}{\partial r_j} + \frac{\partial}{\partial r_j}\left(D \frac{\partial C}{\partial r_j}\right) + S \tag{2.18}$$

A typical transport equation for solids in multiphase applications may be taken as the following:

$$\frac{\partial C}{\partial t} = -U_j \frac{\partial C}{\partial r_j} + \frac{\partial}{\partial r_j}\left(\frac{\nu_t}{\sigma_s} \frac{\partial C}{\partial r_j}\right) + V_s \frac{\partial C}{\partial z_k} + S \tag{2.19}$$

where C represents the transported solids concentration treated as a scalar and U_j the velocity field solved from the main fluid momentum equations. Note that ν_t/σ_s represents eddy solids diffusivity but can be often neglected.

The new term, V_s, represents the settling velocity of the suspended solids in the direction of gravity (z). V_s is an empirical term representing fluid momentum coupling (which results in hindered settling). The simplest model for estimating this velocity within the sedimentation domain, and until recently the most-widely used, has been the exponential model attributed to Vesilind (1968):

$$V_s = V_o e^{-kC} \tag{2.20}$$

where V_o represents a maximal settling velocity which would be found in the case of unhindered settling and the parameter k represents the hindering effect of suspended solids. In CFD modelling for sedimentation Zhou and McCorquodale (1992a, 1992b) used different values of σ_t to modify the assumption of isotropic eddy diffusivity in different regions of the sedimentation tank. They used two different empirical values for σ_t: in the longitudinal direction $\sigma_t = 1.0$; in the vertical direction $\sigma_t = 0.5$ in the bottom sludge layers and $\sigma_t = 1.0$ in the upper layers of the tank. A more detailed treatment of the problem of sedimentation is presented in Chapter 6.

2.2.4.2 Heat (temperature) transport

In several applications in wastewater treatment heat transport can be important because of its effect on buoyancy but also because of temperature effects on biological and chemical reaction rates. For example, Griborio (2004) included the effect of temperature transport and heat transfer through the surface of a clarifier. Energy follows the general conservation form with conduction replacing diffusion (here expressed in the case of an incompressible fluid with constant C_p):

$$\rho C_p \left(\frac{\partial T}{\partial t} + U_j \frac{\partial T}{\partial r_j} \right) = \frac{\partial}{\partial r_j} \left(k \frac{\partial T}{\partial r_j} \right) + S \tag{2.21}$$

where k is the thermal conductivity ($W.m^{-1}.K^{-1}$) and C_p is the heat capacity of the fluid at constant pressure ($J.kg^{-1}.K^{-1}$). The source term S can include external heat transfer (from other phases, boundaries, or heat exchange elements), radiation, and reactive energy. It is common to omit conduction in fluids, as the convection term dominates. Indeed, this intensive form of the energy balance equation is valid for an incompressible fluid with no work term and no heat of reaction (e.g., water with an infinite dilution assumption). It is valid for most wastewater units where the primary fluid is water. Turbulent flows should also consider eddy diffusion (see Chapter 10) though in most cases flow convection will dominate.

It is important to notice here that Equation (2.21) neglects heat generation by viscous dissipation. Indeed, viscous forces convert a proportion of the kinetic energy in the flow into thermal energy due to internal friction, which stems presumably from inter-molecular interactions at the atomic level, and these do heat the fluid. When modelling highly viscous fluids internal friction due to viscosity becomes a significant energy source. For water and wastewater applications where the viscosity is low, it is usually ignored. For waste activated sludge and digester mixing, where thermal effects are being assessed, it is better to include it (in this case, see Ansys (2011)).

2.2.4.3 Reactive CFD

A primary concern in modelling wastewater treatment tanks is to predict biological and chemical transformations which take place in reactors. A brief history of the development of process modelling for wastewater treatment is provided in the Water Environment Federation *Manual of Practice No. 31* (Water Environment Federation, 2014). A long line of development of biokinetic modelling for suspended growth activated sludge reactors and secondary sedimentation tanks began in the 1970s with work by John Andrews and his students at Clemson University (Andrews, 1971; Bryant, 1972; Busby, 1973; Stenstrom, 1975; Tracy, 1973). This was followed by a great deal of work by international researchers such as that incorporated into the activated sludge models (ASMs) and anaerobic digestion model (ADM) published by the IWA (Batstone *et al.*, 2002; Henze *et al.*, 2006). This work has developed relatively sophisticated models for biological and chemical reactions expressed in ordinary differential equations and solved by numerical methods.

Until recently, these process models were based on highly simplified models of hydraulic transport through reactor tanks. Typically, activated sludge reactors have been modelled using tanks-in-series models and secondary clarifiers modelled by solution of one-dimensional partial differential equations. As discussed in Chapter 1, the availability of higher speed computational hardware has made CFD a possible alternative to provide a more realistic hydraulic base for these reactor models. The obvious choice for incorporation of these models into CFD simulations of wastewater treatment reactors is by incorporating reaction and decay into supplemental transport equations solved simultaneously with or subsequently to solution of the fluid flow equations. Detailed examples of how this is can be accomplished will be provided in Chapters 4 and 8 as well as in Chapter 10.

2.2.4.4 Density couple; equation of state

Transport of solids and heat through a reactor can profoundly affect the hydraulic regime since solids concentration and temperature gradients introduce buoyancy forces into the flow field which can significantly affect the flow. Where these buoyancy effects can be expected, such as in sedimentation tanks or in mixed reactors, this effect can be modelled by incorporating an equation of state relating the transported factor to the local density in the reactor. An example equation of state for solids concentration is the following:

$$\rho = \rho_w + C\left(1 - \frac{\rho_w}{\rho_s}\right) \tag{2.22}$$

where C represents the solids volume fraction and ρ, ρ_w and ρ_s represent, respectively, the density of the mixture, the carrier liquid (water) and the solids. When used with an equation of state in this manner the scalar transport approach is called an 'active' scalar approach, as opposed to a passive scalar approach where the scalar is assumed to have no effect on the fluid properties.

2.2.5 Multiphase models

For simulating clouds of particles or droplets in a mixture, where the volume of the transported particle or droplet cannot be neglected, two approaches have been used: the Lagrangian approach and the Eulerian approach.

2.2.5.1 Lagrangian approach

In the Lagrangian approach, the reference frame is that of the particle, droplet, or fluid packet, and properties are specific to the Lagrangian element (i.e. each particle, droplet, or packet has position, velocity, mass, temperature, composition etc.). In the Eulerian approach, the reference frame is absolute (normally referenced to a dimensional origin), and the control volume is a region within that absolute space (e.g., a Cartesian volume). The mass of particles within the control volume will change as particles move in or out of the control volume. This is therefore a second component (fluid) within the control volume. In wastewater applications (e.g., grit removal), the approach is usually to combine the Euler reference frame for the bulk fluid (water) and Lagrangian approach for the particles. The obvious disadvantage of a Lagrangian approach is that each control volume must be described separately, and the number is not fixed (and potentially enormous) in the case of activated sludge particles or gas bubbles. There are a number of advantages, including only simulating in the region of interest (important for open systems where *inherently*, the whole system cannot be simulated). There are also hybrid approaches, including the use of population balance models (Morchain *et al.*, 2014; Nopens *et al.*, 2015).

We will describe here the Eulerian or two-fluid approach, which has been used for simulation of air and water in activated sludge aeration tanks. The Lagrangian approach is often used for grit removal (Chapter 3) and disinfection (Chapter 7) applications.

2.2.5.2 Euler-Euler approach

In this two-fluid approach separate continuity and momentum equations are written for the dispersed phase, subscript '*d*', and the continuous phase, subscript '*c*'. In these equations the volume fractions of the continuous and dispersed phase are designated α_d and α_c, respectively, where:

The continuity and momentum equations in the Euler-Euler approach in a Cartesian coordinates system are the following (Crowe *et al.*, 2011) (no mass transfer between phases considered here):

$$\begin{aligned} &\frac{\partial \alpha_d \rho_d}{\partial t} + \frac{\partial}{\partial t}(\alpha_d \rho_d v_i) = 0 \\ &\frac{\partial \alpha_c \rho_c}{\partial t} + \frac{\partial}{\partial t}(\alpha_c \rho_c v_i) = 0 \end{aligned} \tag{2.23}$$

$$\begin{aligned} &\frac{\partial}{\partial t}(\alpha_d \rho_d v_i) = -\frac{\partial}{\partial x_j}(\alpha_d \rho_d v_i v_j) - \alpha_d \frac{\partial p}{\partial x_i} + \beta_V (u_i - v_i) + T_{d,i} \\ &\frac{\partial}{\partial t}(\alpha_c \rho_c v_i) = -\frac{\partial}{\partial x_j}(\alpha_c \rho_c v_i v_j) - \alpha_c \frac{\partial p}{\partial x_i} + \beta_V (u_i - v_i) + T_{c,i} \end{aligned} \tag{2.24}$$

The term, $\beta_V(u_i - v_i)$, is a drag force term where u and v represent the velocities of the continuous phase and particle phase, respectively, and β_V is a proportionality coefficient. See Crowe *et al.* (2011) for more detailed explanations.

The two-fluid approach has been used most commonly in wastewater treatment for analysis of interactions of gas and liquid phases in aerated reactors. For analysis of solids transport and sedimentation it has been more common to use an active scalar transport equation rather than a full two-fluid approach to solution. Since the two-fluid approach requires solution of eight partial differential equations (not including turbulence model equations) rather than the five required for solution of the RANS equations plus one transport equation for solids, the scalar transport approach is more computationally efficient than a full multiphase model.

2.2.5.3 Drift flux

The drift flux approach was developed to overcome difficulties in specifying interfacial interactions between the two phases in a two-fluid model. Ishii (1977) cites, as an example of these difficulties, that 'the interaction terms should include first-order time and spatial derivatives under certain conditions.' In addition, the drift flux approach is computationally more efficient, requiring solution of two continuity equations for the mixture and the dispersed phase, but only one set of momentum equations for the mixture. The drift flux equations as presented in Ishii (1977) are as follows:

Mixture continuity equation:

$$\frac{\partial \rho_m}{\partial t} + \nabla(\rho_m \vec{v}_m) = 0 \tag{2.25}$$

Continuity equation for dispersed phase:

$$\frac{\partial \alpha_d \rho_d}{\partial t} + \nabla(\alpha_d \rho_d \vec{v}_m) = -\Gamma_d - \nabla\left(\frac{\alpha_d \rho_d \rho_c}{\rho_m} \vec{v}_{dj}\right) \tag{2.26}$$

Mixture momentum equation:

$$\frac{\partial \rho_m \vec{v}_m}{\partial t} + \nabla(\rho_m \vec{v}_m \vec{v}_m) = -\nabla P_m - \nabla\left(\bar{\bar{\tau}} + \bar{\bar{\tau}}^T - \frac{\alpha_d}{1-\alpha_d} \frac{\rho_d \rho_c}{\rho_m} \vec{v}_{dj} \vec{v}_{dj}\right) + \rho_m \vec{g} \tag{2.27}$$

Here, ∇ is the divergence operator where:

$$\text{div}\, \vec{a} = \nabla(\vec{a}) = a_{i,i} = \frac{\partial a_1}{\partial x_1} + \frac{\partial a_2}{\partial x_2} + \frac{\partial a_3}{\partial x_3} \tag{2.28}$$

In Ishii's notation, α_k and ρ_k are void ratio and density for, respectively, the continuous phase ($k = c$) and the dispersed phase ($k = d$) and $\vec{v}_m$, ρ_m and P_m represent the velocity vector, density, and pressure of the mixture. $\bar{\bar{\tau}}$ and $\bar{\bar{\tau}}^T$ represent viscous stress and turbulent diffusion flux. Γ_d represents a 'mass-transfer term for phase k due to phase changes.' The velocity $\vec{v}_{dj}$ represents the drift velocity of the dispersed phase$=(1-\alpha_d)\vec{v}_r$ where $\vec{v}_r$ is the relative velocity between the phases. See Ishii (1977) for further details.

Brennan (2001) used a drift flux model for analysis of secondary sludge sedimentation. In his work he preferred the drift flux model to a two-fluid model for reasons of computational simplicity. He rejected the scalar transport approach because he assumed that it could not be used to model density-driven flow. He did not appear to consider using an equation of state, however, to capture density gradient effects on the flow in an active scalar method as described above.

2.2.5.4 Volume-of-fluid (VOF) approach

This approach is a Euler-Euler model with a fixed grid (mesh). It is used to simulate free surface flows or bubbly flows. The interface between the phases is tracked. It is suitable in the case of immiscible fluids with a clearly defined interface. Some classical applications are motion of large bubbles in a liquid, motion of liquid after a dam break and, more generally, steady or transient tracking of any liquid-gas interface. In the case of bubbles, the bubble size must be significantly larger than a control volume.

Each cell is assumed to contain only one phase or the interface between them. A single set of momentum equations is solved. Additional source terms are included to account for surface tension and wall adhesion. A volume fraction conservation equation is solved for each phase (equation below for incompressible fluid) while only one set of momentum equations is solved.

$$\frac{\partial \alpha_d}{\partial t} + u_j \frac{\partial \alpha_d}{\partial x_i} = S_{\alpha d} \tag{2.29}$$

This volume fraction can take values with three possible situations:

- 0 if the cell is empty of the secondary fluid;
- 1 if the cell is full;
- Between 0 and 1 if the cell contains the interface.

Usually, the source-term $S_{\alpha d}$ takes a zero value but mass transfer between phases can be modelled by specifying a non-zero value for it.

2.2.5.5 Rheology models

Above a certain concentration, water-particle suspensions may exhibit a non-Newtonian behaviour: there is no direct proportionality between shear rate and shear stress. This is particularly important in processes where biological sludge is present. Mainly, rheological properties will have an impact on (1) sludge pumping, (2) bioreactor hydrodynamics, (3) oxygen transfer, (4) secondary settler hydrodynamics, (5) membrane filtration and (6) sludge dewatering (Ratkovich *et al.*, 2013). Ratkovich *et al.* (2013) provide a comprehensive review of rheological measurement and modelling, from a Good Modelling Practice perspective. Also, rheological properties affecting CFD modelling of unit processes are discussed in the corresponding chapters (Chapter 4: suspended growth, Chapter 6: sedimentation, Chapter 8: anaerobic digestion).

2.3 NUMERICAL METHODS FOR CFD

The equations of fluid flow and transport can be solved analytically for only a limited number of cases and those are not applicable to industrial process flows. For practical purposes these equations must be solved using numerical techniques for approximation based on a discrete analogue of the real process element. After selection of the appropriate governing equations for a given application, the next step is to develop a method for 'discretization' of the fluid domain.

2.3.1 Discretization

Discretization is 'a method of approximating the differential equations by a system of algebraic equations for the variables at some set of discrete locations in space and time' (Ferziger & Perić, 2002). We will discuss three popular methods which have been used in application of CFD to wastewater treatment problems: finite difference, finite element, and finite volume.

2.3.1.1 Finite difference

The finite difference method was the first approach used for numerical solution of partial differential equations. Ferziger and Perić (2002) suggest that this method was developed originally by Euler in the eighteenth century. Since that time numerous researchers in mathematics and engineering have further refined this method for solution of different problems. We will develop some basic concepts for the finite difference method here. For further details see standard references starting with Ferziger and Perić (2002).

The basic idea for the finite difference method is to approximate the partial derivative of a state variable by the slope of the tangent to the function at a local point in a numerical grid.

$$\left(\frac{\partial \phi}{\partial x}\right)_{x_i} = \lim_{\Delta x \to 0} \frac{\phi(x_i + \Delta x) - \phi(x_i)}{\Delta x} \tag{2.30}$$

This example illustrates what is called a forward difference formula where the differential is approximated by the average of the downstream point in the flow direction in the computational grid minus the current point. Backward (or upwind) difference formulas use the point upstream of the current point, while in central difference schemes the average between the downstream and upstream points around the current point is used. These alternatives can be used for spatial as well as time approximations. Finite difference methods can be explicit, where new values for parameters are calculated from known boundary and initial values at a previous time step, or implicit, where values at a new time step are solved in terms of values at the same time level by iteration or matrix methods. Well-known explicit techniques include the (unconditionally unstable!) forward in time central in space (FTCS), Leapfrog (central in time and space) and Dufort-Frankel schemes. An important implicit method is the Crank-Nicholson scheme. For details of these see textbooks in numerical methods, such as Smith (1985), Peyret and Taylor (1983) or Fletcher (1991, 1998).

Harlow and Welch (1965) presented one of the first numerical solutions to an important free-surface incompressible flow problem (the dam-break problem) using finite difference techniques. They developed a numerical grid as indicated in Figure 2.2, where pressure (p) is calculated at the centre of the computational grid lines with velocities (u and v) calculated along the grid lines.

Their solution to the horizontal momentum equation,

$$\frac{\partial u}{\partial t} + \frac{\partial u^2}{\partial x} + \frac{\partial}{\partial y}(uv) + \frac{\partial P}{\partial x} = \nu\left(\frac{\partial^2 u}{\partial x^2} + \frac{\partial^2 u}{\partial y^2}\right) + g_x \tag{2.31}$$

was resolved into finite difference form as follows:

$$\begin{aligned} &\frac{\left(u_{i+1/2,j}^{n+1} - u_{i-1/2,j}^{n}\right)}{\Delta t} + \frac{u_{i,j}^2 - u_{i-1,j}^2}{\Delta x} + \frac{\left(u_{i+1/2,j-1/2}\right)\left(v_{i+1/2,j-1/2}\right) - \left(u_{i-1/2,j+1/2}\right)\left(v_{i-1/2,j+1/2}\right)}{\Delta y} \\ &= \nu\left(\frac{\left(u_{1\text{-}3/2,j} + u_{1\text{-}1/2,j} - 2u_{i+3/2,j}\right)}{\Delta x^2} + \frac{\left(u_{i+1/2,j+1} + u_{i+1/2,j-1} - 2u_{i+1/2,j}\right)}{\Delta y^2}\right) + g_r \\ &+ \frac{P_{i,j} - P_{i+1,j}}{\Delta x} + O\left(\Delta x^2\right) \end{aligned} \tag{2.32}$$

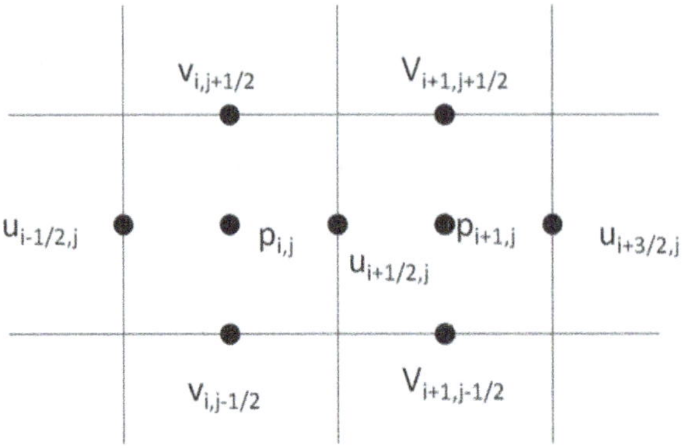

Figure 2.2 Finite difference computational grid (adapted from Harlow and Welch (1965)).

Here the superscript '$n+1$' implies calculation at a new time step and the superscript 'n' indicates values for velocity at the previous time step. Likewise, the subscripts 'i' and 'j' correspond to spatial locations as indicated in Figure 2.1.

Similar techniques were used for solution of the sedimentation problem by Gottardson as reported by Larsen (1977) who used a Du-Fort Frankel scheme and by Imam and McCorquodale (1983) who used a weighted upwind-centred difference approximation for solution of the sedimentation flow in vorticity-stream function form (see below).

2.3.1.2 Finite element

The finite element method has also been used in applications of CFD to wastewater problems. In this technique the velocity and pressure fields are recast into what are called interpolation functions:

$$
\begin{aligned}
u &= \sum_l u_l \varphi_l^u(\xi, n) \\
v &= \sum_l v_l \varphi_l^v(\xi, n) \\
p &= \sum_l p_l \varphi_l^p(\xi, n)
\end{aligned}
\tag{2.33}
$$

which are substituted into the governing equations for continuity:

$$\int\int\left(\frac{\partial u}{\partial x}+\frac{\partial v}{\partial y}\right)\varphi_m^p dxdy = 0 \tag{2.34}$$

and momentum:

$$\int\int\left\{\frac{\partial}{\partial x}\left(u/v^2 + P\right)+\frac{\partial}{\partial y}(uv)\right\}\varphi_m^{u/v} \tag{2.35}$$

and then solved iteratively.

Finite element techniques have the advantage over finite difference methods in that they can be easily adapted to unstructured meshes that can be configured to fit complex geometric elements like sedimentation tank inlets and outlets. For more details on finite element analysis see standard texts, for example, Finlayson (1972, 1980) and Fletcher (1991, 1998). The finite element method was used to develop a CFD model using the Galerkin weighted residual method of a radial flow secondary sedimentation tank by Kleine and Reddy (2005). The numerical results from an 85 by 58 cell grid were compared to field test results from three different activated sludge sedimentation tanks in South Africa with good results.

2.3.1.3 Finite volume

The finite volume method is the discretization technique used in most commercial CFD codes. It is similar to the finite difference method but casts the equations in terms of fluxes into and out of discrete control volumes across the solution domain. A classical description of the technique is included in Patankar (1980). Figures 2.3–2.5 illustrate control volume discretization of the continuity, horizontal momentum, and vertical momentum equations based on Patankar's development of the finite volume technique. This approach is widely used in CFD for wastewater applications using commercial CFD packages.

2.3.1.4 Grid-less methods

Several grid-less methods for CFD have been developed, including the vortex method developed by (Chorin, 1973), and the smoothed particle hydrodynamics (SPH) method developed from astrophysics. Only the SPH method has seen any use in wastewater applications (Meister *et al.*, 2017; Meister & Rauch, 2016; Winkler *et al.*, 2016).

2.3.2 Solution approaches for pressure

A key problem in numerical solution of the Navier-Stokes equation is the problem of solving for the pressure term. Two strategies that have been used successfully in application of CFD to wastewater treatment are the vorticity/stream function approach and the approach developed by Patankar (1980) for solution of semi implicit pressure linked equations (SIMPLE).

2.3.2.1 Vorticity/stream function

In the vorticity/stream function approach pressure is eliminated by transforming the Navier-Stokes and continuity equations into a transport equation for vorticity and a Poisson equation for solution

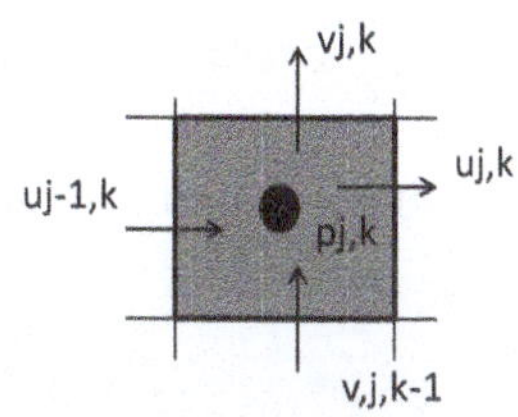

Figure 2.3 Finite volume discretization for the continuity equation.

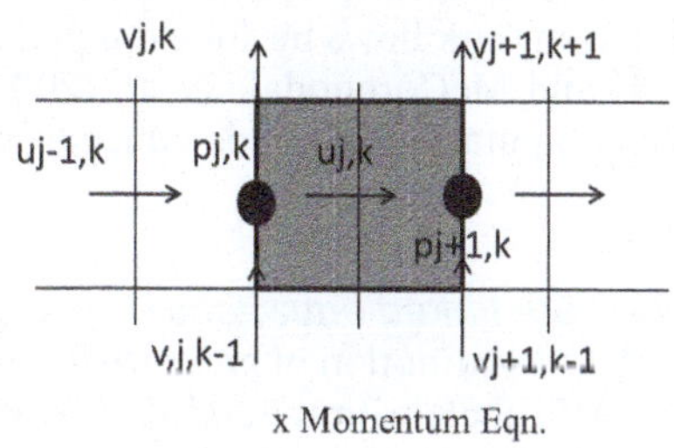

Figure 2.4 Finite volume discretization of the horizontal momentum equation.

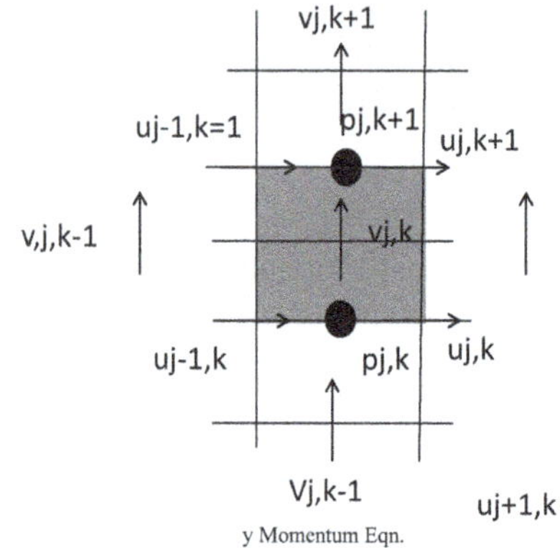

Figure 2.5 Finite volume discretization of the vertical momentum equation.

of the stream function (Roache, 1972). Vorticity, ζ, is defined as the difference between the cross differentials of velocity (in horizontal and vertical directions):

$$\zeta = \frac{\partial u}{\partial y} + \frac{\partial v}{\partial x} \tag{2.36}$$

The transport equation for vorticity is:

$$\frac{\partial \zeta}{\partial t} = -u\frac{\partial \zeta}{\partial x} - v\frac{\partial \zeta}{\partial y} + \nu\left(\frac{\partial^2 \zeta}{\partial x^2} + \frac{\partial^2 \zeta}{\partial y^2}\right) \tag{2.37}$$

The stream function, ψ, is defined (in the horizontal and vertical direction) as follows:

$$u = \frac{\partial \psi}{\partial y} \text{ and } y = \frac{\partial \psi}{\partial x} \tag{2.38}$$

The stream function is then recast from the continuity equation into a Poisson equation:

$$\frac{\partial^2 \psi}{\partial x^2} + \frac{\partial^2 \psi}{\partial y^2} = \zeta \tag{2.39}$$

The vorticity/stream function approach is practically a two-dimensional technique. Extending this to three dimensions turns out to be more complicated than other techniques which use the so-called primitive variables of velocity and pressure directly. The vorticity/stream function approach has been widely used for solution of sedimentation tank flows by, for example, Imam and McCorquodale (1983) and, more recently, Griborio (2004) and McCorquodale *et al.* (2004) whose '*2DC*' model is widely used in consulting engineering for computation of solids concentration profiles in activated sludge clarifiers.

2.3.2.2 SIMPLE – semi implicit pressure linked equations

A second (and widely used) approach for elimination of pressure for use with the primitive variables in the Navier-Stokes equations is the SIMPLE algorithm developed by Patankar (1980). In this approach a staggered grid is used in which calculations for velocity components, pressure and transported scalars are calculated on a finite volume grid as shown in Figures 2.3–2.5.

The solution algorithm proceeds as follows:

(1) Guess the pressure field
(2) Solve the momentum equations
(3) Solve the pressure correction equation
(4) Calculate the pressure by adding the pressure correction to the guessed pressure field
(5) Calculate velocities
(6) Solve for other properties (temperature, solids)
(7) Update the pressure field and iterate to convergence

Patankar's algorithm is widely used in commercial CFD software. It has also been used in custom CFD work by, for example, Zhou and McCorquodale (1992a). Other similar algorithms to SIMPLE, including SIMPLEC, SIMPLER, and PISO are discussed in studies by Patankar (1980) and Ferziger and Perić (2002).

2.3.3 Other topics

This introduction to classical approaches to CFD has been necessarily brief. For a more general overview reference should be made to standard texts, especially Crowe *et al.* (2011), Ferziger and Perić (2002) and Fletcher (1991, 1998). For discussion of the important volume-of-fluid method for free surface flows, start with Hirt and Nichols (1981).

2.4 GOOD MODELLING PRACTICE

This section is partly reprinted from *Water Science and Technology*, volume 73, issue number 5, pages 969–982, with permission from the copyright holders, IWA Publishing.

The overall CFD modelling approach and process is outlined in Figure 2.6. All projects start with identifying the objective, followed by a system analysis to determine the level of physics needed for the problem. In general, the least amount of computational effort should be used to reach the answer to the design or research question being sought from the model results. The modeller must determine if the problem can be defined in two- or three-dimensional space; if the problem is steady state (boundary conditions and solution are time independent) or dynamic; if the problem is single species, single phase with multiple species, or multiphase; if the problem is Newtonian or non-Newtonian; if there is heat transfer, and if additional transport, momentum transfer or mass transfer equations are needed for solutions of secondary species, phases and reactions. Once the modelling approach is determined, model development begins with defining the problem geometry including the key physical details influencing the flow and process being simulated. In many cases, the modeller has to balance complexity with computational efficiency. As an example, small details that do not significantly influence the bulk flow field and are in the order of, or significantly smaller than the computational mesh size are often neglected. This step is followed by meshing, application of boundary conditions, flow solution, and post processing. Each of these activities has minimum requirements to develop quality results, all of which will be discussed in more detail in the following sections.

2.4.1 Approach and key assumptions

Several key choices need to be made at the outset of any CFD modelling problem to guide the direction of the modelling process. These choices include (1) spatial dimensionality; two vs three dimensions; (2) time averaged vs time marching; (3) the number of phases and species; (4) material properties and their interactions when multiple species are present, and (5) if additional transport or transfer models are needed beyond what is already included in the CFD package.

In terms of spatial dimensionality, a three-dimensional (3D) problem may not substantially require more user effort to set up as opposed to a two-dimensional (2D) problem. However, it may have a significant impact on computational requirements due to the increased degrees of freedom. Hence,

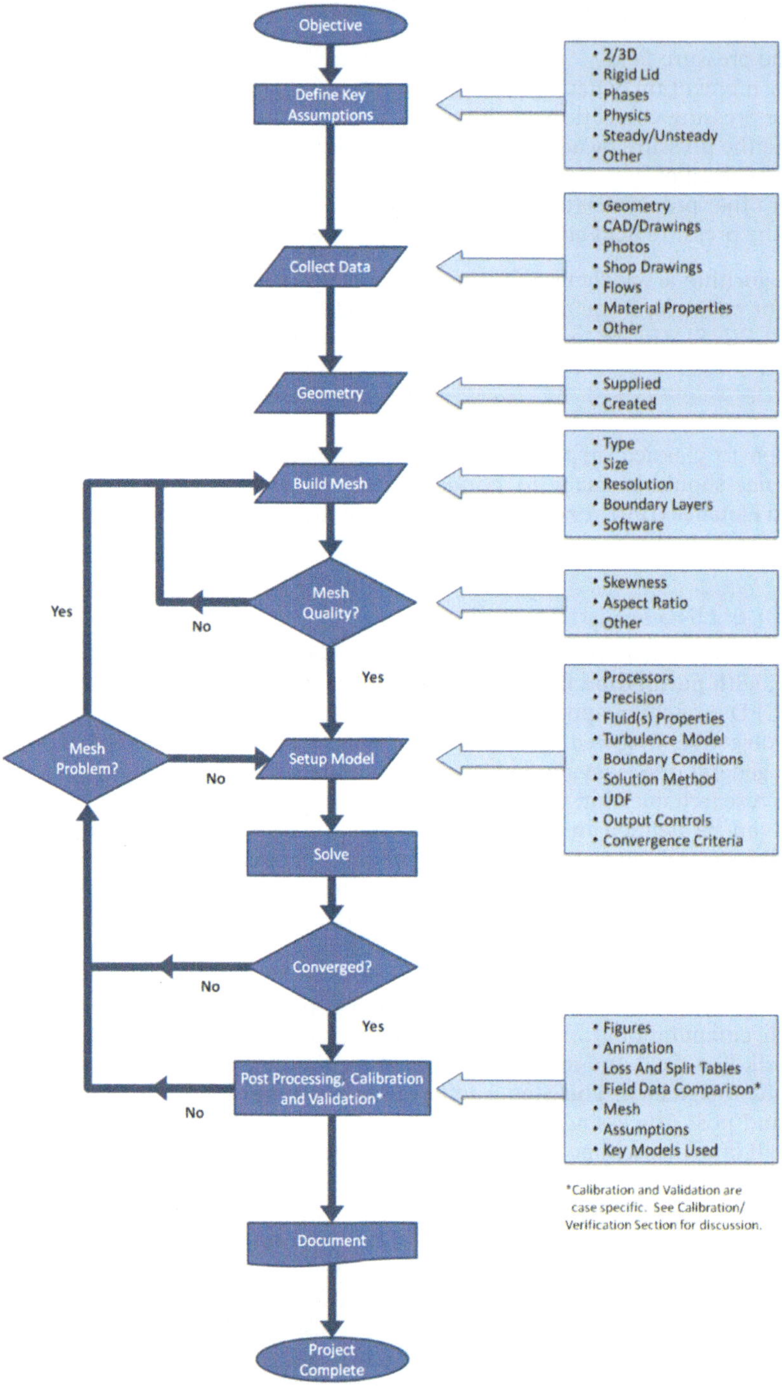

Figure 2.6 Complete flow of a CFD modelling process (Wicklein *et al.*, 2016). (Reprinted from *Water Science and Technology,* 73(5), 969–982, with permission from the copyright holders, IWA Publishing.)

one should check whether the assumptions made when modelling in 2D are appropriate for the overall model objective. If this is the case, a 2D model might be able to provide the answer to the objective in a fraction of the time and cost required to develop and obtain the results in 3D. It should be noted that there are some opportunities to infer 3D outcomes from 2D geometry, for example with areas of axial symmetry in circular tanks or if a tank section is representative of the complete rectangular tank width. In these cases, the inlet and outlet flow distribution should be uniform and two dimensional as well, as non-uniformity or rotating flow features will be lost in the solution with a 2D assumption.

Another consideration is time. For some applications, a steady-state solution of time-averaged conditions is appropriate, particularly when boundary conditions are not changing. For systems with a water surface that is planar and steady, a rigid lid approximation may be used where the water surface is fixed at a known level, with no shear stress (i.e. absence of wind) applied at the plane that would be an interface with air. However, if the flow is gradually or rapidly varied, a free surface model may be required. If boundary conditions or the solution are dynamic, a transient solution approach may be more appropriate. Although transient and free-surface models are more computationally expensive and require appropriate initial conditions, they may be the only approach in solving a characteristic system which cannot be solved implicitly as a steady state.

A third important consideration is the number of species and phases required. Phases are states of matter including gases, liquids, and solids, whereas species are different materials having a common state. Water resource recovery facility (WRRF) primary flows are typically liquid water that may include additional dissolved or fluid species, as well as solid and gas phases. Modelling of species can range from straightforward for passive, non-reacting materials, to very complex with reactions and leading to changes of phase state. Multiphase modelling is generally complex due to the phase interaction. Multispecies and multiphase modelling can be done with either a Lagrangian or Eulerian approach. A Lagrangian model individually tracks the species and/or phases as discrete elements within the bulk fluid, whereas a Eulerian approach treats the species and/or phases as a second (or more) fluid(s) in continuum (Crowe *et al.*, 2011). The material properties for the species and phases being modelled will dictate the approach required.

Material properties can be chosen independently for all species and phases based on known values for steady thermal and non-reacting conditions or set as an active function of other properties. Density and viscosity of the fluid is fundamentally dependent on factors such as solids concentration and temperature. Thermal effects can be included using an energy equation. The impact of solids on density is relatively straightforward to include in Newtonian flows using an additional transfer equation to couple the fluid and solid states in the solution. Recent research indicates that using a fixed density analysis of activated sludge tanks (i.e. assuming density of water) can dramatically over-predict velocities and the degree of mixing (Samstag *et al.*, 2012). The impact of solids concentration on viscosity and density has been studied (De Clercq, 2003; Markis *et al.*, 2014; Ratkovich *et al.*, 2013) and any analysis of sedimentation and mixing should account for these processes. The same may be said for the limit at which the liquid phase may be treated as a Newtonian fluid (Bridgeman, 2012). For solids concentrations above 2%, the fluid becomes non-Newtonian where the physics are more complex and must be accounted for. Stress yield models have been generally applied in the literature (Eshtiaghi *et al.*, 2013; Ratkovich *et al.*, 2013), but there is a lack of consensus, particularly at higher solids concentrations.

The Lagrangian particle tracking method (see Section 2.2.5.1) is most used in WRRF modelling for grit and diffuse secondary species and phases that do not have a significant influence on the bulk flow. Lagrangian models are also commonly used for tracking passive organisms' exposure to UV light intensity (Computational Fluid Dynamics Based Models for Assessing UV Reactor Design and Installation, 2011) when modelling UV disinfection. The Lagrangian approach can become cumbersome when modelling of a significantly large number of particles is required (Crowe *et al.*, 2011) or they have a significant influence on the bulk flow characteristics.

A Eulerian approach is generally used in WRRF modelling for multispecies flows and for higher concentrations of secondary phases and/or when there is strong coupling between the phases. The first Eulerian approach uses a transport model (see Section 2.2.4.1). The governing equations for fluid motion of the primary phase are solved, and the other phases and/or species follow the advective flow and are modelled as separate species or scalars with transport equations that predict local concentrations accounting for both advection and diffusion. Since most flows of importance in wastewater treatment are turbulent, diffusion is usually thought to be dominated by turbulence, rather than by molecular diffusion. When reactions are included (see Section 2.2.4.3), the transported species could be transformed to other phases of matter as solid flocs are formed, or gases are released. The interaction between momentum and mass conservation equations for the primary phase is complicated by making density and viscosity a function of scalar concentration, which could include temperature, salinity, or high biomass concentrations (see Section 2.2.4.4). The approach is computationally efficient as there are a limited number of additional equations to solve.

The second Eulerian approach uses a mixture model (Brennan, 2001; Ishii, 1977) in which the phases are treated as a single continuous phase that is a blend of the discrete phases (see Section 2.2.5.3). The probability of each phase's existence is based on the phase's volume fraction at a point in space. In this approach, a single differential equation is solved for continuity of the mixture. Also, a single differential equation is solved for the momentum of the mixture and tracks the motion of each phase relative to the centre of the mixture mass. This is handled by introducing the concept of a diffusion or drift velocity. A disadvantage of the standard mixture model approach for multiphase transport is that it relies upon input of characteristic sizes for calculation of phase coupling. This can be complicated in wastewater applications where a wide range of particle diameters are present and changing as they flow through a tank (Wicklein & Samstag, 2009) due to mechanisms of flocculation and deflocculation. Further, additional empirical functions must be introduced to evaluate hindered and compression settling regimes that can occur, for example, in secondary clarifiers.

A Euler-Euler approach (see Section 2.2.5.2) is a third option, where momentum and continuity equations are solved separately for each phase. The Euler-Euler approach has been thought to be preferable for liquid-gas flows (Amato & Wicks, 2009) since it can incorporate a range of typical bubble properties, for example, surface tension, agglomeration and break-up. However, it does come at significant computational cost due to the number of equations that must be solved.

A final choice is whether additional models apart from the CFD are required. If conversion of suspended and dissolved species is occurring (and of interest to hydraulics, rheology, or the process itself), conversion terms need to be added to their respective scalar equations (see Chapters 4 and 8). If settling is to be modelled, a settling velocity in the direction of gravity should be introduced (see Chapter 6). If kinetics are to be modelled, source and sink terms are to be included in the respective scalar equations. These terms can easily be recovered from a stoichiometric matrix if the model is available in that form. It will then consist of the vertical summation of the products of stoichiometric terms and process rate expressions. Further details on kinetics can be found in IWA STR No. 9 (Henze *et al.*, 2006).

2.4.2 CFD model development

Once key problem elements have been identified and choices made, the CFD model can be built. CFD models are used to solve a numerical form of the fundamental equations of fluid flow known as the Navier-Stokes equations, which are based on momentum balances. For most practical engineering problems being addressed today, the Reynolds-averaged form of the Navier-Stokes (RANS) equations are used (De Clercq, 2003; Ho *et al.*, 2011; Khan *et al.*, 2006; Liu & García, 2011; McCorquodale *et al.*, 2007). These equations do not form a closed set (ASCE, 1988), owing to the non-linearity of the original Navier-Stokes equations and their temporal averaging. Current CFD models solve additional equations representing the turbulence characteristics of a flow field, which is a key parameter in determining the nature of flow, eddy formation, circulation, flow separation, and flow interaction with surfaces.

In their general form, the RANS equations cannot be solved analytically. For solution of the RANS equations, it is necessary to discretize the flow domain into a computational mesh to define the actual locations where equations of flow will be solved. CFD models approximate the differential equations by numerical methods, that resolve the equations into a set of algebraic equations or time-dependent ordinary differential equations (Lomax *et al.*, 2001). These equations are solved to provide hydraulic information (e.g., velocity, water surface elevation, and pressure) at a finite number of discrete points within the flow domain. Most current CFD models use the finite volume method as this approach allows the use of an unstructured computational mesh; however finite difference and finite element methods have also been also used (see Section 2.33).

There is increased interest in more complete transient versions of Navier-Stokes equations to capture the dynamic nature of turbulence through use of detached eddy simulation (DES) and large eddy simulation (LES) as computational resources continue to improve and increasingly complex problems are being addressed. However, these more advanced models have not found their way into routine use to date and will not be discussed in any significant detail here (Geurts, 2004; Spalart, 2009).

2.4.2.1 Geometry

For CFD models, accurate geometric information is required, as this can have a significant effect on flow behaviour. Furthermore, the model objective often is to assess the impact of geometry on fluid hydraulics. If available, 2D or 3D electronic geometry files from most CAD packages can be imported into most CFD packages using standard electronic formats. If these are not available, it is advisable to spend ample time collecting these data accurately from sources such as 2D paper design, as-built drawings, photographs, and field measurements and verifications. If older design drawings are used, an effort should be made to validate their accuracy through field measurement. The model domain should be sufficiently large so that flow conditions at the model boundaries do not create artificial conditions within the region of interest. A sufficient amount of the upstream geometry must be included to achieve a fully developed flow field prior to the area of interest. Traditionally, the upstream boundary needs to be at least 5–10 times the characteristic length upstream from the area of interest, but should be farther upstream if there are bends, transitions, obstructions, or other changes in flow direction within this region. Figure 2.7 displays an example of a UV system that highlights the influence of providing sufficient upstream geometry to provide fully developed flow conditions at the area of interest. In this example, the area of interest is between two bends, where the reactor is located. Placing the upstream boundary at 2.5 vs 10 diameters upstream from the reactor shows subtle differences in the velocity field at a vertical plane through the lamps. The pressure drop may be similar, but the differences in velocity will have a direct impact on the calculated dose from the lamps.

2.4.2.2 Meshing

A computational mesh is developed within the flow geometry. There are many available approaches to mesh generation and computational cell type. It is important that the mesh is sufficiently refined in the region of interest, and of sufficient quality overall to provide a converging solution. In areas of the domain where solution variables are likely to have steep gradients, for example, a finer mesh is recommended to reduce propagation of numerical errors and promote convergence. The task of mesh generation is generally accomplished through the use of mesh-generating software that allows for definition of the model geometry, computational cell size, and grid density, and provides tools for grid quality analysis. Unstructured computational grids are the most common type, as they allow the greatest flexibility in defining the model domain and meshing properties.

The mesh comprises computational cells that can range from 2D triangles to 3D, 20-faced polyhedral elements. Numerous meshing algorithms have been developed to match the mesh better with the surface and capture flow details. An ideal computational cell has flow entering and exiting perpendicular to the faces, with cell sides having similar edge lengths and equilateral angles between edges. Compound curved surfaces and transitions between various complex surfaces lead to cell

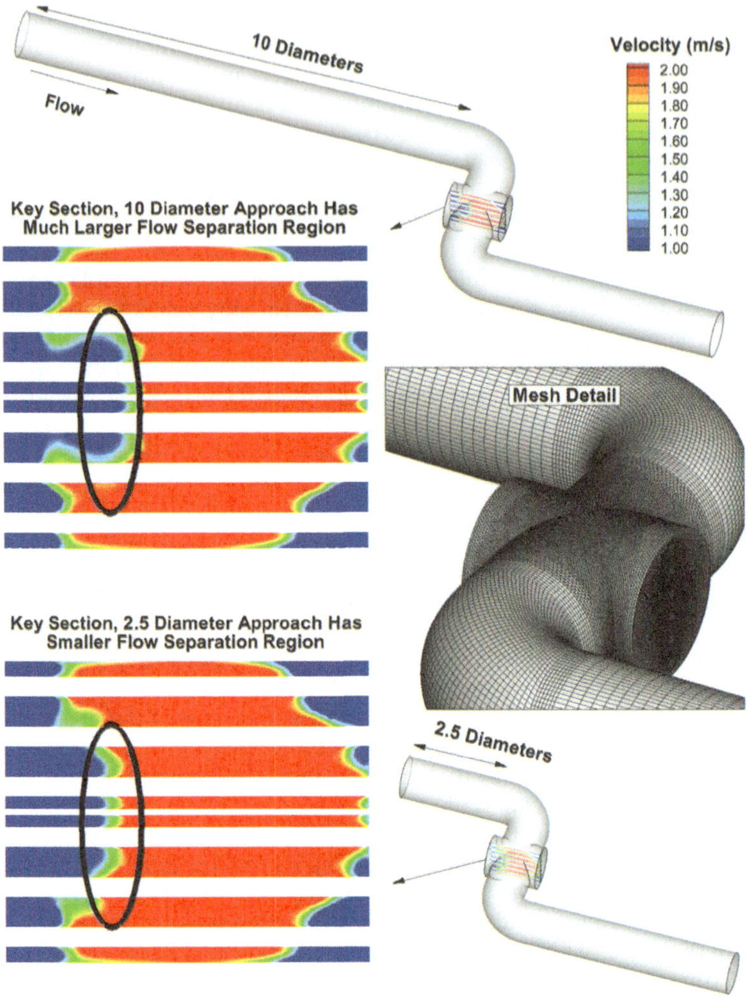

Figure 2.7 Sufficient upstream geometry is required for fully developed flow in the region of interest (Wicklein *et al.*, 2016). (Reprinted from *Water Science and Technology*, 73(5), 969–982, with permission from the copyright holders, IWA Publishing.)

deformations that reduce mesh quality and can lead to numerical diffusion errors. Key general mesh quality metrics include mesh skewness and length ratio, as shown in Figure 2.8. Skewness is a measure of how far the cell faces deviate from equilateral and should generally be below 0.9 for most solvers, particularly in the area of interest. The cell-length ratio is the comparison between the lengths of the sides and should generally be minimized such that all sides are similar in length. As the length ratio increases, solution quality will diminish, particularly if flow is not travelling parallel to the longest cell dimension. There are cases where higher length ratios may be tolerated, such as in the boundary layer region, where flow generally does travel parallel to the longer length sides.

Although modern computers have ever-increasing capabilities, a balance is always needed between overall mesh resolution and computational efficiency to generate a solution in a reasonable amount of time. Mesh independence analysis is generally recommended to verify the solution is not dependent upon the number of nodes used to solve a given problem, that is the model solution changes with

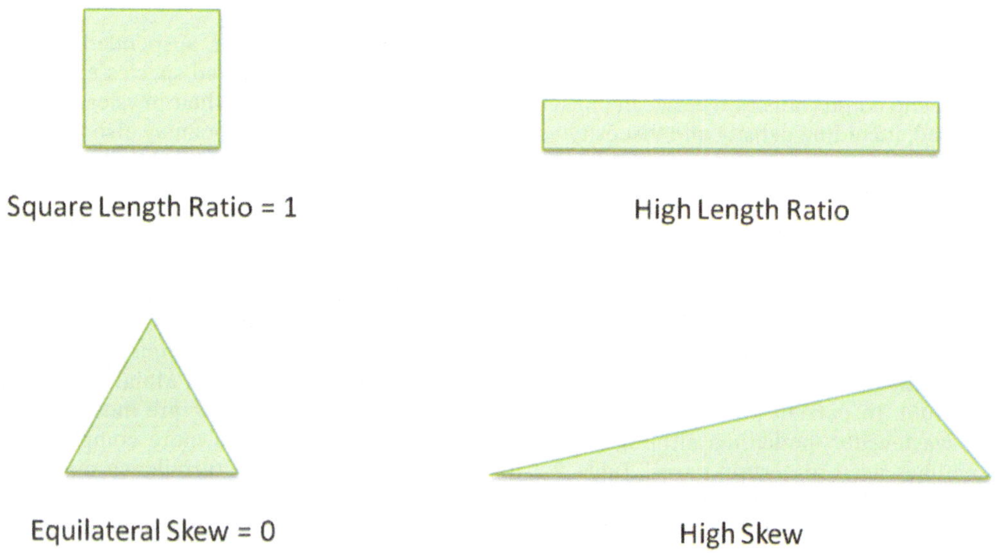

Figure 2.8 Cell quality examples (Wicklein *et al.*, 2016). (Reprinted from *Water Science and Technology*, 73(5), 969–982, with permission from the copyright holders, IWA Publishing.)

reducing the computational cell size (and increasing the overall cell count). A coarse grid may have a lower level of inherent stability (and higher degree of numerical diffusion), while a fine grid will be numerically more stable, but will require increased computational effort, both because of the increased degrees of freedom, and because of an increased model stiffness (different order of magnitude in model process rates). Increasing the fineness of the mesh also imposes the need for smaller time steps to maintain stability for transient simulations. Consideration must be given also to the minimum number of computational cells across flow regions to maintain accuracy of the numerical scheme. For example, five cells as a minimum are required for second order accurate solutions. The mesh should be finer in areas of interest and where sharp gradients may occur (e.g., inlet, outlets, free surface, near agitators…), and smoothly transition between finer and coarser mesh regions. For most wastewater applications the bulk flow region is of primary interest and this is where most wastewater process reactions typically occur. Therefore, the flow details within the boundary layer are usually not of primary interest and consequently should be sufficiently refined for use of wall functions to determine the transition from no-slip at a wall to the free stream flow.

It is recognized that modellers often do not run mesh independence studies for all problems due to resource constraints, particularly where previous or similar representative geometries have been analysed, or reasonable engineering judgment can be utilized to develop the model mesh. Thus, when reporting CFD results, the mesh type should always be noted as well as the overall cell count and a detailed image of the mesh should be provided. This is a prerequisite for judging the quality of the CFD exercise as mesh quality is critical for accurate solutions and subsequent interpretation of the results.

2.4.2.3 Solver setup

Once the mesh is developed, the model solver needs to be set up to calculate the CFD solution. In general, a double precision solver with second order discretization should be used for all wastewater problems. Many of the species modelled occur in low concentrations, implying the need for high numerical accuracy. Details of pressure-velocity coupling schemes and equation discretion can be found in Versteeg and Malalasekera (2007), Lomax *et al.* (2001) and Patankar (1980).

2.4.2.4 Multiphase models

Wastewater treatment (WWT) typically requires several phases: wastewater, suspended solids and gas phase in aerated zones. If biokinetic changes are to be modelled, dissolved species must also be introduced and could transition between phases. The model phases will need their physical properties to be applied, including density and viscosity as a minimum. Additional factors may also be required when supplemental energy (like mixing) is included in the solution. Additional details concerning multiphase models are mentioned in Section 2.2.5.

2.4.2.5 Turbulence models

Almost any CFD problem in wastewater flows will typically require a suitable approach to turbulence (see also Section 2.2.3, p. 3). A wide range of turbulence models are available and routinely used. The selected model should be appropriate for the problems in question. There are numerous references that discuss the details of the various models (Ferziger & Perić, 2002; Versteeg & Malalasekera, 2007; Wilcox, 2006). In current practice, two-equation models, such as k-ε and k-ω, are most commonly used for wastewater modelling, although both computationally simpler and more complex models are sometimes used in certain cases. Table 2.2 displays different types of turbulence models with advantages and disadvantages in their performance. Launder and Spalding (1972) provide an early overview reference on turbulence models. The basic turbulence models commonly used in hydraulic engineering have been reviewed by ASCE (1988) and Rodi (2000).

Table 2.2 Turbulence model selection, advantages and disadvantages.

Model	Advantages	Disadvantages
Two-Equation (standard k-ε, Renormalized Group (RNG) k-ε, Chen Kim k-ε and Realizable k-ε)	Simplest turbulence models that only require initial/boundary conditions Most widely tested models that have performed reasonably well for many types of flow conditions	Have been found to perform poorly for the following cases: • some unconfined flows • flows with curved boundary layers or swirling flows • rotating flows • fully developed flows in non-circular ducts
Second moment (Reynolds Stress Transport Model (RSTM)/ Algebraic Stress Transport Model (ASTM))	Most general of traditional turbulence models that only require initial/boundary conditions More accurate representation of Reynolds stresses that better characterize the turbulent flow properties in: • wall jets • flows with curved boundary layers or swirling flows • rotating flows • fully developed flows in non-circular ducts	Higher computing cost Has not been widely tested compared to the two-equation models Has been found to behave poorly for certain flow conditions: • axis-symmetric jets • unconfined recirculating flows
Detached Eddy Simulation (DES)	A hybrid model that blends a two-equation model near the walls with the LES model in the free stream More accurate for dynamic flow phenomenon than the simpler turbulence models	Has not been as widely tested as the two-equation models Higher computing cost
Large Eddy Simulation (LES)	Appropriate for complex free shear flows Used for turbulent boundary layers with high grid resolution at low Re	Has not been widely tested compared to the two-equation models Significantly higher computing cost

2.4.2.6 Boundary conditions

The flow field computed by the CFD model is a direct function of the flow conditions applied at the domain boundaries, known as boundary conditions. Typical fluid flow boundaries include inlet boundaries, outlet boundaries, pressure boundaries, symmetry boundaries, and wall boundaries. Inlet boundaries provide a velocity or mass flow rate in the three vector components into or out of the model domain, as well as turbulence characteristics. Pressure boundaries have constant pressure and turbulence characteristics, and flow can move in or out of the domain. Outlet boundaries only allow flow to travel out of the domain and may have no pressure constraints or turbulence constraints. Symmetry boundaries allow no vector component normal to the boundary. Wall boundaries are considered solid with no flow through the boundary. The wall boundary type can either be no slip (zero velocity) with a roughness component, or a slip wall with no roughness component. Typically, a wall function is used to approximate the transition from the zero velocity at the boundary through the boundary layer into the free stream. These wall functions model the effective drag from the roughness at the wall, without requiring the large number of computational elements required to resolve the flow field within the boundary layer.

For wall boundary conditions, the no-slip condition (i.e. velocities equal to zero at the wall) is applied to all solid surfaces. At very small distances from the solid wall, a viscous sub-layer exists and is followed by an intermediate layer and turbulent core. In the viscous sub-layer, the flow is influenced by viscous forces and does not depend on free stream turbulent parameters. The velocity in this sub-layer only depends on the distance normal to the wall, fluid density, viscosity, and the wall shear stress. Typically, the viscous sub-layer is too thin to discretize and is therefore not included as a significant region in turbulence models in WWT reactors. However, the intermediate region, which includes effects of both the viscous sub-layer and the turbulent core, is more significant in size and requires more care in predicting the velocity and turbulence.

The intermediate sub-layer is bridged by utilizing empirical wall functions to provide near-wall boundary conditions for the mean-flow and turbulence transport equations. The purpose of these empirical functions is to connect the wall shear stress to the dependent variables at the near-wall grid node. This grid node must lie outside this sub-layer and reside in the fully turbulent zone. There are two types of wall function provided in commercial CFD codes: equilibrium log-law wall functions and non-equilibrium log-law wall functions. Non-equilibrium log-law wall functions should be used when there is turbulent transport of heat, and species at a reattachment point.

Reactions, energy, biokinetics and multiphase flows require additional boundary condition information. The additional parameters are typically added at the flow and wall boundary conditions and could include the wall and fluid temperature, species concentrations, or phase information.

2.4.2.7 Steady vs dynamic

The selection of whether to perform a steady-state or dynamic (transient, time-dependent) CFD simulation is a key stage in the process. A steady-state simulation uses an iterative scheme to progress to convergence, but persistent oscillations in the convergence residuals may indicate dynamic flow (transient) characteristic and the model may need to be solved using a time-dependent approach. Dynamic simulations also take much longer to compute, so the initial choice of a steady simulation is generally the most practical option, unless dynamic conditions are part of the problem objective, such as when the input variables are simply changing (e.g., diurnal flow, a cascading flow, or flows with changing salinity, temperature or influent composition). A dynamic simulation approach can also be used to converge to the equivalent steady-state solution over time. Applying transient simulations can be very useful for multiphase applications, particularly biokinetic modelling when the interaction between multiple phases is complex.

2.4.3 Convergence

Numerical solution methods are typically solved iteratively, the convergence error being the difference between the calculated solution and exact solution (Wilcox, 2006). Unfortunately, the model will typically not reach the exact solution; therefore, metrics are needed to judge when the model is sufficiently converged. Commonly, the residual error, the sum of the differences between the currently iterated solution and the previously iterated solution for each cell and its nearest neighbours (or next-nearest neighbours in the case of second order schemes) is the first measure. This difference should reduce over time, while satisfying the main criteria that the mass flow should remain continuous. Finally, the solution at discrete points should be stable (Wilcox, 2006).

2.4.4 Calibration and validation

The core CFD model, if set up properly, should not require calibration to solve the fundamental fluid flow equations accurately for a single species, single phase flow with appropriate boundary conditions and an adequate mesh, as the parameters used are based on physics or known properties, and hence are not contestable. However, there are uncertainties in many typical wastewater applications that may require some level of calibration. Sub-models for scalar quantities (e.g., solids) often require empirical fits that are subject to many factors that cannot be isolated for a given problem without some local data. For example, settling velocities can be measured on site. Solids profiles can be measured under one condition and the model compared to the measured data. In some cases, modifications in the fit parameters for the empirical models can improve the prediction of local site performance of the model. In general, however, the focus of data-model comparison for CFD purposes is model validation rather than parameter estimation or calibration, as it is preferable to measure parameters for empirical model development through dedicated controlled experiments. The methods and extent of validation required for a proper CFD study are discussed extensively in Chapter 9. Particularly, the use of tracer tests and the concept of residence time distribution (RTD) are discussed in Chapter 10.

REFERENCES

Abdel-Gawad S. M. and McCorquodale J. A. (1984a). Hydrodynamic of circular primary clarifiers. *Canadian Journal of Civil Engineering*, **11**(2), 299–307, https://doi.org/10.1139/l84-043

Abdel-Gawad S. M. and McCorquodale J. A. (1984b). Strip integral method applied to settling tanks. *Journal of Hydraulic Engineering*, **110**(1), 1–17, https://doi.org/10.1061/(ASCE)0733-9429(1984)110:1(1)

Abdel-Gawad S. M. and McCorquodale J. A. (1985). Numerical simulation of rectangular settling tanks. *Journal of Hydraulic Research*, **23**(2), 85–100, https://doi.org/10.1080/00221688509499358

Amato T. and Wicks J. (2009). The practical application of computational fluid dynamics to dissolved air flotation, water treatment plant operation, design and development. *Journal of Water Supply: Research and Technology – AQUA*, **58**(1), 65–73, https://doi.org/10.2166/aqua.2009.003

Andersson B., Andersson R., Hakansson L., Mortensen M., Sudiyo R. and van Wachem B. (2011). Computational Fluid Dynamics for Engineers. Cambridge University Press, Cambridge.

Andrews J. F. (1971). Kinetic models of biological waste treatment processes. *Biotechnology Bioengineering Symposium*, **2**(5), 5–33.

Ansys I. (2011). ANSYS FLUENT Theory Guide. Southpointe, 275 Technology Drive, Canonsburg, PA 15317, USA.

ASCE, T.C. on T.M. in H.C. (1988). Turbulence modeling of surface water flow and transport: part I, II, III, IV and V. *Journal of Hydraulic Engineering*, **114**(9), 970–991, https://doi.org/10.1061/(ASCE)0733-9429(1988)114:9(970)

Batstone D. J., Keller J., Angelidaki I., Kalyuzhnyi S. V., Pavlostathis S. G., Rozzi A., Sanders W. T. M., Siegrist H. and Vavilin V. A. (2002). The IWA anaerobic digestion model No 1 (ADM1). *Water Science and Technology*, **45**, 65–73, https://doi.org/10.2166/wst.2002.0292

Brennan D. (2001). The Numerical Simulation of Two-Phase Flows in Settling Tanks. PhD Thesis, University of London.

Bridgeman J. (2012). Computational fluid dynamics modelling of sewage sludge mixing in an anaerobic digester. *Advances in Engineering Software*, CIVIL-COMP **44**(1), 54–62, https://doi.org/10.1016/j.advengsoft.2011.05.037

Bryant J. O. (1972). Continuous Time Simulation of the Conventional Activated Sludge Wastewater Renovation System. PhD Thesis, Clemson University, Clemson, South California.

Busby J. B. (1973). Dynamic Modelling and Control Strategies for the Activated Sludge Process. Clemson University, Clemson, South California.

Chorin A. J. (1973). Numerical study of slightly viscous flow. *Journal of Fluid Mechanics*, **57**(4), 785–796.

Computational Fluid Dynamics Based Models for Assessing UV Reactor Design and Installation (2011). Denver, CO, USA.

Crowe C. T., Schwarzkopf J. D., Sommerfeld M. and Tsuji Y. (2011). Multiphase Flows with Droplets and Particles, 2nd ed. Taylor & Francis, Boca Raton, FL, USA.

De Clercq B. (2003). Computational Fluid Dynamics Of Settling Tanks: Development Of Experiments And Rheological, Settling, And Scraper Submodels. PhD Thesis, Ghent University, Ghent, Belgium.

Eshtiaghi N., Markis F., Yap S. D., Baudez J.-C. and Slatter P. (2013). Rheological characterisation of municipal sludge: A review. *Water Research*, **47**(15), 5493–5510, https://doi.org/10.1016/j.watres.2013.07.001

Ferziger J. H. and Perić M. (2002). Computational Methods for Fluid Dynamics, 3rd ed. Springer, Berlin, Heidelberg.

Finlayson B. A. (1972). Existence of variational principles for the Navier-Stokes equation. *The Physics of Fluids*, **15**(6), 963–967.

Finlayson B. A. (1980). Nonlinear Analysis in Chemical Engineering. McGraw-Hill, New-York, NY, USA.

Fletcher C. A. J. (1991). Computational Techniques for Fluid Dynamics 2: Specific Techniques for Different Flow Categories. Springer, Berlin Heidelberg.

Fletcher C. A. J. (1998). Computational Techniques for Fluid Dynamics 1: Fundamental and General Techniques. Springer, Berlin Heidelberg.

Geurts B. (2004). Elements of Direct and Large-Eddy Simulation. R.T. Edwards, Inc., Philadelphia.

Griborio A. G. (2004). Secondary clarifier modeling: A multi-process approach. PhD Thesis, University of New Orleans.

Hangos K. M. and Cameron I. T. (2001). Process Modelling and Model Analysis, 1st ed. Academic Press, San Diego, CA, USA.

Harlow F. H. and Welch J. E. (1965). Numerical calculation of time-dependent viscous incompressible flow of fluid with free surface. *Physics of Fluids*, **8**(12), 2182–, https://doi.org/10.1063/1.1761178

Henze M., Gujer W., Mino T. and van Loosdrecht M. (2006). Activated Sludge Models ASM1, ASM2, ASM2d and ASM3. IWA Publishing, London, UK.

Hinze J. O. (1959). Turbulence. An introduction to its Mechanism and Theory. McGraw-Hill, New-York, NY, USA.

Hirt C. W. and Nichols B. D. (1981). Volume of fluid (VOF) method for the dynamics of free boundaries. *Journal of Computational Physics*, **39**(1), 201–225, https://doi.org/10.1016/0021-9991(81)90145-5

Ho C. K., Khalsa S. S., Wright H. B., Wicklein E. A. and Khalsa S. S. (2011). Computational Fluid Dynamics Based Models for Assessing UV Reactor Design and Installation. Water Research Foundation, Denver, CO, USA.

Hossain M. S. (1980). Mathematische Modellierung von turbulenten Auftriebsströmungen (Mathematical modelling of turbulent, increasing currents). PhD Thesis, University of Karlsruhe, Germany.

Imam E. and McCorquodale J. A. (1983). Simulation of flow in rectangular clarifiers. *Journal of Environmental Engineering*, **109**(3), 713–730, https://doi.org/10.1061/(ASCE)0733-9372(1983)109:3(713)

Ishii M. (1977). One-dimensional Drift-Flux Model and Constitutive Equations for Relative Motion Between Phases in various two-Phase Flow Regimes (No. ANL-77-47). Argonne National Lab., Ill. (USA).

Karpinska A. M. and Bridgeman J. (2016). CFD-aided modelling of activated sludge systems – A critical review. *Water Research*, **88**, 861–879, https://doi.org/10.1016/j.watres.2015.11.008

Khan L. A., Wicklein E. A. and Teixeira E. C. (2006). Validation of a three-dimensional computational fluid dynamics model of a contact tank. *Journal of Hydraulic Engineering*, **132**(7), 741–746, https://doi.org/10.1061/(ASCE)0733-9429(2006)132:7(741)

Kleine D. and Reddy B. D. (2005). Finite element analysis of flows in secondary settling tanks. *International Journal for Numerical Methods in Engineering*, **64**(7), 849–876, https://doi.org/10.1002/nme.1373

Larsen P. (1977). On the Hydraulics of Rectangular Settling Basins Experimental and Theoretical Studies. Department of Water Resources Engineering, Lund Institute of Technology, University of Lund, Lund, Sweden.

Launder B. E. and Spalding D. B. (1972). Lectures in Mathematical Models of Turbulence. Academic Press, New-York, NY, USA.

Launder B. E. and Spalding D. B. (1974). The numerical computation of turbulent flows. *Computer Methods in Applied Mechanics and Engineering*, **3**(2), 269–289.

Laurent J., Samstag R. W., Ducoste J. M., Griborio A., Nopens I., Batstone D. J., Wicks J. D., Saunders S. and Potier O. (2014). A protocol for the use of computational fluid dynamics as a supportive tool for wastewater treatment plant modelling. *Water Science and Technology*, **70**(10), 1575–1584, https://doi.org/10.2166/wst.2014.425

Liu X. and García M. H. (2011). Computational fluid dynamics modeling for the design of large primary settling tanks. *Journal of Hydraulic Engineering*, **137**(3), 343–355, https://doi.org/10.1061/(ASCE)HY.1943-7900.0000313

Lomax H., Pulliam T. H., Zingg D. W., Pulliam T. H. and Zingg D. W. (2001). Fundamentals of computational fluid dynamics (Vol. 246). Berlin: Springer.

Markis F., Baudez J. C., Parthasarathy R., Slatter P. and Eshtiaghi N. (2014). Rheological characterisation of primary and secondary sludge: impact of solids concentration. *Chemical Engineering Journal*, **253**, 526–537, https://doi.org/10.1016/j.cej.2014.05.085

McCorquodale J. A., La Motta E. J., Griborio A., Homes J. and Georgiou I. (2004). Development of software for modeling activated sludge clarifier systems. A Technology Transfer Report, Department of Civil and Environmental Engineering, University of New Orleans, LA, 70148.

McCorquodale J. A., Griborio A., Li J., Horneck H. and Biswas N. (2007). Modeling a retention treatment basin for CSO. *Journal of Environmental Engineering*, **133**(3), 263–270, https://doi.org/10.1061/(ASCE)0733-9372(2007)133:3(263)

Meister M. and Rauch W. (2016). Smoothed particle hydrodynamics – an innovative method for solving fluid dynamics problems in urban water management. In: Proceedings of the IWA World Water Congress, Brisbane, Australia: International Water Association.

Meister M., Winkler D., Rezavand M. and Rauch W. (2017). Integrating hydrodynamics and biokinetics in wastewater treatment modelling by using smoothed particle hydrodynamics. *Computers & Chemical Engineering*, **99**, 1–12, https://doi.org/10.1016/j.compchemeng.2016.12.020

Morchain J., Gabelle J. C. and Cockx A. (2014). A coupled population balance model and CFD approach for the simulation of mixing issues in lab-scale and industrial bioreactors. *AIChE Journal*, **60**(1), 27–40, https://doi.org/10.1002/aic.14238

Munk W. H. and Anderson E. R. (1948). Otes on a theory of the thermocline. *Journal of Marine Research*, **7**, 276–295.

Nopens I., Torfs E., Ducoste J., Vanrolleghem P. A. and Gernaey K. V. (2015). Population balance models: A useful complementary modelling framework for future WWTP modelling. *Water Science and Technology*, **71**(2), 159–167, https://doi.org/10.2166/wst.2014.500

Patankar S. V. (1980). Numerical Heat Transfer and Fluid Flow. Taylor & Francis, Boca Raton, FL, USA.

Peyret R. and Taylor T. D. (1983). Computational Methods for Fluid Flow. Springer, Berlin, Heidelberg.

Ratkovich N., Horn W., Helmus F. P., Rosenberger S., Naessens W., Nopens I. and Bentzen T. R. (2013). Activated sludge rheology: A critical review on data collection and modelling. *Water Research*, **47**(2), 463–482, https://doi.org/10.1016/j.watres.2012.11.021

Roache P. J. (1972). Computational Fluid Dynamics. Hermosa Publishers, Albuquerque.

Rodi W. (2000). Turbulence Models and Their Application in Hydraulics: A State-of-the-art Review, 3rd ed. CRC Press, London, UK.

Samstag R. W., Wicklein E. A., Reardon R. D., Leetch R. J., Parks R. M. and Groff C. D. (2012). Field and CFD analysis of Jet aeration and mixing. *Proceedings of the Water Environment Federation*, **2012**(12), 4113–4139, https://doi.org/10.2175/193864712811708301

Smith G. D. (1985). Numerical Solution of Partial Differential Equations: Finite Difference Methods. Oxford Clarendon Press, Oxford, U.K.

Spalart P. R. (2009). Detached-Eddy simulation. *Annual Review of Fluid Mechanics*, **41**(1), 181–202, https://doi.org/10.1146/annurev.fluid.010908.165130

Stenstrom M. K. (1975). A Dynamic Model and Computer Compatible Control Strategies for Wastewater Treatment Plants. PhD Thesis, Clemson University, Clemson, South California.

Tracy K. (1973). Mathematical Modeling of Unsteady-State Thickening. PhD Thesis.

Versteeg H. K. and Malalasekera W. (2007). An Introduction to Computational Fluid Dynamics: The Finite Volume Method, 2nd ed. Pearson Education Limited, London.

Vesilind P. A. (1968). Design of prototype thickeners from batch settling tests. *Water Sewage Works*, **115**(7), 302–307.

Water Environment Federation (2013). Wastewater Treatment Process Modeling, 2nd ed. WEF Manual Of Practice 31. Water Environment Federation, Alexandria, VA 22314-1304.

Wicklein E. A. and Samstag R. W. (2009). Comparing commercial and transport CFD models for secondary sedimentation. *Proceedings of the Water Environment Federation*, **2009**(10), 6066–6081, https://doi.org/10.2175/193864709793952765

Wicklein E., Batstone D. J., Ducoste J., Laurent J., Griborio A., Wicks J., Saunders S., Samstag R., Potier O. and Nopens I. (2016). Good modelling practice in applying computational fluid dynamics for WWTP modelling. *Water Science and Technology*, **73**(5), 969–982, https://doi.org/10.2166/wst.2015.565

Wilcox D. C. (2006). Turbulence Modeling for CFD, 3rd ed, DCW industries, Inc., La Canada, CA, USA.

Winkler D., Meister M., Rezavand M. and Rauch W. (2016). SPHASE – Smoothed Particle Hydrodynamics in Wastewater Treatment. World Environmental and Water Resources Congress 2016, Reston, VA, American Society of Civil Engineers, pp. 303–311.

Zhou S. and McCorquodale J. A. (1992a). Modeling of rectangular settling tanks. *Journal of Hydraulic Engineering*, **118**(10), 1391–1405, https://doi.org/10.1061/(ASCE)0733-9429(1992)118:10(1391)

Zhou S. and McCorquodale J. A. (1992b). Mathematical modelling of a circular clarifier. *Canadian Journal of Civil Engineering*, **19**(3), 365–374, https://doi.org/10.1139/l92-044

doi: 10.2166/9781780409030_0037

Chapter 3

Hydraulic analysis and headworks

Edward Wicklein

Carollo Engineers, Seattle, WA, USA; Email: ewicklein@carollo.com

3.1 INTRODUCTION

The most basic and commonly used computational fluid dynamics (CFD) application in water and wastewater treatment is hydraulic evaluation, as the flow field is the basis for all advanced process evaluations. In this context, CFD is used for complex hydraulic analysis that is difficult or impossible to solve with analytical methods or not fully represented by less complex models. The CFD model is able to capture the influence of flow distribution on losses, manage very complex geometries, and capture turbulent effects. Some treatment processes benefit from plug flow with uniform approach conditions, and some require a high level of mixing in a continuous stirred-tank reactor (CSTR) mode, the CFD model can solve the hydraulics of both and provide insights on further improvement.

The CFD model is used both for closed conduit flows with a wall boundary all around the fluid, as well as open channel applications where the water surface has a boundary with air. Open channel problems can be solved two ways, depending on the area being modeled and required results. The most common approach is use of a rigid lid approximation, which assumes a fixed water level for the solution and is used for areas where the water level is known and stable such as in a tank or sub-critical channel. A zero-shear boundary is used at the surface in these cases. Free surface modeling is used for gradually and rapidly varying flow problems where the surface level may not be explicitly known or becomes an important component of the overall solution. Typical closed conduit CFD modeling has been experimentally validated in Ducoste and Linden (2006) for UV piping systems. Khan *et al.* (2006) provide experimental validation of an open channel contact tank using a rigid lid approximation, and Li *et al.* (2004) validated a rigid lid model of pump intake. A free surface UV system was validated by Peng *et al.* (2012). Akoz *et al.* (2009) validated a sluice gate in 2D using a free surface method and Zeng *et al.* (2017) validated 3D free surface models for spillways and weirs.

3.2 HYDRAULIC ANALYSIS

3.2.1 Flow distribution

The CFD model can be used to study flow features within a system such as flow distribution, which contributes to different aspects of system performance. Figure 3.1, adapted from Wicklein *et al.* (2016) shows an example of how flow separation at an elbow changes the velocity profile downstream. Common applications where flow distribution is important are some types of flow measurement

Editor(s): Julien Laurent, Randal Samstag, Jim Wicks and Ingmar Nopens

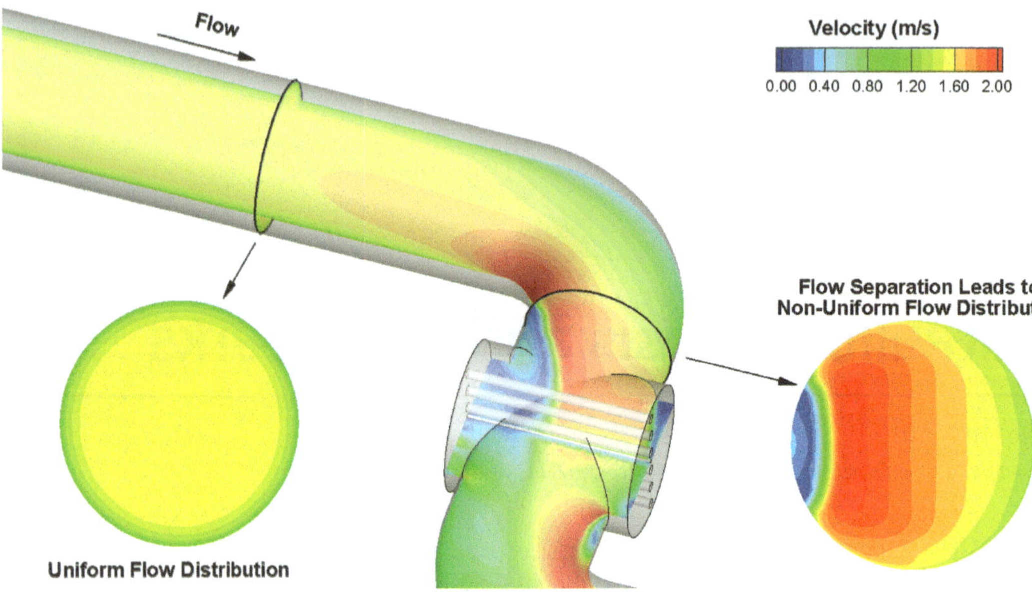

Figure 3.1 Velocity profile in closed conduit flow (source: Wicklein, 2016).

devices, and chemical addition or plug flow type reactors such as contact tanks or UV disinfection. Flow distribution is also a very important approach for pumps, which are discussed in more detail in Section 3.2.4.

A contact tank needs to provide sufficient contact time for all of the particles of water moving through the tank. Flow separation leads to a downstream non-uniform flow distribution which can lead to more rapid paths through tank than the design travel time, leading to decreased hydraulic efficiency. A comparison study of tank baffling by Hannoun and Boulos (1997) shows that changes in internal geometry of tank baffling altered the flow separation and could be used to optimize hydraulic efficiency. The hydraulic efficiency of baffling was further studied by Crozes *et al.* (1999) who found that computational domains with a relatively coarse grid could predict baffling factors generally within 15% to the conservative side of the curve. Within a few years others, such as Khan *et al.* (2006) and Wols *et al.* (2010), showed that CFD models were accurately calculating the velocity profiles within the contact tanks and tracer curves through tanks (see also Chapters 9 and 10).

UV disinfection systems rely on good flow distribution, as each flow path needs sufficient exposure to the UV light source. Ducoste and Linden (2006) evaluated hydraulics through closed-vessel UV reactors, comparing CFD model calculated values using many two-equation turbulence models with particle image velocimetry (PIV) measurements. Their work showed the flow field within the reactor is very complex, and no single two-equation turbulence model was superior, but all generally captured the flow field. Further work by Ho *et al.* (2011) using the *k*-epsilon realizable model generally found good agreement between head loss and dose delivery for several reactors, indicating the velocity field was accurately captured by the model.

3.2.2 Flow splitting analysis

Flow splitting is a critical unit operation that enables balanced flow to multiple units across a range of operations. Complex flow splitting problems are commonly evaluated using CFD analysis. The three-dimensional CFD model is an ideal tool for flow split analysis as the losses are often not easy to quantify due to the complexities of the energy losses that occur when dividing and combining flows. In addition, traditional loss coefficients assume uniform approach conditions to a respective fitting, which is often not the case in the compact footprint utilized for treatment facilities. The CFD model can accurately capture these losses through the turbulence models.

CFD models are used for both closed condition and open-channel flow splitting problems. Open-channel flow splitting can be divided into problems that can be resolved with a steady-state rigid lid approximation, and those that require simulation of the free surface. A rigid lid can be used for flows that remain subcritical through the area of interest where surface roughness and form losses govern the flow split, but the system may have a common downstream level control weir. Problems which have complex approach conditions near flow splitting weirs must be resolved using a free surface model as the approach direction and water level may vary near weirs such as would be observed in a confined weir splitter box.

3.2.2.1 Closed conduit

Manifolds are commonly used to split and divide flow between unit process, as well as to combine flows following the unit process. The dividing and combining flow losses can be complex. Flow passively traveling through the system will typically not be evenly balanced between the reactors. There are multiple flow paths available through the manifold system. Each flow path has a different head loss that is the sum of many different head loss components. These head loss sources can include different path lengths, dividing flows, combing flows, and approach conditions to fittings and other cross-sectional changes in the flow path. The head loss from the start of the manifold system to the end of the system is equal. The head loss is a function of velocity along the path, and therefore, the flow (velocity) will trend toward a balanced head loss.

As the overall head loss is increased across the manifold, the flow will trend toward a more balanced condition as the relative difference between the head loss along each path is diminished. The flow split can be actively balanced with valves by increasing head losses along selective flow paths.

Hassan *et al.* (2014) present results from CFD analysis on a tapered longitudinal manifold and a uniform longitudinal manifold. The two manifold configurations were tested at different inlet flows, and results suggested that the tapered manifold provided relatively equal flow distribution compared with the uniform longitudinal manifold. Tong *et al.* (2009) present results from an investigation on balancing flow through a manifold. CFD analysis of a wide range of manifold geometries was carried out. Results showed that optimal flow uniformity could be achieved by modifications simulated by the CFD analysis. These modifications were: expansion of the cross-sectional area, linear or non-linear tapering of the distribution manifold or varying the cross-sectional area of the outflow channels. Knatz (2005) indicated that the influent direction relative to the channel may impact the results.

3.2.2.2 Open channel

3.2.2.2.1 Fixed water surface (rigid lid)

For uniform flow and gradually varied flow conditions, a rigid lid is often appropriate for simulation as the water level is generally known and not unduly influencing the flow conditions within the water column. Wicklein *et al.* (2014) present results from CFD analysis on three baffle wall configurations on the inlet channel of parallel train UV channels (Figure 3.2). CFD results from the base condition showed poor flow split based on evaluation of flows in the different channels, which was improved by the addition of a baffle wall in the inlet channel. The best flow split results were obtained with a baffle acting as a tapering wall in the inlet channel (Knatz, 2005).

3.2.2.2.2 Free surface

Rapidly varied flow conditions or when free surface effects are important or unknown require use of a free surface model. Examples include weirs, spillways, high Froude number flows, and

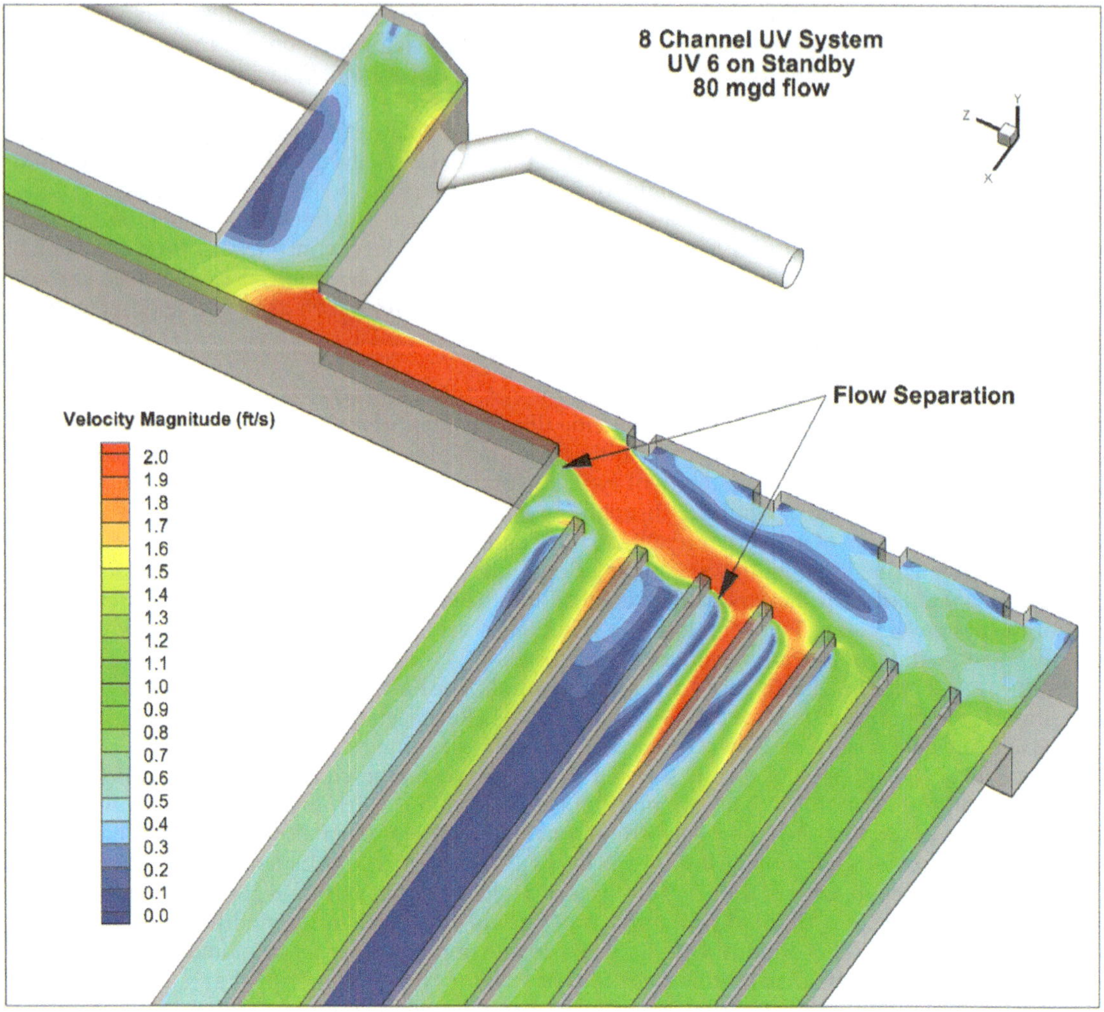

Figure 3.2 Horizontal section view of velocity showing inlet condition variability (source: Wicklein *et al.*, 2014).

complex open channel problems where the water level may not be well known, such as through UV equipment. CFD has been used extensively in engineering practice for analysis of flow splitting between process units and for estimation of complex head losses in hydraulic profiles. However, not very much of this practical work has been reported in the literature. One example of complex free surface systems with known flow splitting issues is the study by Wicklein *et al.* (2019b), as shown in Figure 3.3. CFD modeling was used to identify locations that contributed to poor flow split leading to reduced hydraulic capacity. Figure 3.4 presents a visualization of a 3D CFD flow splitter simulation developed at a recent workshop sponsored by the CFD working group (Marques, 2015).

A second example from unpublished work shows free surface water level variability approaching weirs (Figure 3.5). The flow in this example is oscillating between the three outlets (as shown in Figure 3.6), but on average shows a flow imbalance.

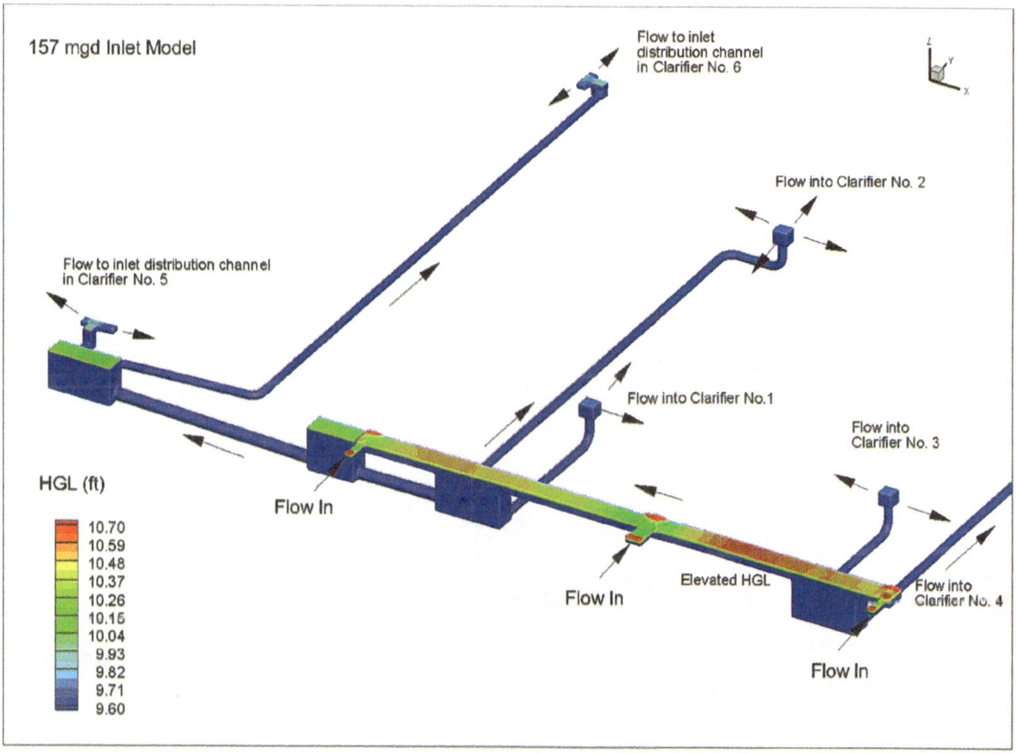

Figure 3.3 Effluent channel WSE at 157 mgd (HGL: hydraulic grade line, WSE: water surface elevation) (source: Wicklein *et al.*, 2019a).

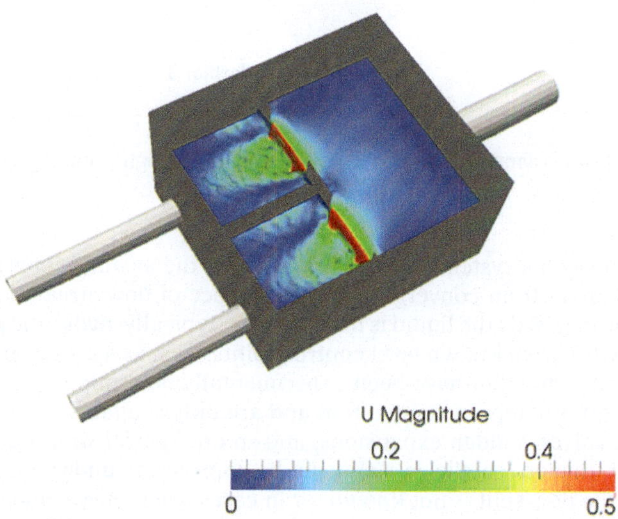

Figure 3.4 Surface velocity profile for a flow splitter from open source CFD software (source: Marques, 2015).

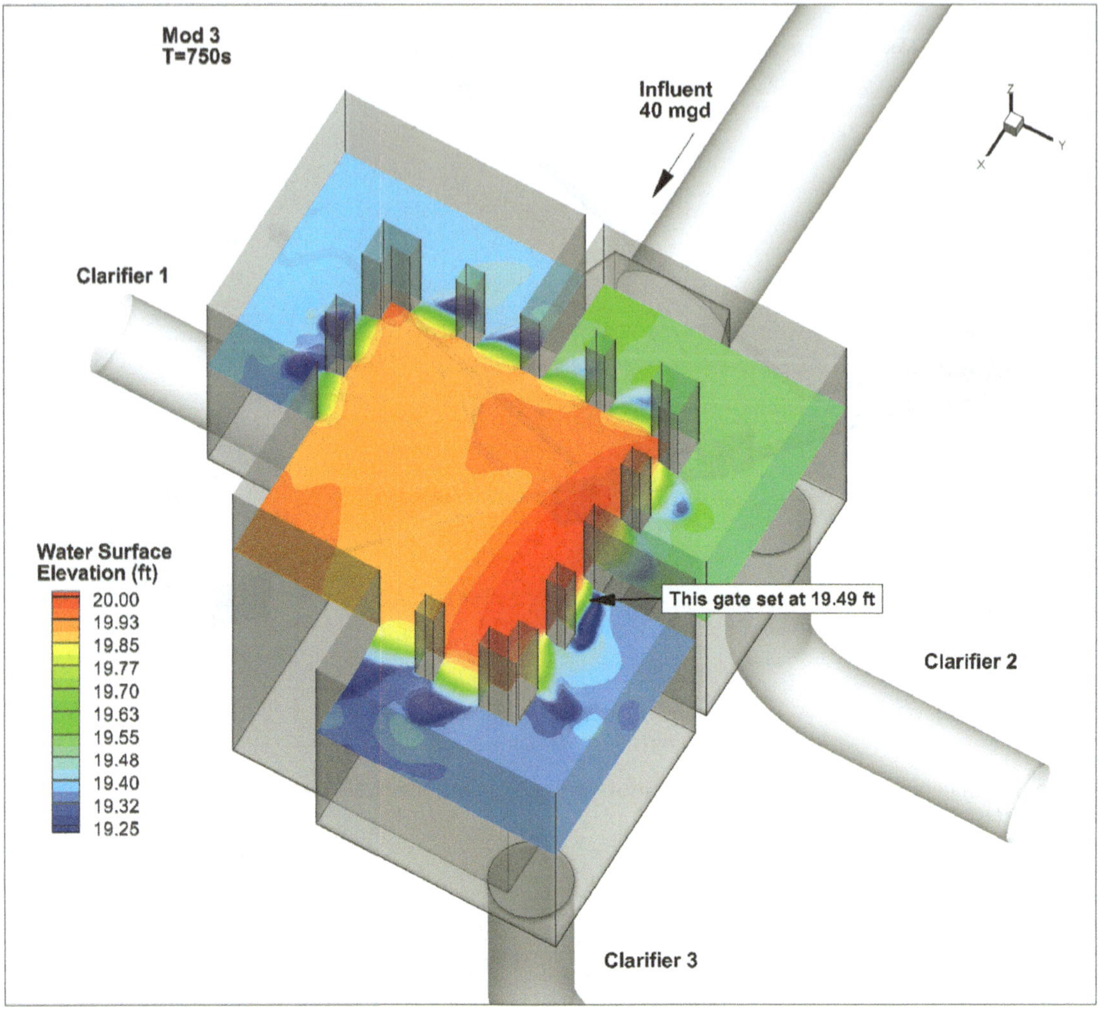

Figure 3.5 The CFD models the asymmetric design which contributes to high water level near some weirs.

3.2.3 Hydraulic profile

The hydraulic profile through a system is a balance between the available energy and losses that are a combination of form losses from convergence or divergence of flow streamlines, and losses due to surface roughness of the materials the liquid is flowing over. Typically, hydraulic profiles are calculated backward through a facility from known level control points, such as a weir or pump station set point, using standard loss coefficients that have been experimentally determined for common fittings. The coefficients assume a uniform approach condition and are only available for features that have been tested, such as elbows, valves, sudden expansions, and so on. Typical values are readily available in the literature. However, the loss coefficients may not be appropriate under non-uniform or turbulent flow conditions, when the flow split is not known, or in cases where new geometries such as complex equipment or custom transitions are being used. For these applications CFD models can be employed to determine the losses.

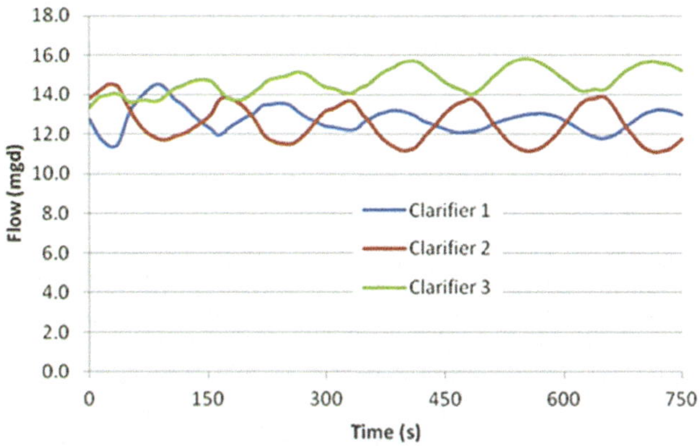

Figure 3.6 Monitoring the flow split shows oscillations, but also a general imbalance.

The hydraulic profile through an open channel UV system was investigated by Peng *et al.* (2012) using a free-surface CFD model. Their model showed close agreement in water surface through channel contractions and the UV equipment. Moreover, they were able to obtain more accurate UV performance results with the correctly calculated water level.

Figure 3.7 shows an example of a flow splitting system, where flow is diverted through two ports in a dividing structure, from an unpublished project. Conventional orifice loss coefficients predicted a driving head required for the desired flow spilt. However, flow approaches the orifices at an angle, leading to higher losses than predicted, requiring a greater upstream hydraulic grade line (HGL). A CFD model was used to both identify the problem and test modifications of the structure to reduce the head loss thereby lowering the upstream HGL.

3.2.4 Pump intakes

Pump impellers are generally designed for uniform stable flow. The wet well geometry coupled with flow conditions can create hydraulic conditions such as air entrainment, vortex formation, swirl, turbulence, and velocity imbalances that lead to pump performance problems, as identified by the Hydraulic Institute (ANSI/HI 9.8, 2018). Further, wastewater influent pump stations may have large

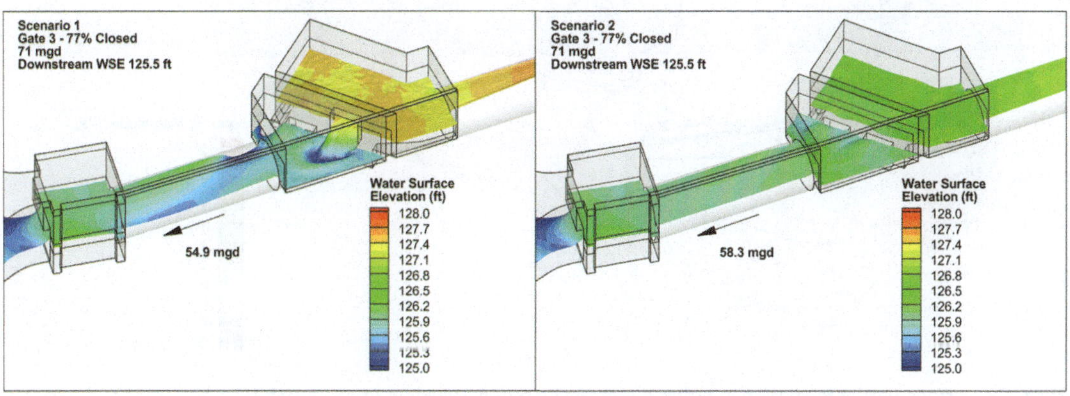

Figure 3.7 Water surface profile through complex structures (WSE: water surface elevation).

and highly variable solids loads that need to be moved through the system in addition to the liquid flows. The hydraulics of large pump stations, greater than 144 mgd, are frequently evaluated with scale physical modeling during design to create more robust stations following HI 9.8 guidelines. Most water and wastewater pump stations are much smaller than 144 mgd, the recommended threshold for scale physical modeling. Several standard wet well designs have emerged that generally work well, but assumptions on approach conditions are often built into the layouts. CFD has been extensively used to evaluate typical water and wastewater pump stations.

Early CFD modeling of pump hydraulics was evaluated and compared with laboratory tests by Rajendran *et al.* (1999), who found that the CFD model could reasonably predict the formation and location of experimentally measured vortices. Work by Okamura *et al.* (2007) demonstrated that the CFD model is able to accurately calculate the velocity distribution at the pump bell entrance, comparing the CFD results with experimental PIV data. Work by Lin and Take (2017) comparing flow details from CFD models of complex pump stations with scale model data show the CFD accurately captures features such as velocity distribution and swirl. Beyond the laboratory, CFD modeling is a proven tool for analysis of pump intake hydraulics (Wicklein *et al.*, 2006, 2019b). The CFD models are able to identify potently problematic intake conditions and were used to develop remediation measures leading to pumping facilities that operate well. Figures 3.8 and 3.9 (Wicklein *et al.*, 2006) show the CFD model is able to capture key hydraulic details and improve the pump wet well design. In this case there were surface vortices present (Figure 3.8) and flow separation in suction piping (left image of Figure 3.9). The wet well design was improved by making it deeper thereby reducing risk of surface vortex formation and allowing use of long radius elbows that reduced flow separation and improved pump intake conditions as shown in the right image of Figure 3.9.

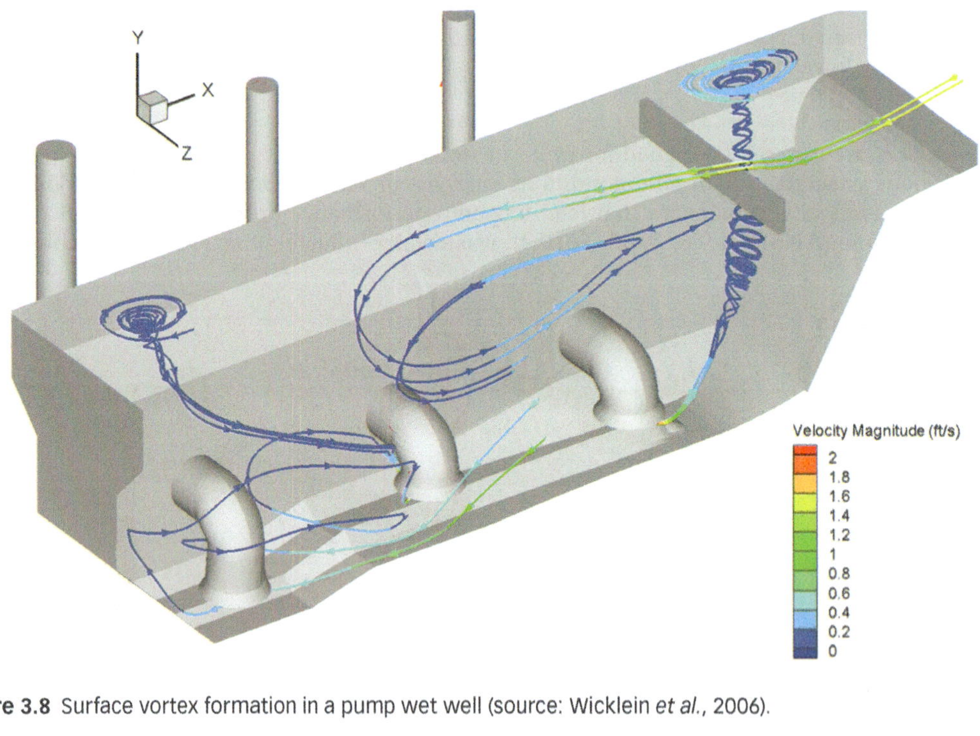

Figure 3.8 Surface vortex formation in a pump wet well (source: Wicklein *et al.*, 2006).

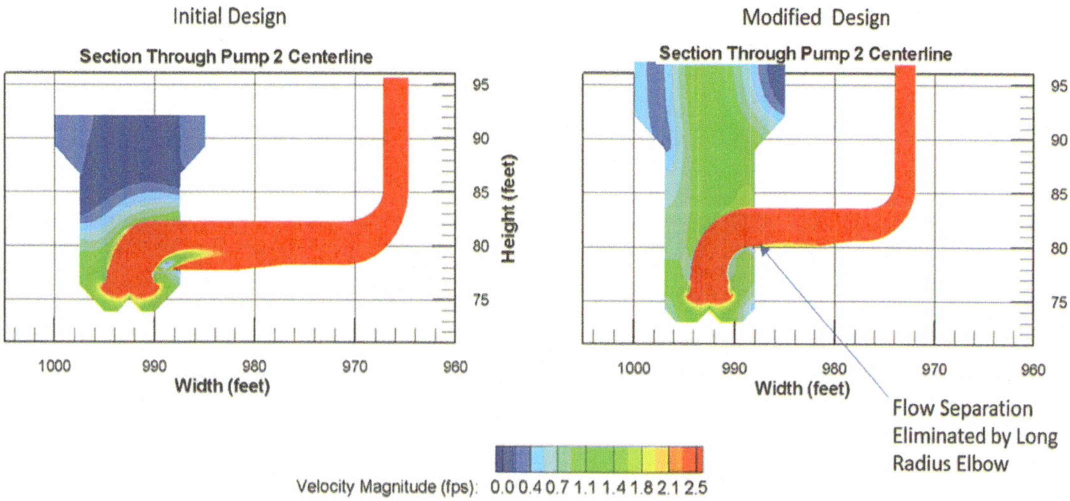

Figure 3.9 Section views of velocity through pump suction piping before and after modifying the design (source: Wicklein *et al.*, 2006).

3.2.5 Outfalls

Treated water is often returned to the environment via an outfall. The purpose of the outfall is to rapidly mix effluent with the receiving water stream to minimize any environmental impacts due to differences in properties or constituents between the two water sources. Outfalls have been modeled with CFD. Modeling by Kheirkhah Gildeh *et al.* (2015) showed that CFD models are accurately able to simulate environmental mixing hydraulics of outfall diffusers.

3.3 HEADWORKS

Headworks of most wastewater treatment facilities include equipment to physically remove larger materials from the raw influent flow. Screens are used to capture larger debris that is often neutrally buoyant, such as larger plastic pieces and fiber type materials. The hydraulic conditions such as flow splitting and flow distribution are both applicable within headworks where parallel physical treatment components such as screens and grit removal systems are present (Figure 3.10).

3.3.1 Screening

Screens remove larger, often neutrally or positively buoyant solids from the wastewater stream. The screens can be modelled simply as a porous loss. To better capture the flow distribution, straightening effects as well as losses in the system, the approach taken is more often to model the bars.

3.3.2 Grit removal

Grit chambers use gravity or centrifugal forced sedimentation to separate large, dense particles from raw wastewater in the plant headworks. While the principles are relatively simple, the multiphase nature, and multiple mechanisms can make analysis complex.

Analytical evaluation of grit removal goes back to the work of Camp (1945) who developed a rational approach for sizing of grit removal channels based on ideal settling behavior. Much of the

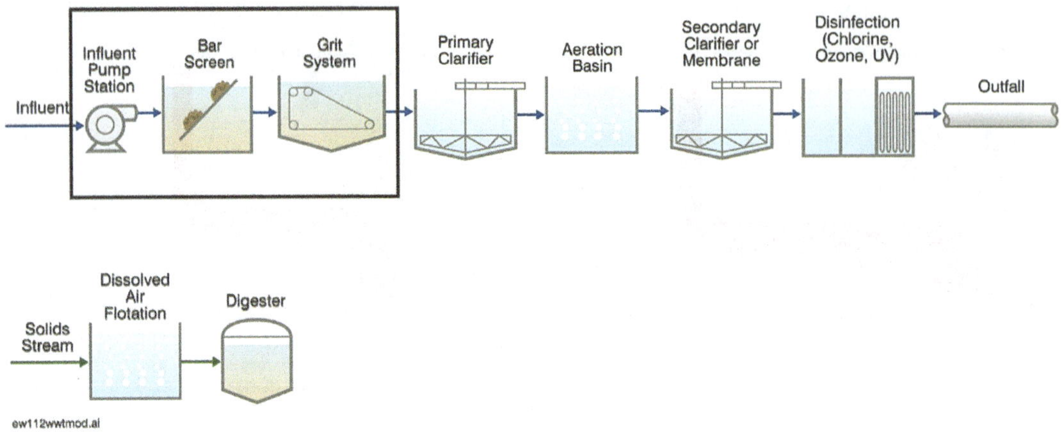

Figure 3.10 Placement of the anaerobic digestion in a wastewater resources recovery facility.

current practice, however, relies on manufacturer's recommendations without an analytical basis. CFD provides an alternative to reliance on manufacturer's claims, particularly where the hydraulic configuration mixes gravity and centrifugal sedimentation, such as in a vortex separator.

McNamara *et al.* (2012) present a comprehensive evaluation of three different types of grit removal tank geometries: forced vortex, detritor and lamella. Grit particles of nine different diameters were converted to a continuous distribution equation and modelled with specific gravities of 2.65 based on silica sand and 1.5 based on lab samples from field installations, and a sphericity ratio of 0.65. Multiphase (water and air) simulations were carried out to steady state prior to injection of grit particles into the inlet stream.

Major process factors for consideration in the design and operation of grit removal tanks mentioned in the literature include:

- Type
- Shape
  - Gross shape: rectangular, square or circular
  - Depth to diameter or length ratio
  - Length to width ratio
- Inlet configuration
- Outlet configuration
- Grit and scum withdrawal systems
- Solids settleability and flocculation
- Aeration
- Hydraulic regime
- Significance of biological activity
- Rheology

3.3.2.1 Case study

A typical coarse screen and intake pump station was presented by Wicklein *et al.* (2019a). The facility shown in Figure 3.11 included a sharp 90-degree bend in the sewer entering the headworks, parallel coarse screens, and parallel self-cleaning wet wells (only one wet well included in the model). A range of flows were simulated, the velocity was relatively uniform approaching the screens as shown

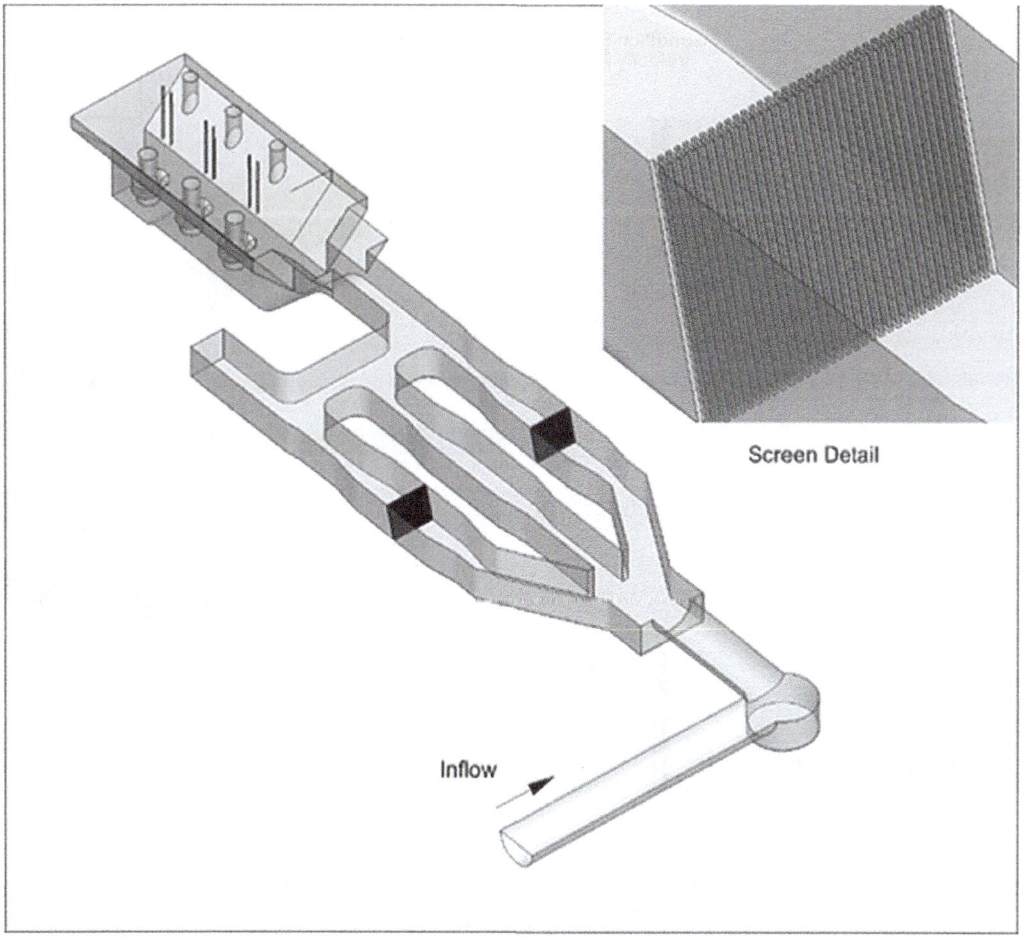

Figure 3.11 Headworks model with screens and pumps (source: Wicklein *et al.*, 2019b).

in Figure 3.12. The maximum flow split found between the screens was 7% difference. The model was also used to simulate the self-cleaning pump down of the wet well. Figure 3.13 shows the free surface results near the end of the pump down. Flow accelerates down to the ogee inlet, creating high velocity supercritical flow conditions that scour solids and grit from the floor, carrying them to the downstream pump and later grit removal process. A hydraulic jump forms that entrains any floatable material and maintains sufficient submergence on the downstream pumps to pump out all the solid materials.

3.4 RESEARCH NEEDS

Research work has focused on general flow distribution, as fluid flow manifolds are ubiquitous in engineering practice, from fuel cells to chemical reactors. However, there is very little peer-reviewed research that is specific to flow splitting in water and wastewater treatment processes. A key limitation

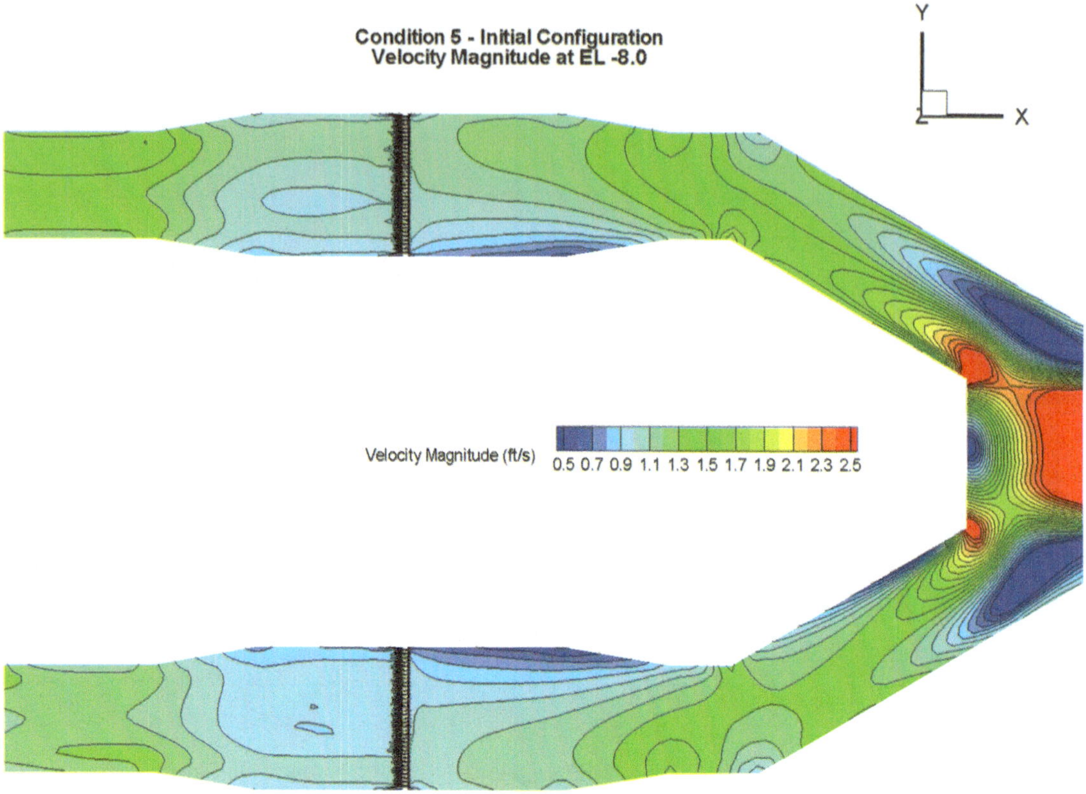

Figure 3.12 Approach velocity distribution to the screens (source: Wicklein *et al.*, 2019b).

that has not been addressed in the literature is the potential impact of solid and gas phases. Solids may impact splitters by differential movement of particulates within the splitter, which may increase viscosity or density and hence induce differential flow (see also Chapters 4, 6 and 8). The gas phase, particularly in aeration or used in splitter mixing systems intended to keep solids in suspension, can have a strong impact on fluid density and momentum (where momentum is imparted from an aerator, e.g.). This is critically important where a splitter has a particularly turbulent inlet cell, or where splitters are used to distribute flows from an aerobic unit to multiple clarifiers, particularly where centrifugal or laterally directed aeration is used.

Recent work on grit removal has included many desirable elements of good CFD analysis: efforts at calibration or verification, 3D approach and multiphase analysis. The best work appears to have included analysis of solids transport and discrete settling. It is not so clear that underflow concentrations have been monitored, however. It has generally been assumed that grit settlement is not density-coupled. A particular opportunity is the use of approaches from other unit operations such as population balances (Nopens, 2005; Nopens *et al.*, 2002) to describe performance under dynamic conditions. While sedimentation in the primary unit is generally unhindered, there may also be links to other sedimentation approaches in primary and secondary units, particularly hindered settling in solids collection zones.

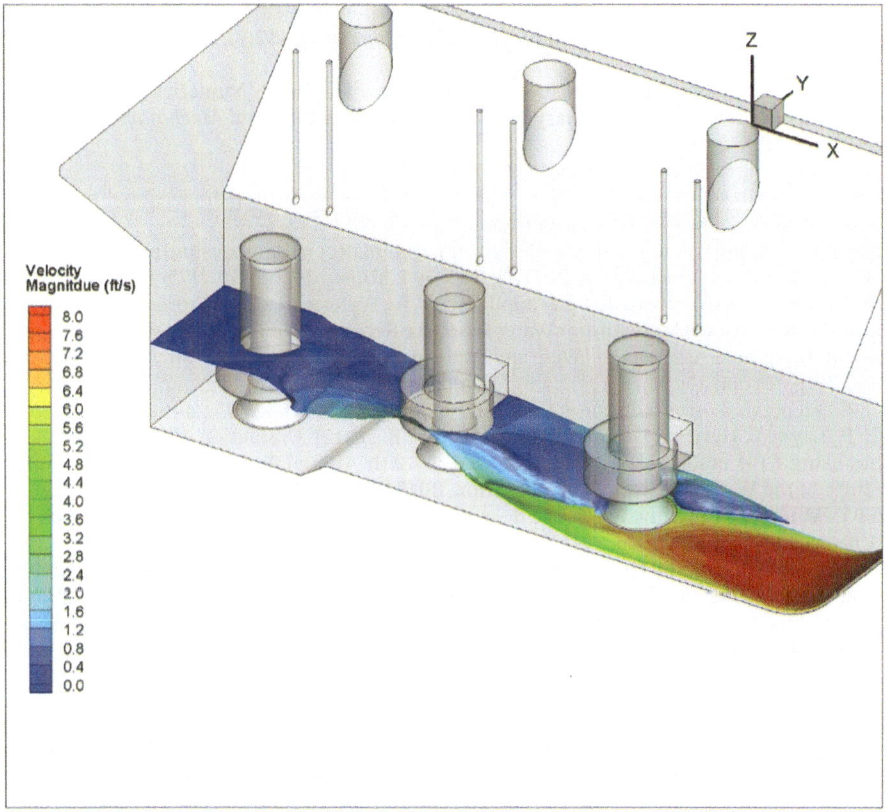

Figure 3.13 Water surface during a self-cleaning pump down of a headworks pump station (source: Wicklein *et al.*, 2019b).

REFERENCES

Akoz M. S., Kirkgoz M. S. and Oner A. A. (2009). Experimental and numerical modeling of a sluice gate flow. *Journal of Hydraulic Research*, **47**(2), 167–176, https://doi.org/10.3826/jhr.2009.3349

ANSI/HI 9.8. (2018). American National Standard for Pump Intake Design. American National Standards Institute, Washington D.C., USA.

Camp T. R. (1945). Sedimentation and the design of settling tanks. *Transactions, ASCE*, **111**(1), 895–936, https://doi.org/10.1061/TACEAT.0005912

Crozes G. F., Hagstrom J. P., Clark M. M., Ducoste J. and Burns C. (1999). Improving Clearwell Design for CT Compliance. AWWA Research Foundation, Denver, CO, USA.

Ducoste J. J. and Linden K. J. (2006). Hydrodynamic Characterization of UV Reactors. American Water Works Association Research Foundation, Denver, CO, USA.

Hannoun I. A. and Boulos P. F. (1997). Optimizing distribution storage water quality: a hydrodynamic approach. *Applied Mathematical Modelling*, **21**(8), 495–502, https://doi.org/10.1016/S0307-904X(97)00043-7

Hassan J. M., Mohamed T. A., Mohammed W. S. and Alawee W. H. (2014). Modeling the uniformity of manifold with various configurations. *Journal of Fluids*, **2014**, 1 8, https://doi.org/10.1155/2014/325259

Ho C. K., Khalsa S. S., Wright H. B., Wicklein E. A. and Khalsa S. S. (2011). Computational Fluid Dynamics Based Models for Assessing UV Reactor Design and Installation, Water Research Foundation. Denver, CO, USA.

Khan L. A., Wicklein E. A. and Teixeira E. C. (2006). Validation of a three-dimensional computational fluid dynamics model of a contact tank. *Journal of Hydraulic Engineering*, **132**(7), 741–746, https://doi.org/10.1061/(ASCE)0733-9429(2006)132:7(741)

Kheirkhah Gildeh H., Mohammadian A., Nistor I. and Qiblawey H. (2015). Numerical modeling of 30° and 45° inclined dense turbulent jets in stationary ambient. *Environmental Fluid Mechanics*, **15**, 537–562, https://doi.org/10.1007/s10652-014-9372-1

Knatz C. (2005). Modeling with CFD. *Public Works*, **136**(3), 50–55.

Li S., Lai Y., Weber L., Silva J. M. and Patel V. C. (2004). Validation of a three-dimensional numerical model for water-pump intakes. *Journal of Hydraulic Research*, **42**(3), 282–292.

Lin F. and Take J. (2017). Use of CFD modeling for designing permanent canal closures and pump stations. *Proceedings of the Water Environment Federation*, **2017**(2), 368–394, https://doi.org/10.2175/193864717821494961

Marques N. (2015). OpenFOAM® in wastewater applications. In: Watermatex 2015 Workshop: Computational Fluid Dynamics – Hands-on CFD Modeling for Water Resource Recovery Facilities. 9th IWA Symposium on Systems Analysis and Integrated Assessment (Watermatex) from 14 to 17th June 2015. Surfers Paradise, Queensland, Australia. https://rsamstag.wordpress.com/downloads-2/iwa-cfd-working-group-documents/?frame-nonce=729478071b&preview=true&iframe=true&calypso_token=65d8e63c-675c-4480-9bc7-81586328dbd4

McNamara B. F., Layne J., Hyre M., Kinnear D. J. and Bott C. B. (2012). Evaluation of three full-scale grit removal processes using CFD modeling. In: WEFTEC 2012 – 85th Annual Technical Exhibition and Conference. Proceedings of the Water Environment Federation, **2012**(9), 6008–6030.

Nopens I. (2005). Modelling the activated sludge flocculation process: a population balance approach. PhD thesis, Ghent University, Ghent, Belgium.

Nopens I., Biggs C. A., De Clercq B., Govoreanu R., Wilén B. M., Lant P. and Vanrolleghem P. A. (2002). Modelling the activated sludge flocculation process combining laser light diffraction particle sizing and population balance modelling (PBM). *Water Science and Technology*, **45**(6), 41–49, https://doi.org/10.2166/wst.2002.0092

Okamura T., Kamemoto K. and Matsui J. (2007). Cfd Prediction and Model Experiment on Suction Vortices in Pump Sump. The 9th Asian Conference on Fluid Machinery, Jeju, Korea.

Peng B., Santoro D., Sasges M., Petri B. and Olson D. (2012). Free surface effects and UV disinfection in open channel reactors. International Water Association Conference, México City, Mexico.

Rajendran V. P., Constantinescu S. G. and Patel V. C. (1999). Experimental validation of numerical model of flow in pump-intake bays. *Journal of Hydraulic Engineering*, **125**(11), 1119–1125, https://doi.org/10.1061/(ASCE)0733-9429(1999)125:11(1119)

Tong J. C. K., Sparrow E. M. and Abraham J. P. (2009). Geometric strategies for attainment of identical outflows through all of the exit ports of a distribution manifold in a manifold system. *Applied Thermal Engineering*, **29**(17–18), 3552–3560, https://doi.org/10.1016/j.applthermaleng.2009.06.010

Wicklein E. A. (2016). Aspects of modeling UV systems. WEFTEC 2016 Workshop – CFD for WRRF, WEF, New Orleans.

Wicklein E., Sweeney C., Senon C., Hattersley D., Schultz B. and Naef R. (2006). Computation fluid dynamic modeling of a proposed influent pump station. *Proceedings of the Water Environment Federation*, **2006**(5), 7094–7114, https://doi.org/10.2175/193864706783761356

Wicklein E., Hadler K. and Summers A. (2014). Using CFD modeling to balance flow for a new 80-mgd uv disinfection system retrofit into an existing chlorine contact Basin. *Proceedings of the Water Environment Federation*, **2014**(18), 7199–7213, https://doi.org/10.2175/193864714815942936

Wicklein E., Batstone D. J., Ducoste J., Laurent J., Griborio A., Wicks J., Saunders S., Samstag R., Potier O. and Nopens I. (2016). Good modelling practice in applying computational fluid dynamics for WWTP modelling. *Water Science and Technology*, **73**(5), 969–981, https://doi.org/10.2166/wst.2015.565

Wicklein E., Stone E. and Kreidler D. (2019a). Getting more out of your treatment system with improved flow splitting. Florida Water Resources Conference, Tampa, FL.

Wicklein E. A., Wickstrom J. and Huff D. (2019b). Minimizing facility size while improving screening and self cleaning pump stations with CFD modeling. *Proceedings of the Water Environment Federation*, **2019**(7), 5085–5093

Wols B. A., Hofman J. A. M. H., Uijttewaal W. S. J., Rietveld L. C. and van Dijk J. C. (2010). Evaluation of different disinfection calculation methods using CFD. *Environmental Modelling & Software*, **25**(4), 573–582, https://doi.org/10.1016/j.envsoft.2009.09.007

Zeng J., Zhang L., Ansar M., Damisse E. and González-Castro J. A. (2017). Applications of computational fluid dynamics to flow ratings at prototype spillways and weirs. II: framework for planning, data assessment, and flow rating. *Journal of Irrigation and Drainage Engineering*, **143**(1), 04016073, https://doi.org/10.1061/(ASCE)IR.1943-4774.0001113

doi: 10.2166/9781780409030_0051

Chapter 4
Suspended growth tanks

Anna M. Karpinska Portela[1], Usman Rehman[2] and Jim Wicks[3]

[1]Southern Water, Brighton, UK; Email: anna.karpinska-portela@southernwater.co.uk
[2]AM-TEAM, Gent, Belgium; Email: Usman.Rehman@AM-TEAM.com
[3]The Fluid Group, Oxford, UK; Email: jim.wicks@thefluidgroup.com

4.1 INTRODUCTION

Since its first full scale introduction into the UK over a hundred years ago (Ardern & Lockett, 1914), the activated sludge (AS) process has become the most common suspended growth method of secondary treatment (Figure 4.1), in use at thousands of wastewater resource recovery facilities (WRRFs) worldwide. By designing a sequence of dedicated (un)aerated basins to carefully promote the growth of biological floc and combining these with secondary clarifiers to recirculate biomass and separate out treated water, a rich tapestry of carbonaceous, nitrifying and phosphorous removal systems can be created which meet a broad range of effluent compliance targets.

Aeration basins are typically compartmentalized, have large sludge and nitrate recycle flows and are critically dependent on good hydraulics, high levels of inter-phase contact time and careful biomass management. The treatment performance relies on an interdependence of a variety of complex power systems: air blowers, mixers and pumps, with aeration by far the most energy-intensive aspect of the process, typically accounting for 75% of the total energy consumption at an activated sludge facility.

Accordingly, there are numerous potential benefits offered by computational fluid dynamics (CFD). From an extended insight into the hydrodynamics (such as residence time, short circuiting and adequate mixing), understanding the gas/liquid mass transfer and optimizing the biological functionality of the AS tanks, detailed numerical modelling can improve process efficiency, reduce the impact on the environment and costs of construction, and lower operating expenditure.

However, the physics of a typical aeration basin are complex due to the presence of solid, liquid and gaseous phases, a large length scale variation between suspended sludge floc, air bubbles and the external tank geometry, and the relative velocities of the continuous and dispersed phases due to the energy imparted by aerators and mixers. The overall (macro) mixing phenomena must be determined accurately, as these govern the distribution of substrate concentrations within the tank, and the intensity and frequency of turbulent interactions between individual phases (micro-mixing) have a significant impact on substrate conversion rates, which ultimately determine plant performance.

Editor(s): Julien Laurent, Randal Samstag, Jim Wicks and Ingmar Nopens

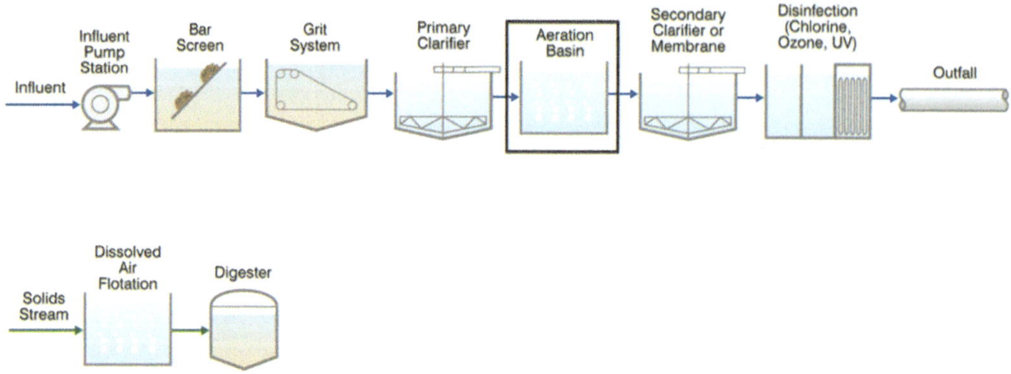

Figure 4.1 Placement of the suspended growth tank in a wastewater resources recovery facility.

In short, the accurate prediction of air/liquid/solid, multiphase, biokinetic phenomena in the turbulent environment of a suspended growth tank represents a major challenge for computational fluid dynamics.

4.2 LITERATURE REVIEW

Over the last two decades, the search for improved performance, greater efficiency and reduced cost in the activated sludge process has led to an increased interest in the application of CFD. In recent work, Samstag *et al.* (2016) provided a comprehensive overview of fluid dynamics to model different unit wastewater treatment processes, including flow splitting, primary and secondary treatment, anaerobic digestion and disinfection. Similarly, Karpinska and Bridgeman (2016, 2018) focused on modelling an aeration tank in isolation, comparing hydrodynamics using both a steady and unsteady Reynolds Averaged Navier-Stokes simulation (RANS and URANS); modelling turbulence using different $k-\varepsilon$ and $k-\omega$ models and the Reynolds Stress Model (RSM); conducting multiphase flow simulations using Eulerian, slip mixture, volume of fluid (VOF) and Lagrangian models, and compared active and passive scalar solids transport. Both authors acknowledged that a complete, dynamic, CFD simulation of AS tanks involving complex multiphase flow and biokinetic reactions remained a challenge, due to the computational costs required. It was also emphasized that although an unequivocal protocol on CFD methodology had not yet been agreed, the most computationally inexpensive, but still efficient scenario, namely RANS/URANS closed by a Standard $k-\varepsilon$ turbulence model, has been adopted as the standard scheme for the modelling of AS tanks.

4.2.1 Gas/liquid transfer

The focus of most early CFD work was on a better understanding of the mass transfer mechanism in aerated and agitated AS tanks. Cockx *et al.* (2001) and Do-Quang *et al.* (1998) simulated a two-dimensional (2D) model of the full-scale closed-loop aeration tank using an Eulerian two-fluid model where the turbulence was predicted using a Standard $k-\varepsilon$ model and assuming a constant bubble diameter. To simplify numerical simulations, a fixed value for the liquid velocity field was set in the impeller zone. The oxygen mass transfer was obtained via introduction of an experimentally verified value of the global volumetric mass transfer coefficient, K_La, as a transport source term. The simulation attempted to optimize the air and liquid velocities within the tank to allow longer contact times between the phases, hence improving oxygen mass transfer. It was also found that an increase in K_La in the closed-loop agitated tanks depended to a large extent on the fluid circulation. The authors concluded that a modelling approach based on 2D simulation was insufficient to adequately represent the three-dimensional (3D) flow regime inherent within a tank. Nevertheless, despite the lack of experimental validation and a model which ignored the presence of the solid phase, this work provided an early valuable research framework for future numerical studies on aeration basins.

Further attempts to develop an improved numerical procedure to predict oxygen mass transfer in aeration tanks have been described in Fayolle *et al.* (2007). The 3D models of several pilot-scale and full-size oxidation ditches aerated with diffusers were simulated using an Eulerian two-fluid model for water and air. The Standard $k-\varepsilon$ model was used to predict turbulence of the continuous phase and modelling of the slow-speed mixers was simplified by fixing the velocity within a specified sub-volume. The results of the simulations showed that the oxygen mass transfer in these AS tanks depended on the type, number, and spatial distribution of the flow boosters. The use of an incrementally more complex modelling framework accounted for the impact of axial fluid velocities on the bubble size and gas hold-up was able to reproduce experimental values of K_La directly. It was also highlighted that the accuracy of their overall solution depended on an assumed value of the bubble size, hence either measurement of the *in situ* bubble diameter or use of a numerical sub-model estimating bubble sizes at the diffuser level would be necessary in future models. A limitation of the modelling framework was the use of clean water and air, which therefore discounted any inertial forces which would affect mixing due to the local solids concentration or density gradients.

Littleton *et al.* (2007a) created a 3D CFD model to simulate hydrodynamics in an Orbal ditch operating as a biological nutrient removal (BNR) process. Water was defined as the working fluid and turbulence was modelled using the Standard $k-\varepsilon$ model. An oxygen source term was introduced in the disk aerator's volume, while a simplified oxygen sink term was implemented throughout the fluid volume to mimic the uptake of oxygen by the biomass. The velocity field imparted by disk aerators was modelled using the fixed velocity method. The CFD results were validated from field trial data obtained for a pilot scale and full-scale tank. The main features of recirculating flow, secondary flow currents and non-uniform oxygen distribution, were captured by the model, suggesting alternating anaerobic and aerobic zones formed lengthwise and across the width of the tank - a strong indication that simultaneous nitrification and denitrification processes were occurring. While the robust BNR performance of the Orbal ditch was associated with partial mixing which ensured spatially varying oxygenation conditions, the impact of the solids on the hydrodynamics and resulting dissolved oxygen (DO) profile was not examined in detail.

Yang *et al.* (2011) focused on the simulation of the hydrodynamics and oxygen mass transfer in a multichannel carousel ditch equipped with horizontal rotors and submerged mixers. The CFD studies were aimed at optimizing the existing operational conditions by promoting formation of larger DO gradients for enhanced denitrification. The flow was simulated using an Eulerian mixture approach, with additional transport equations to model transfer of oxygen introduced by the rotors and a sink term to model oxygen consumption for biochemical oxygen demand (BOD) removal. The surface aerators were simulated using a moving wall model, while submerged mixers were modelled as a momentum source. While the study presented an energy efficient operating scenario, this approach was based on average flow, neglecting the presence of solids and assuming a constant influent quality.

4.2.2 The importance of solids

As seen from several literature examples presented above, modelling of the AS tanks based on neutral density simulations, therefore not accounting for the presence of solids, has become standard practice for both aeration and mixing as it facilitates a relatively fast setup of the model, and usually involves the use of water and air only.

Nevertheless, one of the most relevant phenomena occurring in suspended growth systems, although not sufficiently well understood and explored via CFD analysis, is sludge flocculation (Samstag *et al.*, 2016). In early work Parker *et al.* (1970, 1971) identified factors affecting flocculation in activated sludge tanks which may lead to poor settling and sludge bulking. These have a direct impact on biokinetic conversion rates and, as a result, final effluent quality (e.g., turbidity). Parker *et al.* (1972)

developed a simple mathematical model for floc breakup and agglomeration, where both phenomena have been related to the root mean velocity gradient. Whilst the proposed model has been later implemented in CFD models of sedimentation tanks, further experimental studies by Ducoste and Clark (1998) showed that floc breakup and agglomeration depend on the turbulence intensity rather than the velocity. The floc-forming capacity of the activated sludge, and its settling and compacting properties are linked to the type of biomass (zoogleal and filamentous). The relative amount of the microorganisms present in the AS tank and growth depends on the organic loading rates, local dissolved oxygen concentrations and sludge age. Moreover, flocculation depends on floc properties, namely size and structure, both influenced by local shear stresses induced by the flow regime, sludge concentration, and ionic strength. Consequently, Nopens *et al.* (2002) and Nopens (2005) focused on the implementation of a population balance model (PBM) approach to study activated sludge flocculation *in situ*. The authors emphasized the ongoing need to gather detailed information about the flocculation process to achieve a robust description of sludge settling for future CFD models and proposed a calibration methodology for PBMs based on experimental size distribution data obtained online during flocculation tests.

The second most important aspect of solids is the inertial dampening effect their inherent mass has on mixing and substrate distribution. Without mixers in place (aerated or physical), solids will settle to the base of an aeration tank. Where mixing levels are low, they accumulate (e.g., close to the inner walls of an oxidation ditch). Piles of solids may contribute anaerobically to treatment efficiency (of interest itself in more detailed process calculations), but to neglect the presence of solids assumes that mixing is universally uniform and comprehensive, or that solids are neutrally buoyant with water and density (or concentration) gradients throughout the tank negligible.

Fan *et al.* (2010) simulated an oxidation ditch aerated with an inversed umbrella surface aerator. The multiphase flow was simulated using a solid-liquid Eulerian two-phase model and the aerators were modelled using a multiple reference frame (MRF) approach. The 'solid' phase properties were adapted to simulate activated sludge floc. The studies aimed at generating a detailed description of the mixing regime for selection of the optimal aerator speed ranges, ensuring a uniform distribution of the solids within the tank and minimizing the risk of solids settling. Similarly, Xie *et al.* (2014) studied mixing in a full-scale carousel ditch. The 3D flow was simulated using a solid-liquid mixture model and the sludge settling was coupled through the slip velocity. The surface rotors and submerged mixers were simulated using moving wall and fan models. The resulting flow patterns and solids profiles within the ditch allowed identification of the near-zero velocity regions, where biomass settling would occur.

Samstag *et al.* (2012) tested a density-coupled 3D CFD model incorporating solids transport and settling parameters to assess mixing in a full-scale sequencing batch reactor (SBR). The work focused on evaluating a jet aeration system designed to keep solids suspended and determined the power consumption for two primary operating modes, either mixing with or without air. The multiphase flow was simulated using a slip mixture approach enhanced with an active scalar transport model for solids, calibrated from field data. The authors found that a simulation based on water and air did lead to an over-prediction of the degree of mixing (as it would when mixers are tested in clean water baths, then moved into sludge tanks, even with relatively low solids content). Accordingly, a reliable assessment of the mixing in a suspended growth tank requires an accurate prediction of local density gradients, due to the impact of the flow regime on solids transport.

4.2.3 Including biokinetics

Over the last decade, with increasing computational power available and an interest in multiphase flow, there has been extensive research carried out on the development of the *complete* CFD model to simulate actual physical-chemical-biological processes in different AS systems.

Such numerical approaches provide more valuable output than a standalone IWA Activated Sludge Model (ASM) (Henze *et al.*, 2006) which hydraulically contains a relatively simple, completely mixed 'tanks-in-series' flow assumption. Attention has been focused on whether 'real' tanks at larger scale have much greater substrate concentration variation and internal DO profile distribution across the length, width and depth (particularly true for circular geometries, where the flow velocity is spatially non-uniform) – a concept relevant for understanding biological nutrient removal where the location and relative volume of poor and fully-aerated regions are key (Pereira *et al.*, 2009).

The coupling of hydrodynamics with biokinetics is achieved through embedding a selected model of the ASM series in a CFD model, using additional species of transport with source terms which account for biochemical reaction rates. The resulting coupled CFD-ASM model applied to study suspended growth tanks enables prediction of the transport and interactions between the water-air, water-biomass and air-biomass phases, accounting for the multiphase velocity field, shear forces and turbulent mixing, oxygen mass transfer, growth, decay and metabolic activity of the AS. Despite the complexity of the approach, CFD-ASM integration allows assessment of the hydrodynamics and its impact on the reaction conversion rates, represented by the concentration profiles within the simulated AS system. Such 'complete' prediction of the AS system dynamics provides key information required for troubleshooting and optimization of the aeration, mixing and hydraulic design, leading to enhanced process efficiencies achievable at reduced operating costs.

One of the early works on CFD-ASM coupling reported in the literature (Glover *et al.*, 2006), considered a 3D simulation of a full-scale oxidation ditch, where multiphase flow was simulated using a Eulerian two-fluid model (solids were excluded), assuming a fixed bubble size. The study demonstrated a practical use of the CFD-ASM1 approach by identifying poor performance in a diffused aeration system, highlighting potential low nitrification efficiencies.

Laursen (2006) provided a comprehensive study of the hydrodynamics and biokinetics of several full-scale wastewater treatment tanks. Preliminary stage research focused on modelling individual sub-processes, that is aeration (bottom mounted diffusers) and mixing (banana blade, or hydrofoil, slow speed mixers). Consequently, biokinetics (based on ASM3), sedimentation and rheology models were integrated with the hydrodynamic model. Three case studies were presented, where the basic model was gradually expanded to cover (i) simulation of the sludge settling in the AS tank using a simple two-phase model, accounting for sedimentation and rheology; (ii) modelling of an integrated aeration tank by extending the hydrodynamic model to three-phase flow to account for air, mixing and bulk microbial oxygen uptake; and (iii) a hydrodynamic model integrated with the ASM3 model to simulate biological processes in the tank. A few additional modelling scenarios were explored in this study, namely, density coupling via an active scalar approach and a calibrated viscosity model were enabled to account for the presence of the solids; turbulence of the liquid phase was simulated using both $k-\varepsilon$ and $k-\omega$ shear stress transport (SST) models; and the mixers were simulated using a sliding mesh model and a momentum source term. The author provided discussion on the strengths and weaknesses of each of the sub-models used and proposed future extensions to the improved process tank model.

Le Moullec *et al.* (2010a) modelled an AS tank using a Eulerian two-fluid model coupled to ASM1. The simulations considered a channel AS tank aerated by a porous tube. A compromise between the complexity of the hydrodynamic-biokinetic model which could be run in a realistic time frame, and the range of simplifications required to achieve this resulted in a few assumptions: the channel reactor was in a pseudo steady-state; the biomass content in the reactor was constant; the air phase was modelled with a constant bubble diameter; and the AS flocs were soluble in the liquid phase. These assumptions prompted a series of errors in simulation outcomes, namely discrepancies between the measured and predicted K_La, underestimation of the nitrate concentration and overestimation of the ammonia in the effluent. The importance of the bubble size estimation for the accurate prediction of KLa was also highlighted.

Gresch *et al.* (2011) provided further analysis of the flow field induced by diffusers in a conventional activated sludge plug-flow tank. The CFD approach proposed in this work consisted of a gas-liquid Eulerian model using a SST $k-\omega$ turbulence model, where the physical properties of the continuous phase were approximated to those of activated sludge via implementation of the bulk density. While the studies did not account for modelling of the oxygen transfer, hydrodynamic simulations were enhanced with additional transport equations representing a nitrification process. The CFD output data were validated experimentally in a full-scale aeration tank. The studies provided a comprehensive analysis of the flow patterns generated by changing the diffuser layout, and its impact on the gas hold-up and ammonia removal.

Sobremisana *et al.* (2012) focused on simulation of hydrodynamics and biokinetics in a baffled flow-through reactor. The approach involved a 2D macro-scale model of the activated sludge system accounting for micro-scale processes occurring in the floc through the simulation of floc aggregation and breakup using a quadrature method of moments (QMOM) and included the biokinetic mechanism of ASM1 implemented via user defined function (UDF) sub-routines. While lacking experimental validation, the proposed approach based on the CFD/QMOM/ASM1 integration provided a more comprehensive insight into the mechanisms of wastewater treatment in flocculant tanks through prediction of the impacts of mixing and species diffusion (carbon, nitrogen, dissolved oxygen) on the floc size as well as on the reactions occurring within the floc.

Lei and Ni (2014) simulated hydrodynamics, oxygen mass transfer, biodegradation, nitrification and denitrification in a carousel oxidation ditch equipped with mechanical aerators and mixers. The three-phase flow was simulated using the slip mixture approach. The CFD outcomes were in good agreement with experimental data (velocity, DO, chemical oxygen demand (COD) and nitrogen compounds), moreover, pseudo-solid properties imposed on the AS floc phase allowed the prediction of physical kinematics of sludge settling.

Nopens *et al.* (2017) and Rehman *et al.* (2014, 2017) presented results of an extensive, integrated hydrodynamic, biokinetic model to study the velocity patterns, DO and ammonium profiles in the outer ring (anoxic/aerobic) of a full-scale concentric, closed-loop reactor in Eindhoven (Netherlands) (see also case study presented later in this chapter, Section 4.3.1). The 3D hydrodynamics were modelled using the Eulerian mixture model. Bulk density, determined from field mixed liquor suspended solids (MLSS) measurements, was included as part of the liquid phase combined with a drift flux model. A Realizable $k-\varepsilon$ model was used to predict the effects of turbulence, while the dispersion of the air phase was modelled via slip velocity. Mixers were modelled as momentum sources using the fixed velocity method. Biokinetics was modelled using direct CFD-ASM1 coupling and an experimental value of k_La was used to model the oxygen mass transfer in the tank. Results, shown in Figure 4.2, were in good agreement with field sampling data. The study investigated the impact of different operating scenarios on the mixing, concentration profiles within the reactor and, ultimately, on the overall treatment performance. This CFD model is shown in greater detail within this chapter, highlighting the various steps in its construction.

As seen from the few examples presented above, a fully coupled CFD-ASM simulation has the potential to provide a more detailed result than a single ASM model, by creating a snapshot of the three-dimensional substrate and dissolved oxygen concentration profiles within a tank. As a result, CFD-ASM output data are more suitable for the design and scale-up of bioreactors. However, in addition to the expensive computational costs involved, an emerging downside of the CFD-ASM modelling approach is its limited viability due to strict convergence criteria and (typically) the equilibrium, or single fixed influent solution it generates (Nopens & Wicks, 2012).

As a result, less computationally expensive procedures for coupling CFD data with ASM simulations have also been intensively studied (Glover *et al.*, 2006; Le Moullec *et al.*, 2010b; Pereira *et al.*, 2012; Portela, 2013). Residence time distribution (RTD) curves, hydraulic retention time (HRT), recycle ratio, local velocities and other key hydrodynamic parameters obtained from CFD simulations can be used to generate an improved chemical reactor model, for example in terms of a number of continuous stirred-tank reactors (CSTRs) in series, recirculation rates and the flow pattern between each of the reactors, where a faster one (or two) dimensional biokinetic model can be implemented. The combination of RTD/virtual tracer tests approaches with CFD is discussed in Chapter 10.

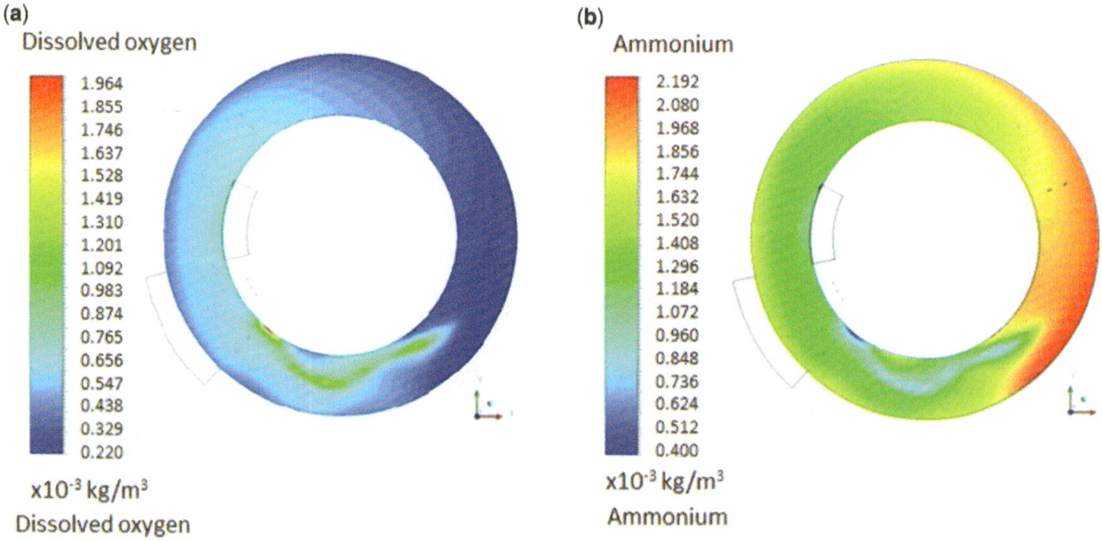

Figure 4.2 (a) Oxygen and (b) ammonium concentration plots in a closed-loop reactor (source: Rehman *et al.*, 2017). (Reprinted from Water Science and Technology, volume 76, issue number 8, pages 1950–1965, with permission from the copyright holders, IWA Publishing).

An example study was described by Pereira *et al.* (2012), where a 3D single phase CFD simulation of an oxidation ditch configuration was implemented to determine RTD characteristics from first principles. The hydrodynamics was simulated using three different turbulence models: RANS with the Standard $k-\varepsilon$ model; unsteady RANS (URANS); and a large eddy simulation (LES) with Smagorinsky's subgrid-scale model. Once the flow field was solved, CSTR and plug-flow reactor models incorporating ASM1 models were fitted to the RTD characteristics identified in the CFD simulation.

4.2.4 Hybrid systems and rheology

A review of the literature related to individual biokinetic and filtration models for submerged membrane bioreactor (MBR) systems with integrated hydrodynamic and biokinetic models was provided by Naessens *et al.* (2012a, 2012b). Ratkovich *et al.* (2013) also provided a critical review on data collection and modelling of sludge rheology (i.e., shear stress) in MBRs, which is vital from the point of view of pumping, hydrodynamics, mass transfer rates, sludge–water separation, and fouling management. Sludge characteristics were also covered in Boyle-Gotla *et al.* (2014) including their effect upon process performance and operating costs. Brannock *et al.* (2010) studied the impact of the hydraulic design, that is membrane configuration, and sludge rheology on hydrodynamics and pollutant removal in two full-scale MBRs. Their complete 3D model, which was validated experimentally, accounted for aeration, biokinetics (ASM1) and a density coupled solids concentration, while the hydrodynamics and reactor performance were characterized using RTD simulation.

Table 4.1 presents a summary of important contributions to the literature.

4.3 CASE STUDIES

Two case studies are presented here which describe the CFD modelling process for suspended growth systems in detail. The modelling parameters and boundary conditions are included.

Table 4.1 A summary of published CFD models with biological processes in WRRFs.

Reference	Solution Technique	Dim.	Turbulence Model	Multiphase Approach	Mesh/CFD Implementation	Density Coupling?
Cockx *et al.* (2001); Do-Quang *et al.* (1998)	FVM	2D, 3D	RANS + Standard $k-\varepsilon$	Gas-liquid Eulerian; species transport (oxygen mass transfer)	6.7 k (2D), 62 k (3D)/Astrid	No
Glover *et al.* (2006)	FVM	3D	RANS + Standard $k-\varepsilon$	Gas-liquid Eulerian; species transport with bioreaction (oxygen mass transfer; complete biokinetics – ASM1); Lagrangian-particle tracking (RTD)	- /Fluent	No
Laursen (2006)	FEM/FVM	3D	URANS + Standard $k-\varepsilon$; URANS + SST $k-\omega$	Gas-liquid Eulerian; species transport with reaction (oxygen mass transfer; complete biokinetics – ASM3)	150 k (case 1), 480 k (case 2), 340 k (case 3)/ CFX	Yes
Littleton *et al.* (2007a, 2007b)	FEM/FVM	3D	RANS + Standard $k-\varepsilon$	Single phase; species transport with reaction (oxygen mass transfer)	-/CFX	No
Fayolle *et al.* (2007)	FVM	3D	URANS + Standard $k-\varepsilon$	Gas-liquid Eulerian; species transport (oxygen mass transfer)	29 k–452 k/ Fluent	No
Le Moullec *et al.* (2008)	FVM	3D	RANS + Standard $k-\varepsilon$/RSM	Gas-liquid Eulerian; Lagrangian-particle tracking (RTD); species transport (RTD)	350 k/Fluent	No
Le Moullec *et al.* (2010a, 2010b, 2011)	FVM	3D	RANS + Standard $k-\varepsilon$	Gas-liquid Eulerian, species transport with bioreaction (oxygen mass transfer; complete biokinetics- ASM1)	50 k/Fluent	No
Fan *et al.* (2010)	FVM	3D	RANS + Standard $k-\varepsilon$	Solid-liquid Eulerian	70.3 k/Fluent	No
Brannock *et al.* (2010)	FVM	3D	URANS + Standard $k-\varepsilon$	Gas-liquid/solid mixture; species transport (solids settling and transport; RTD); species transport with reaction (oxygen mass transfer; complete biokinetics – ASM1)	240 k/Fluent	Yes
Gresch *et al.* (2011)	FEM/FVM	3D	URANS + SST $k-\omega$	Gas-liquid Eulerian; species transport with bioreaction (biokinetics-NH_4 removal)	400 k/CFX	No
Yang *et al.* (2011)	FVM	3D	RANS + Standard $k-\varepsilon$	Gas-liquid mixture; species transport with bioreaction (oxygen mass transfer; biokinetics – BOD removal)	1.66 M/Fluent	No

Sobrenisana *et al.* (2012)	FEA	2D	RANS + Standard $k-\varepsilon$	Solid-liquid; PMB (sludge flocs formation and breakup – QMOM); species transport with bioreaction (oxygen mass transfer, complete biokinetics – ASM1)	-/Comsol	No
Alvarado *et al.* (2012)	FVM	2D	RANS + Standard $k-\varepsilon$	Single phase; species transport (RTD)	75 k/Fluent	No
Samstag *et al.* (2012)	FVM	3D	URANS + Standard $k-\varepsilon$	Gas-liquid mixture (species transport, solids settling and transport)	415 k/Fluent	Yes
Karpinska *et al.* (2015); Pereira *et al.* (2012)	FVM	3D	RANS/URANS + Standard $k-\varepsilon$; LES + Smagorinsky's SGS	Lagrangian-particle tracking (RTD)	600 k/Fluent	No
Lei and Ni (2014)	FVM	3D	RANS + Standard $k-\varepsilon$	Gas-liquid-solid Eulerian; species transport with bioreaction (oxygen mass transfer; complete biokinetics – ASM1)	162 k/Fluent	No
Xie *et al.* (2014)	FVM	3D	URANS + Standard $k-\varepsilon$	Solid-liquid mixture	1.5 M/Fluent	No
Rehman *et al.* (2014, 2017)	FVM	3D	RANS + Realizable $k-\varepsilon$	Gas-liquid mixture; species transport with reaction (oxygen mass transfer; complete biokinetics – ASM1)	700 k/Fluent	Yes
Karpinska and Bridgeman (2017)	FVM	3D	URANS + Standard $k-\varepsilon$; URANS + SST $k-\omega$	Gas-liquid Eulerian; IAC (bubble coalescence and breakup); species transport with reaction (oxygen mass transfer)	2 M/Fluent	No
Karpinska and Bridgeman (2018)	FVM	3D	URANS + SST $k-\omega$	Gas-liquid/sludge Eulerian; IAC (bubble coalescence and breakup); species transport with reaction (oxygen mass transfer); species transport (RTD)	2.2 M/Fluent	No
Sánchez *et al.* (2018)	FVM	3D	RANS + Standard $k-\varepsilon$	Gas-liquid mixture; species transport with reaction (oxygen mass transfer; complete biokinetics – ASM1)	2.4 M/ OpenFOAM	No
Climent *et al.* (2018)	FEM/FVM	3D	SST $k-\omega$	Single phase biokinetics – ASM1	240 k/CFX	No

FVM: finite volume method; FEM: finite element method; SGS: subgrid scale; IAC: interfacial area concentration.

4.3.1 CFD modelling of a bioreactor at Eindhoven WRRF

The full-scale plant modelled in this study is the Eindhoven WRRF in The Netherlands (Rehman, 2016). This plant treats the wastewater of 750 000 population equivalent (PE) with a design load of 136 g COD/day/PE. The incoming wastewater is treated in three parallel lines, each containing a primary settler, a biological tank and four secondary settlers. The plant is a modified UCT (University Cape Town) configuration (biological COD, N and P removal), as shown in Figure 4.3. It is a carousel-type reactor in which the circular inner ring (an anaerobic tank) is a plug-flow reactor, consisting of four compartments in series, with no aeration and minimal recycled nitrate (the purpose of the recycle is to bring back activated sludge). The middle ring is anoxic from where the mixed liquor enters the outer ring, which is a facultative aerobic/anoxic carousel ring. It contains membrane plate aerators installed at the bottom in certain locations.

Two zones of aerators can be distinguished: the ‘summer package’, which is always active and has its airflow controlled by a NH_4-DO cascade controller accounting for the load, and the ‘winter package’, which can be switched on during the winter or rain events for additional nitrification capacity. Three recycles are active: one to return sludge from the anoxic middle to the anaerobic inner ring, a second to recycle nitrate from the outer ring to the anoxic middle ring, and a third to return settled secondary sludge to the anoxic tank.

In this study, only the outer ring of one of the three biological reactors has been modelled. The outer ring consists of one inlet (flow coming from the middle ring) and four outlets (1 outflow and 3 recycle pumps). There are three pairs of propellers installed along the length of the bioreactor to keep the flow in one direction which provides extra energy to improve mixing. The two aeration packages (summer and winter) installed at the bottom of the bioreactor comprise a total of 196 plate diffusers (168 in summer and 28 in winter). These plate diffusers are spread uniformly along the width and length of the reactor.

An internal view of the reactor is shown in Figure 4.4.

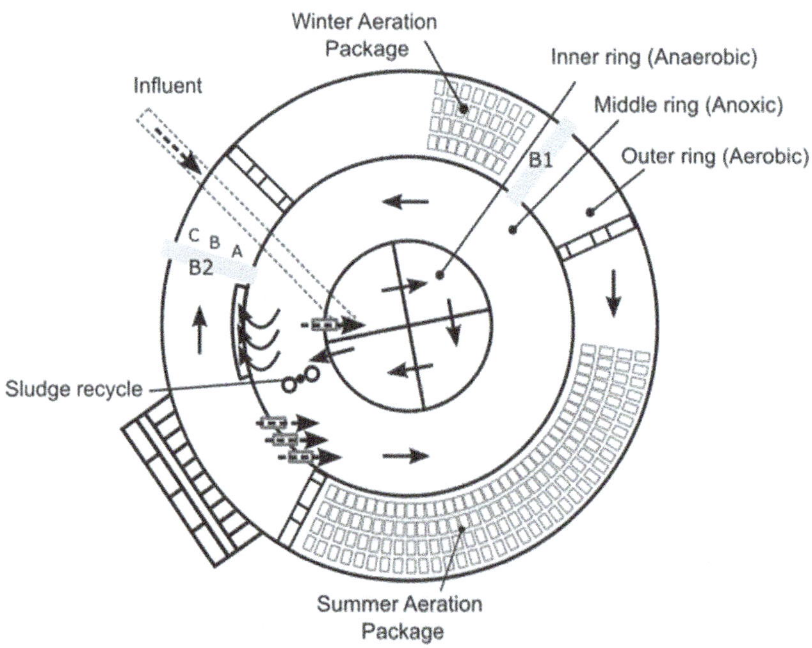

Figure 4.3 Configuration of the bioreactor at the Eindhoven WRRF.

Figure 4.4 The biological reactor at Eindhoven WRRF.

4.3.1.1 Computational fluid dynamics modelling

The stepwise approach provided in Laurent *et al.* (2014) and Wicklein *et al.* (2016) (see also Chapter 2) was followed to conduct the CFD modelling.

The first stage was to develop a 3D geometrical model of the reactor. Secondly, a detailed mesh was created. Thirdly, appropriate boundary conditions were defined, and turbulence and multiphase models were chosen. The simulation was then started, and relevant convergence criteria were applied to assess when the solution had successfully completed. Finally, once convergence had been achieved, post-processing of the data could commence – which included evaluation of the modelling results, and comparison with available experimental results.

4.3.1.1.1 Geometry development

The three-dimensional model of the bioreactor was developed with the help of a commercial tool (AutoCAD Siemens NX 9.0). All the necessary details that would impact the hydrodynamics were included, such as the location and sizes of propellers, inlets, outlets and recycle pumps.

One of the most important features of the model is the aeration package. Fine bubble diffused aerators were included by calculating the total flow volume of air based on the orifice diameters (provided by the manufacturers) and back pressure. It was foreseen that a slip mixture model (the standard CFD model for modelling dense bubbly flows) would be used, where individual bubbles would not be modelled and tracked, but rather the fluid domain would be simulated as a mixture of both phases. Therefore, instead of including multiple individual nozzles, a single aerator equal to the orifice area was constructed, maintaining geometrical similarity with the real diffusers (Figure 4.5). This ensured the inlet velocity of the gas phase remained realistic.

The propellers were also simplified. Individual propeller blades were not modelled (due to the high computational requirements for modelling moving parts) and instead, propellers were simulated as thin cylindrical vessels equal to the size of real propellers and fixed as a momentum source.

It is good modelling practice to simulate the flow boundaries (inlets and outlets) starting a distance away from their actual locations, to achieve a fully-developed flow regime (Versteeg & Malalasekera, 2007). Therefore, the inlet, outlet and recycle pump pipework was extended out from the model.

4.3.1.1.2 Meshing

In the next step, meshing using a commercial mesh/grid generator tool (ICEM CFD, ANSYS) was performed.

Different meshes (coarse (C), fine (F) and very fine (VF)) were generated to check for grid independence. The coarse mesh comprised 0.7 million cells, the fine mesh 1.5 million cells and the VF mesh, 2.5 million cells. The fine and VF meshes provided the best match with experimental values

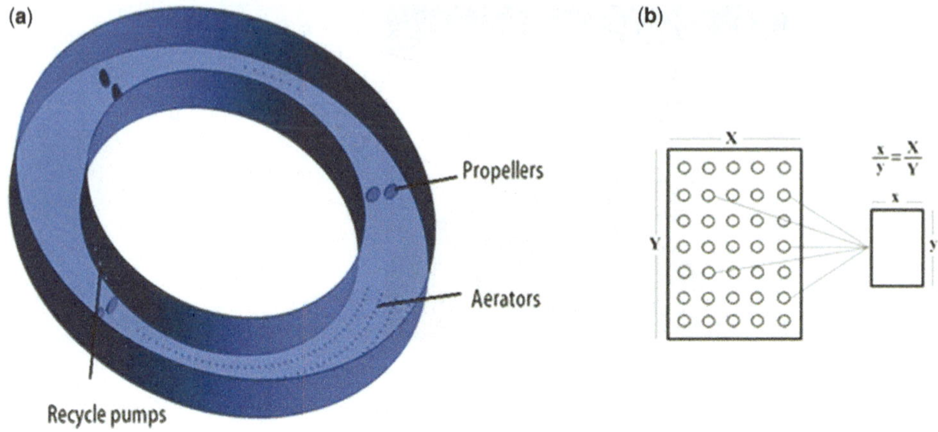

Figure 4.5 (a) The 3D model (geometry) of the bioreactor, and (b) the plate diffuser simplification.

(velocity measurements) and the fine mesh was chosen for the simulation due to its lower computational demand. The meshing includes different sizes and shapes of cells, but was kept structured as much as possible. A total of 90% of the cells were hexahedral which incur fewer numerical instabilities (Andersson *et al.*, 2011) and the mesh ratios are shown in Figure 4.6. The mesh resolution was increased in the region near the diffusers (i.e., smaller than the size of diffusers), and made coarser in the volume where small velocity gradients were expected (i.e., regions of bulk flow). These refinements were made during the grid independence check. The different views of the resulting mesh are shown in Figures 4.7 and 4.8.

The bioreactor was divided into several zones to apply different bulk conditions. The reactor surface was separated from the main reactor to implement a de-gassing boundary condition (termed the 'de-gassing zone'). The reactor height was 6.7 m and an additional layer 200 mm high was added on top of it (this layer is visible in Figure 4.7). The height of 200 mm was chosen so that the mesh inside this region was neither too coarse nor too fine, with an average size equal to the rest of the reactor's mesh. In addition, the propellers were also modelled as separate zones to apply a momentum source which mimicked the real propellers.

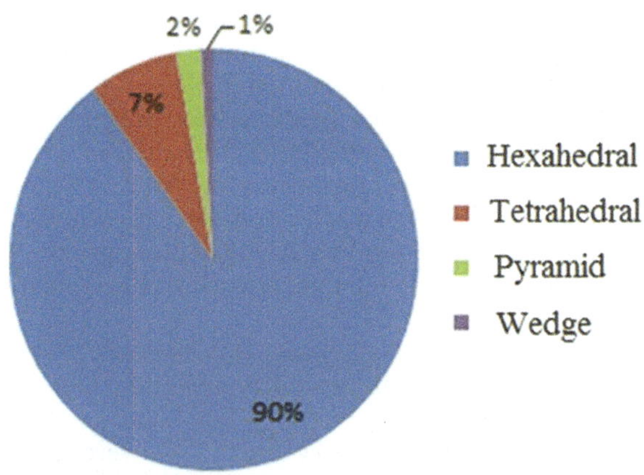

Figure 4.6 The distribution of the fine mesh cell types at Eindhoven WRRF.

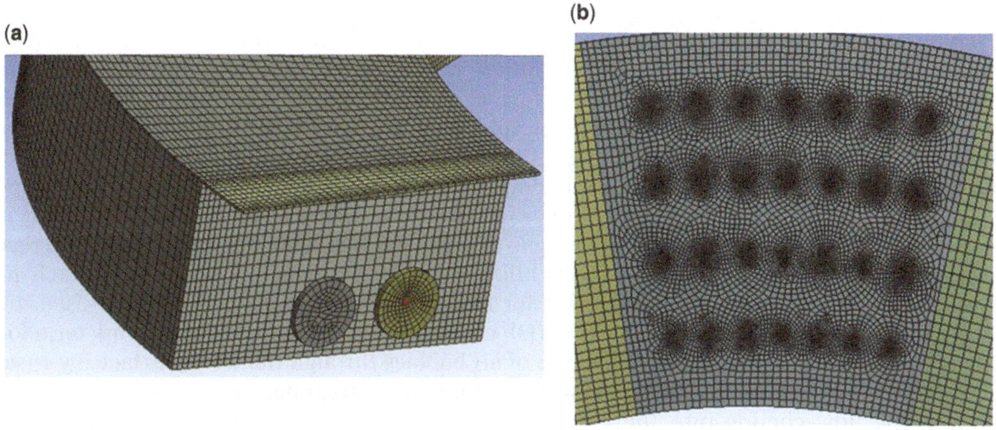

Figure 4.7 (a) An inside view and (b) bottom view of the mesh.

4.3.1.1.3 Boundary conditions

The boundaries of the system were characterized individually.

There were two types of inlet boundaries: one for the liquid and another for the gas (diffusers). The inlet boundary for liquid flow was simply named *Inlet*, whereas the inlet boundary for gas was further characterized as *Summer* and *Winter* to distinguish between the two operational aeration packages. All the diffusers in each package were grouped as one boundary condition (i.e., uniform flow through each diffuser was assumed).

Next, there were four outlets from the reactor, one for the main liquid outflow and three for the recycle pumps. The liquid outflow was simply named *Outlet* and the three pumps were named *P1, P2* and *P3*. If needed, different flow rates could be evaluated with alternate recycle rates. The top of the reactor was a frictionless free surface (named *Top*). The remaining boundaries were physical walls of the reactor and named *Walls*.

Figure 4.8 (a) Top and (b) side view of the mesh at Eindhoven WRRF.

Boundary condition flows are always set based on the given knowledge of the system. As the flow coming into the bioreactor, aeration rate and recycle ratios are known, velocities can be easily calculated. The 'velocity inlet' boundary condition was used for the *Inlet*, *Summer* and *Winter* boundaries. The recycle pumps *P1*, *P2* and *P3* were also set as a 'velocity inlet', but with the direction vector pointing out of the system (a negative flow velocity inlet). The *Outlet* was set with the boundary condition of 'outflow' and calculated by the CFD model during the simulation, by conservation of mass. It should be noted that the outlet on site does not have a pump, therefore flow entirely depends upon the hydrostatic pressure and inertial forces in the reactor. The *Top* boundary was slightly more complex because it is a free surface that allows gas to leave, but liquid must remain within the reactor. The software used (ANSYS Fluent) does not allow separate boundary conditions for each phase and therefore a symmetry boundary condition for free surfaces (Witt *et al.*, 2012) was used. In the de-gassing zone (the top layer of the reactor) a UDF was implemented which acts as a sink term for the secondary phase (air) and mimics the departure of air bubbles through the water surface, by ensuring the secondary phase disappears completely whenever it enters this zone.

Although the flow coming into the bioreactor is dynamic, only steady state simulations were performed. Therefore, selected flow rates which captured snap shots of the dynamics of the system were chosen.

The flow into the outer ring was based on the average daily inflow and the respective recycle ratios. The average lower bound of the liquid flowrate into the outer ring was 1867 m^3/h ('L') and the average higher bound was approximately double this, that is 3734 m^3/h ('2L'). Similarly, the average lower bound of the aeration rate was 2000 Nm^3/h ('G') and the higher bound was approximately triple this, that is 6000 Nm^3/h ('3G').

The winter package was kept turned off because it can only be turned on when the summer package reaches its full capacity, which is not the case under these conditions. The corresponding boundary conditions for the base case defined for CFD modelling are given in Table 4.2.

The propeller zones were fixed with a momentum source to simulate the propeller behaviour. The momentum source used was the thrust provided by the mechanical motion of the propeller in the direction of flow and can be correlated to the power of the propeller in Equation (4.1) (Huang & Li, 2013; Jiang *et al.*, 2010; Yang *et al.*, 2011):

$$P = 2\pi N_r \int F_\theta \, dV \tag{4.1}$$

where P is the power of the impeller, N_r is the rotational speed, F_θ is the tangential momentum source and V is the volume in each cell in the propeller zone.

The power of a propeller can be calculated from the power number correlation as in Equation (4.2):

$$P = N_p \rho N_r^3 D^5 \tag{4.2}$$

where N_p is the power number, ρ is the fluid density and D is the diameter of the impeller.

Table 4.2 The boundary conditions for the base case ('L-G') CFD model at Eindhoven WRRF.

No.	Boundary Condition	Type	Value/Description
1	Inlet	Velocity inlet	1867 m^3/h
2	Summer package	Velocity inlet	2000 Nm^3/h
3	Winter package	Velocity inlet	0
4	P1, P2, P3	Negative velocity inlet	332 m^3/h
5	Outlet	Outflow	To be calculated from the CFD
6	Walls	Wall	No slip condition
7	Top	Symmetry	

4.3.1.1.4 Selection of suitable models

All the CFD simulations were performed in Fluent v14.5 (ANSYS Inc., 2011). With aeration present, it was important to model all phases (i.e., liquid, gas and solids), therefore a suitable multiphase model was needed to accurately capture the gas dispersion within the liquid in the system, along with a second model to capture the density fluctuation of the solids. The flow inside the reactor was clearly turbulent due to the typical flow velocities and aeration processes. Therefore, a fully developed turbulence model was needed which captured all the important features of the fluid flow.

4.3.1.1.5 Multiphase modelling

There are several multiphase models available in the ANSYS Fluent solver to choose from. The choice of the multiphase model depends on the types of the phases (liquid, solid and gas) involved and the interaction between the different phases. In this study, all three phases were present in the bioreactor, that is wastewater being the liquid phase, biomass and particulates, the solid phase and air bubbles (due to aeration), the gas phase. Although solids are important to model the rheology and inertia of the system, the added computational demand for explicitly modelling three phases was too high. Therefore, only liquid and gas flows were modelled as separate phases. With the integration of a biokinetics model, the impact of solids was considered by calculating the sludge density as a function of total suspended solids (TSS) concentration and linking it to this system using a scalar.

The Eindhoven case closely resembles a bubbly flow system, where liquid is considered as the primary (continuous) phase and air as a secondary (dispersed) medium. The slip mixture model was therefore the most suitable model for such a system where individual dispersed phase particle (bubble) tracking was not needed (Andersson *et al.*, 2011). It solves one momentum equation for the mixture and estimates the viscosity of the mixture. This model also calculates the relative velocities of the different phases using buoyancy, drag and other external forces. It also provides a good level of historical accuracy with lower computational demand.

4.3.1.1.6 Turbulence modelling

For these simulations, the Realizable $k-\varepsilon$ model was used, which models jet flows and swirling flows realistically. Standard wall functions were applied.

4.3.1.1.7 Additional model considerations and convergence

The finite volume CFD solver stores values at the cell centres of the grid and, hence an algorithm is required to interpolate these values to the cell faces to calculate the convective flux through each cell face. There are various numerical algorithms/schemes available in ANSYS Fluent for such spatial discretization. These schemes are commonly called upwind schemes, as they interpolate the face values from the downstream cell centre. It is good modelling practice to use a higher order scheme to solve the momentum and turbulence equations (Andersson *et al.*, 2011), therefore, in this study, a second-order upwind scheme was applied.

After selecting the models and numerical algorithm, the next step was to set a criterion for convergence.

As solving the partial differential equations requires an iterative method, the solution must be initialized with values taken from the inlet boundary conditions. During each iteration each cell computes a new value for each variable (e.g., the x-coordinate of velocity, i.e., v_x). After each iteration the difference between the old and the new value is computed and this is commonly called a residual. Part of the convergence criterion is that an acceptable residual value is set, such that when the calculated residual is less than the set residual value, the iteration stops. If the convergence criterion is not met, a new value is assumed and the iterative procedure is repeated until the residuals fall below the set values. The residual criterion for all variables is set separately according to their magnitudes. The residual values for the different variables in the current study are given in Table 4.3.

Table 4.3 The residual values for the convergence criteria at Eindhoven WRRF.

Variable	Residuals
Continuity	10^{-5}
x-coordinate of velocity	10^{-5}
y-coordinate of velocity	10^{-5}
z-coordinate of velocity	10^{-5}
Volume fraction (secondary phase)	10^{-4}
Turbulent kinetic energy	10^{-5}
Turbulent dissipation energy	10^{-5}

Another way of checking the convergence is to compute the mass imbalance of the solution. This calculates the difference between the flux coming in and leaving a cell, and ideally it should be zero. Plus, we can also ensure that the total (integrated) air volume within the simulation is constant, when the simulation has converged.

4.3.1.2 Biokinetic modelling

To predict aerobic carbon and nitrogen removal, the hydrodynamic model was coupled with a standard biokinetic model: the Activated Sludge Model No. 1 (ASM1) (Henze *et al.*, 1987, 2006). This integration was performed by adding 13 user defined scalars (UDSs) representing each of the 13 components in the ASM1 model. ANSYS FLUENT solves these transport equations at the same time as the flow and turbulent models. Each transport equation for a scalar $\varnothing$ can generally be written as:

$$\frac{\partial \alpha_l \rho_l \varnothing_l^k}{\partial t} + \nabla.\left(\alpha_l \rho_l \vec{u}_l \varnothing_l^k - \alpha_l \Gamma_l^k \nabla \varnothing_l^k\right) = S_l^k \quad k = 1,\ldots,N \tag{4.3}$$

where α_l, ρ_l and $\vec{u}_l$ are the volume fraction, density and relative velocity of the liquid phase; Γ and S are the diffusion coefficient and the source term for scalar $\varnothing$; k is the corresponding scalar and N is the total number of scalars.

The first term in the equation is the unsteady term, which in this case is zero as only steady-state simulations are performed. The source terms are introduced by means of UDFs based on the mass balance equations for each of the ASM1 components. The recommended standard values for all the stoichiometric and kinetic parameters are used (Henze *et al.*, 1987).

In addition to all the standard processes of ASM1, an extra term for the oxygen mass balance is included describing the mass transfer from the gas to the liquid phase, as given by Fayolle *et al.* (2007).

The bubble coalescence and breakage phenomena were not considered and a constant bubble size of 3.0 mm was assumed. This value, representative for a wide range of membrane diffusers, was used to calculate the interfacial area and the mass transfer coefficient.

For species transport, it is important to provide a good estimate of turbulent diffusivity. In the current study it was calculated based on a Schmidt number of 0.7 which is a safe assumption for such systems and has been used previously by Le Moullec *et al.* (2010b) (see also Chapter 10).

Given the importance of density in reactive flows (Samstag *et al.*, 2012; Samstag & Wicklein, 2014), bulk density was defined as a function of local suspended solids concentration (De Clercq, 2003) and expressed as:

$$\rho_b = \frac{\rho_l}{1 - \Phi_S(1 - (\rho_l/\rho_s))} \tag{4.4}$$

Table 4.4 Overview of the biokinetic boundary conditions at Eindhoven WRRF.

No.	ASM1 Components	Input Values (mg/L)
1	S_I	42.70
2	S_S	41.00
3	X_I	1955.85
4	X_S	37.78
5	$X_{B,H}$	500.00
6	$X_{B,A}$	98.90
7	X_P	0
8	S_O	0.10
9	S_{NO}	15.20
10	S_{NH}	4.00
11	S_{ND}	0.44
12	X_{ND}	40.63
13	S_{ALK}	30.00

where Φ_S denotes local mass fraction of suspended solids, whereas ρ_b, ρ_l and ρ_s are bulk, liquid and solids density.

A dry solids density of 1450 kg/m³ has been used (Lyn *et al.*, 1992; Stamou *et al.*, 2009). The standard value for water viscosity was used because in highly turbulent flows (with a high Reynolds number) the impact of the change in molecular viscosity was expected to be negligible (Wilcox, 1988). However, the impact of sludge viscosity is subject to further investigation, though debate on the correct measuring method and modelling is still ongoing (Ratkovich *et al.*, 2013).

The simulations for biokinetics were performed on top of a 'frozen' converged steady state hydrodynamic solution to maintain a reasonable convergence time. Computational requirements to achieve convergence increased significantly from 4 hours to 10 hours after including all the biokinetic equations for the steady-state solution using eight parallel processors. It is important to note that the convergence time for subsequent simulations was significantly reduced when the solution was initialized from a previously converged model. It is also recommended to initialize the solution with non-zero soluble substrate (S_S), oxygen (S_O) and ammonium (S_{NH}) concentrations to achieve a smooth convergence.

The influent flow rate and composition were collected from the available plant historical data to use as an inlet boundary condition and these were provided in the form of the ASM1 components in Table 4.4. Not all the data were available and missing values, which included X_I, S_I, X_{ND}, X_S and X_P, were derived from an Eindhoven WRRF plant-wide process model developed by Amerlinck (2015).

4.3.1.3 Simulation setup

Several simulations were carried out to investigate the impact of different combinations of liquid and air flowrates on the local distribution of substrate and thus process performance. The combination of two influent liquid flow rates and two air flow rates have been simulated to produce a total of four scenarios. The low inflow rate was selected to be Eindhoven's WRRF average dry weather flow (1876 m³/h referred to as scenario 'L'), while the high inflow rate was double that (scenario '2L'). Similarly, the low gas flowrate was chosen to be the process base case (2000 Nm³/h referred to 'G') and the high gas flowrate was three times that ('3G') representing the aeration rate at high peak load conditions.

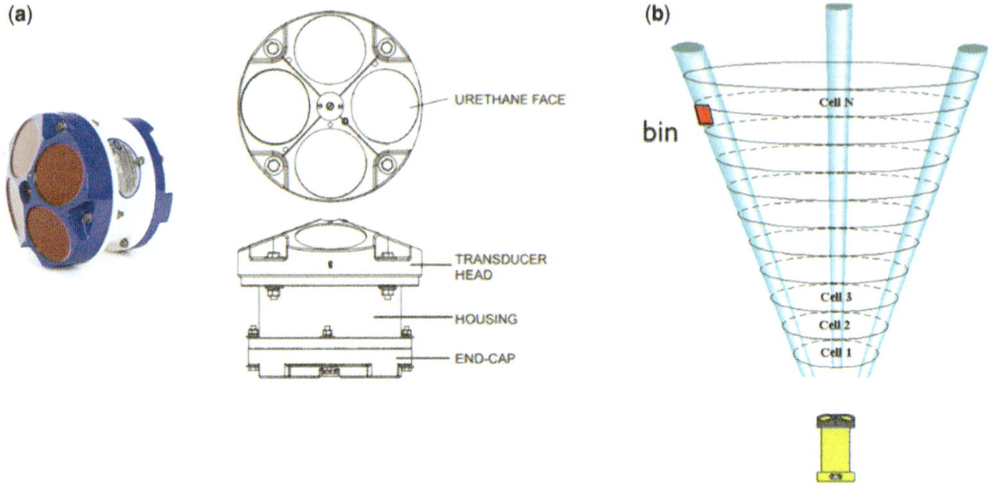

Figure 4.9 (a) Schematic diagram of an Acoustic Doppler Current Profiler (ADCP) and (b) the principle behind its design (Torfs, 2015).

4.3.1.4 Validation of the velocity field

To validate the hydrodynamic output from the model, velocity measurements with the help of a non-invasive Acoustic Doppler Current Profiler (ADCP) Workhorse Rio Grande (Teledyne RD instruments), shown in Figure 4.9, were conducted at the Eindhoven site.

The working principle of the ADCP is based on the Doppler effect. The device has four transducers which transmit sound at a constant frequency into the water. These sound waves travel through the water column and are partially scattered back by particles suspended in the moving water. The ADCP records these echoes at precise time intervals corresponding to fixed depths which can be determined based on the travelling time of the signal. The water column is thus divided into successive vertical elements (called bins) along the axis of the beam (Figure 4.9). The ADCP measures the change in wavelength between the emitted signal and the received signal coming from successive bins (Doppler frequency shift). This allows the determination of the particle velocity over a range of depths (Cobb, 1993; Kinnear & Deines, 2012; Mullison *et al.*, 2011).

The measurements in the current study were performed at two bridges (B1 and B2) (see Figure 4.3). Measurements at the third bridge were too disturbed by the aeration process and thus could not be completed. At each bridge, time-averaged radial, vertical and tangential velocity components at different depths were measured at three radial locations (A, B and C). These locations were chosen because the ADCP needs to be away from possible hindrances such as walls and propellers. At each location, for every measurement, 15 minutes of data were collected, and each measurement was repeated three times to reduce the variations created by the dynamic inflow.

4.3.1.5 Results and discussion

4.3.1.5.1 Comparison between measurements and CFD simulations

The velocity magnitudes derived from the ADCP-measured tangential and axial fluid velocity components along with the corresponding CFD predictions at the B1 and B2 locations are shown in Figure 4.10. At bridge B1, the CFD velocity profiles at locations B and C were close to the measured values. However, at location A the CFD values were over-predicted. At bridge B2, velocity profiles at locations B and C were over-predicted and at location A the CFD values were close to the measured values.

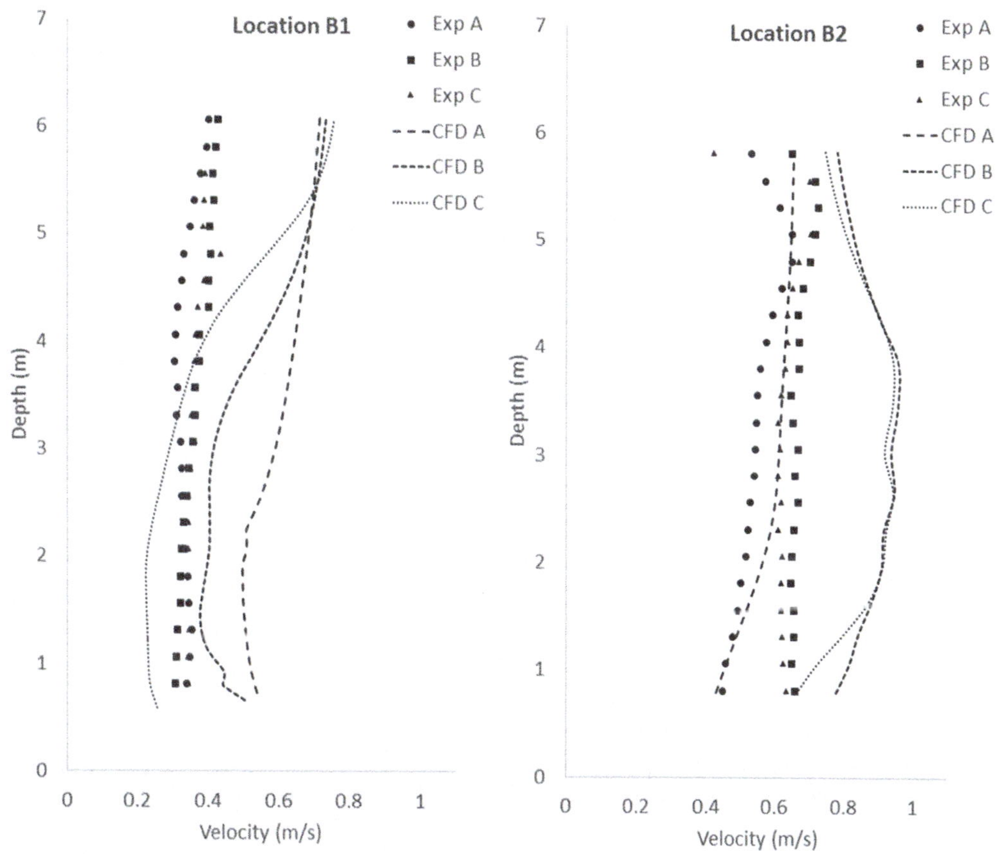

Figure 4.10 Comparison between the CFD results and measured velocities at locations B1 and B2 (A: near the inner wall, B: in the middle and C: near the outer wall) at Eindhoven WRRF. Average measurement error is ±0.0315 m/s.

The underlying cause for these over-predictions is hard to determine. It is important to note that the velocity data were collected under dynamic hydraulic conditions, whereas the model was run at steady-state conditions using the average inflow rate. Also, an approximation (the momentum source method) was made for the propellers, and this type of mixer typically creates swirling flows, which have not been included in the model.

The most likely cause, however, for the over-predictions in the first run of the model, was expected to be that a neutral density approach was used. Water had been defined as a continuous phase and the impact of sludge density had not been considered.

The use of a sludge density correlation (Equation (4.4)) resulted in different predictions of velocities, as compared with the use of a constant water density. The comparison between the new set of CFD velocities and measured values is provided in Figure 4.11. The difference between the CFD and measured velocities was reduced to approximately 10–15%. Therefore, bulk density should always be included when simulating activated sludge systems.

4.3.1.5.2 Base case results

In Figure 4.12, liquid velocities and air volume fraction (gas hold-up) are shown for horizontal cross sections in the reactor, taken at a depth of 3.45 m in the model.

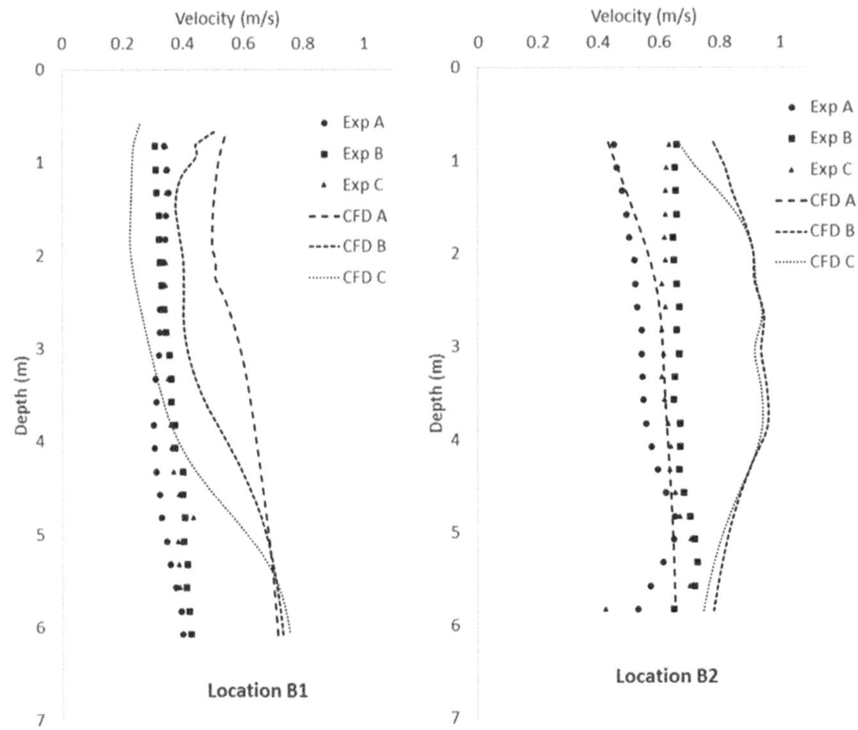

Figure 4.11 Comparison between the CFD results (with sludge density included) and measured velocities at locations B1 and B2 (A: near the inner wall, B: in the middle and C: near the outer wall) at Eindhoven WRRF.

Non-ideal mixing in the reactor can be observed, even in areas close to the propellers and the aeration package. In the aerated region, the gas hold-up is higher near the inner wall and lower near the outer walls. This could be a result of the aeration plume being pushed towards the inner wall due to the bulk flow. That can be further investigated by looking at the gas hold-up distribution at different depths in the reactor (Figure 4.13).

In Figure 4.13, it can be observed that the distribution of air is more even at the top than the bottom. This shows that the air from each diffuser does not rise perpendicularly in the model, instead it is pushed towards the wall, which hints at possible recirculating flow patterns. These findings are important because this will eventually impact the mass transfer between the phases.

In order to investigate the flow patterns in the reactor in greater detail, velocity streamlines are plotted in Figure 4.14. Clearly, the flow does not follow a linear path around the reactor, instead it takes a swirling, helical route. The swirling flow pattern originates directly from the reactor's circular layout.

Velocity vectors are plotted for vertical cross sections within the reactor at different locations in Figure 4.15. In each plot, the left-hand side is the inner wall of the reactor, and the right-hand side is the outer wall, denoted by 'I' and 'O' respectively.

Throughout the reactor a swirling flow prevails. In the beginning of the aerated region, a recirculating flow can be seen, but the magnitude of arrows suggests that the intensity of the recirculation current is small. In the middle of the aerated region, the magnitude of swirling flow increases and two recirculation zones, of unequal sizes, develop. At the end of the aerated region, these two recirculation zones are almost equal in magnitude, and at the outlet (moving downstream from the aerated region),

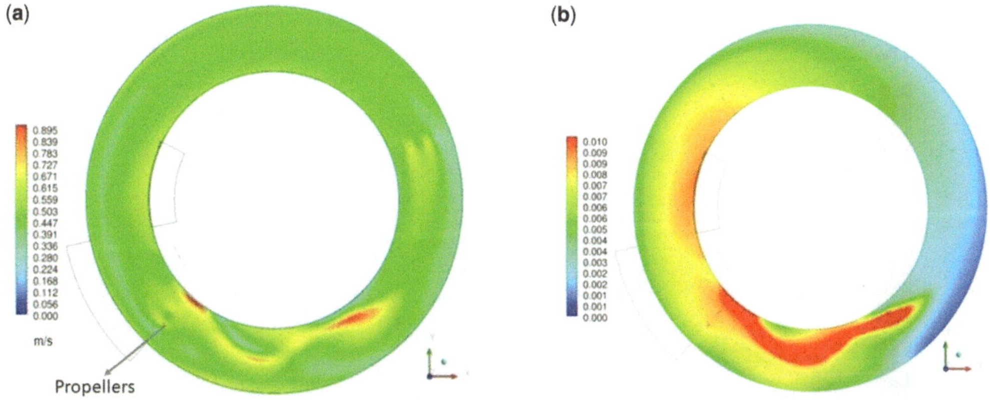

Figure 4.12 (a) Velocity, and (b) gas hold-up (volume fraction of the gas phase) contour plots, at 3.45 m depth in the reactor, at Eindhoven WRRF.

these recirculation zones merge into a single recirculation zone again. The impact of the recirculation zones on the vertical distribution of gas hold-up is shown in Figure 4.16.

Clearly, a complex internal flow regime has developed from a simple circular design for the reactor, and if the air concentration is far from uniform throughout the system, then its treatment performance will be similarly variable.

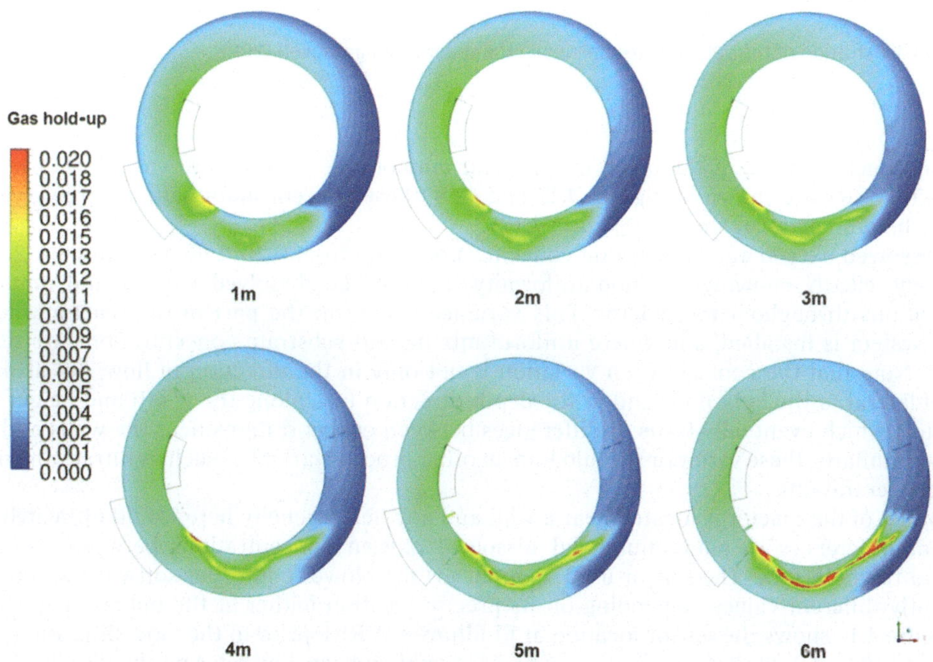

Figure 4.13 Contour plots of gas hold-up at horizontal cross sections at different depths in the reactor (1.0 m is near the top and 6.0 m is near the bottom) at Eindhoven WRRF.

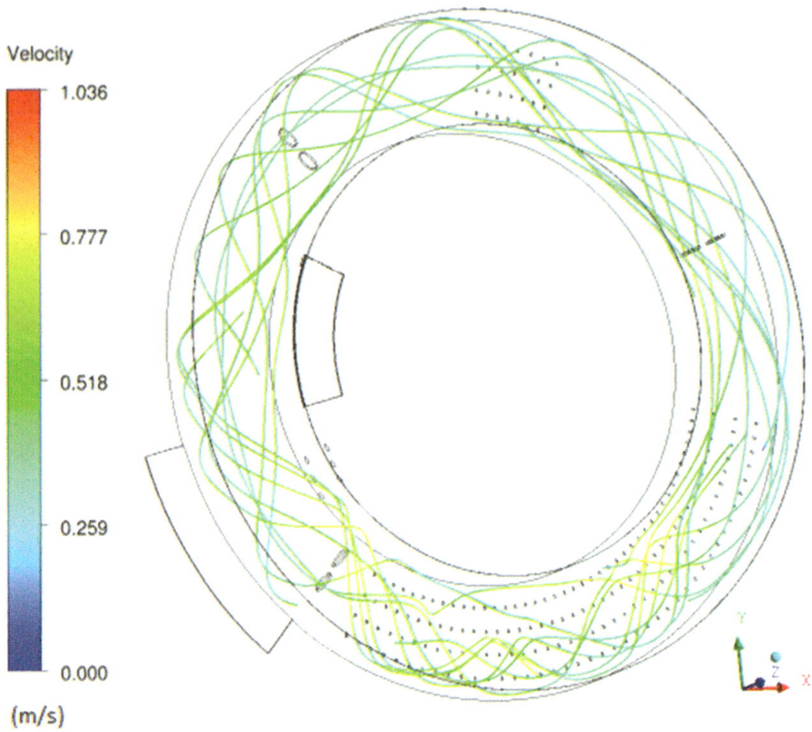

Figure 4.14 Illustration of the velocity streamlines in the reactor at Eindhoven WRRF.

4.3.1.5.3 CFD-ASM1 results

The local concentrations of dissolved oxygen and ammonium for the base case ('L-G') model of Eindhoven WRRF are shown in Figures 4.17 and 4.18 (these concentration patterns are very similar to the gas hold-up patterns shown in Figure 4.12).

The dissolved oxygen and ammonium concentrations vary from 0.20–1.96 mg/l and 0.4–2.2 mg/l, respectively, clearly showing the non-uniformity in both the dissolved oxygen and ammonium concentrations throughout the reactor. This variance occurs in the part of the reactor where the aeration system is installed, and where uniform mixing and substrate concentrations are typically assumed. Note that the concentration variation is not only in the direction of flow, but also in the direction lateral to the bulk flow, and in the depth direction (i.e., along the width and height within the reactor) which eventually leads to differences between concentrations near the walls and in the bulk flow. Similarly, these variations would impact other process variables, such as nitrate and organic carbon concentrations.

The outlet of the reactor is located near a wall and any heterogeneity here would ultimately cause significant differences in ammonium and dissolved oxygen concentrations between the reactor outlet and the bulk flow. The sensor used as input for the blower controller software would record significantly different values, depending on its precise location (either at the outlet, or in the bulk flow). Figure 4.17 shows the sensor location at Eindhoven WRRF prior to the modelling study.

Figures 4.19 and 4.20 show the impact of higher liquid and gas flow rates on the dissolved oxygen and ammonium concentrations in a vertical cross section (the same location as in Figure 4.18).

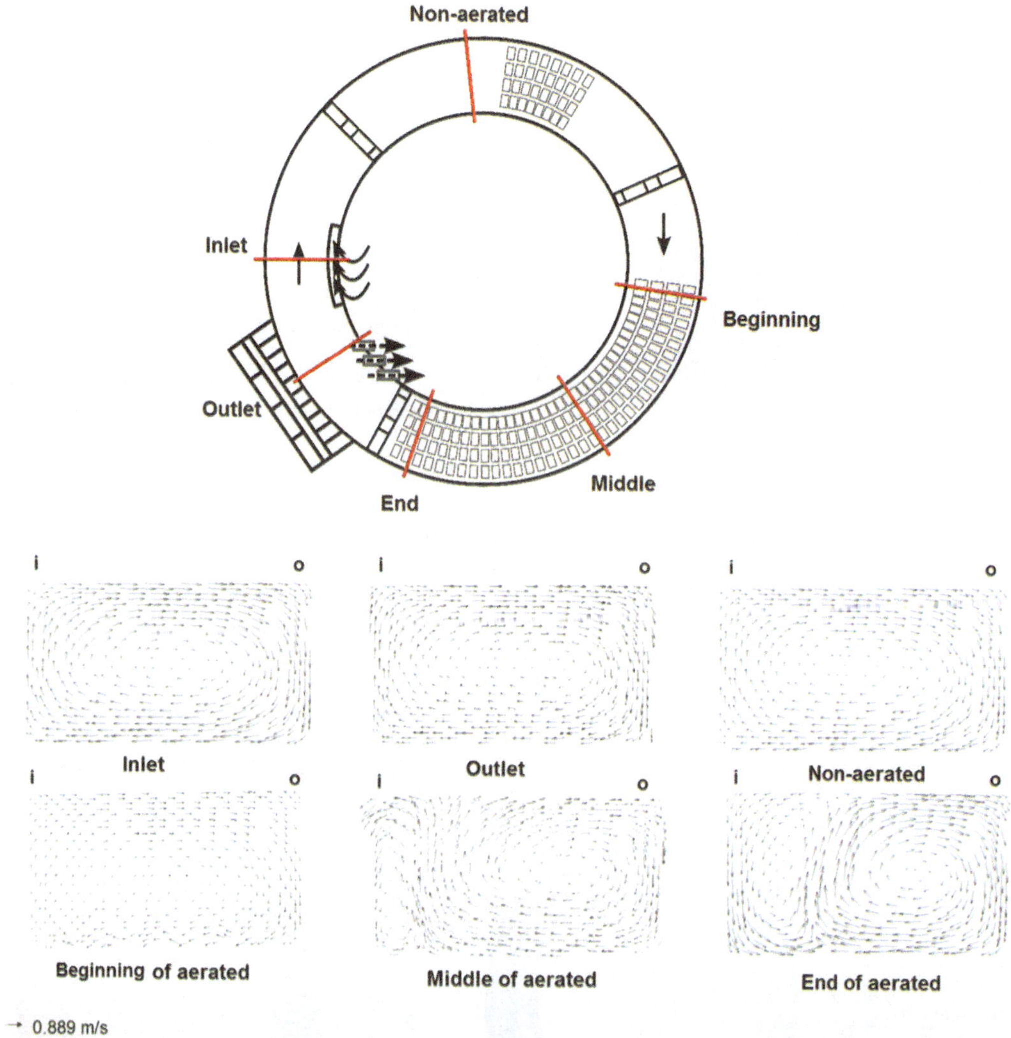

Figure 4.15 (top) Plot locations for the vertical cross sections in the reactor, and (bottom) velocity vector plots at these locations in the model of Eindhoven WRRF.

Doubling the liquid flow rate, while keeping a constant air flow of 2000 Nm^3/h, would not improve the mixing in the reactor. Significant changes were visible in terms of ammonium reduction, originating from the limited availability of air at the doubled influent flow. Tripling the gas flow brought about significant increases in the dissolved oxygen concentration (with consequent decreases in ammonium concentration). At higher liquid flows, non-uniformity in the mixing clearly prevailed throughout, which would not have been easy to predict without the application of CFD.

These models are important for several reasons.

In order to make appropriate decisions in terms of sensor location for both process monitoring and control, knowledge of spatial heterogeneity in dissolved oxygen and ammonium concentration is vital.

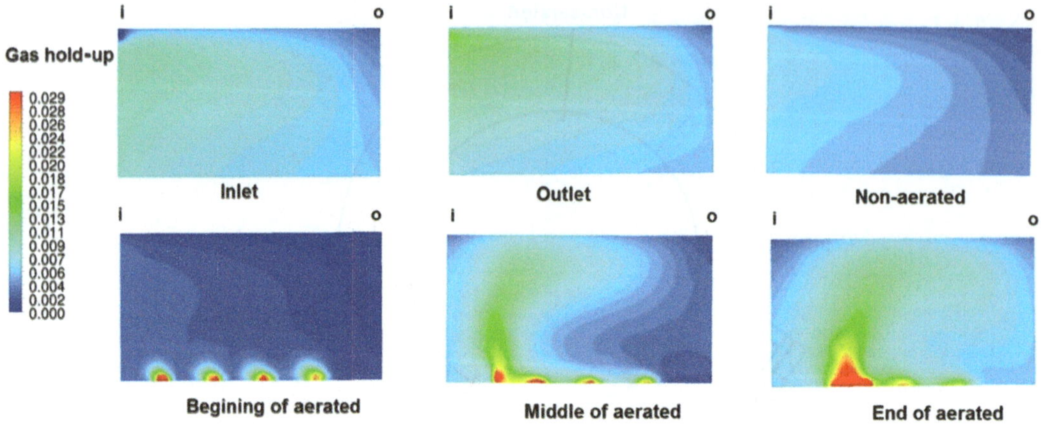

Figure 4.16 Gas hold-up in the vertical cross section at different locations in the reactor at Eindhoven WRRF.

Aerated zones are commonly perceived as homogeneously mixed, and sensor locations are therefore normally dictated by maintenance and operational convenience. Here, it is apparent that different locations along the width and depth of the reactor will bring about significantly different concentration measurements. Therefore, knowledge of the internal mixing pattern in an aeration reactor such as this, should be a minimum requirement when developing an accurate treatment plant control system.

This study has shown that a detailed process design, evaluation and optimization study should consider the effect of local mixing, and its primary effect on process performance. Conventional process engineering does not generally address spatial heterogeneity in the dissolved oxygen and substrate

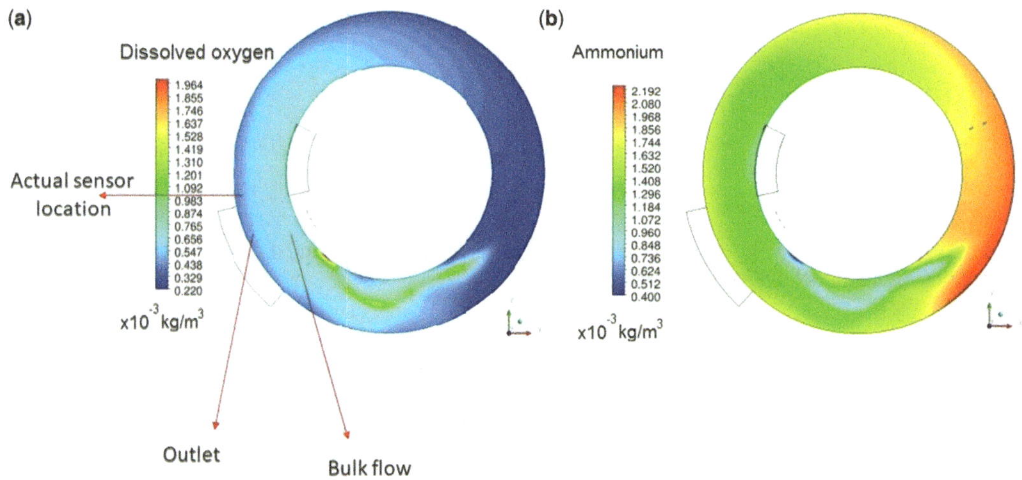

Figure 4.17 (a) Dissolved oxygen and (b) ammonium concentrations at 3.45 m depth in the bioreactor for the L-G case, at Eindhoven WRRF (source: Rehman *et al.*, 2017). (Reprinted from Water Science and Technology, volume 76, issue number 8, pages 1950–1965, with permission from the copyright holders, IWA Publishing).

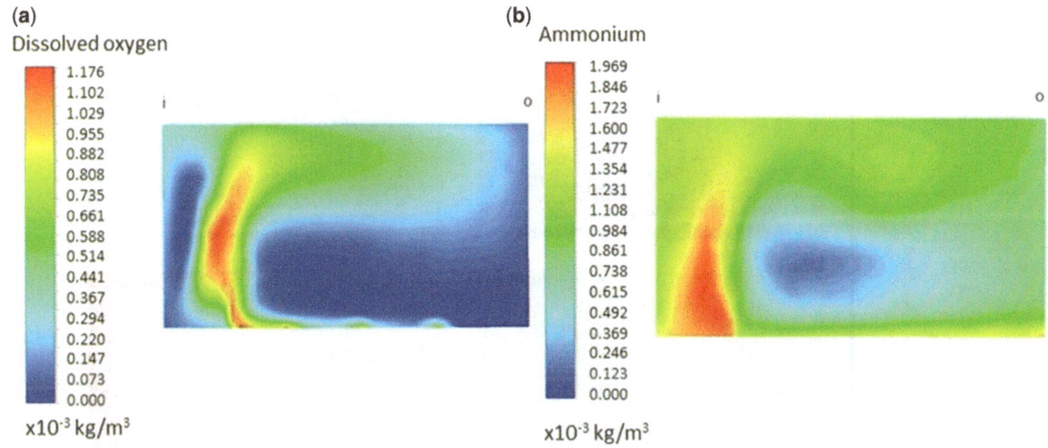

figure 4.18 dissolved oxygen (a) and ammonium (b) concentrations at a vertical cross section in the middle of the aerated region for the l-g case, at eindhoven wrrf (source: rehman *et al.*, 2017). (reprinted from water science and technology, volume 76, issue number 8, pages 1950–1965, with permission from the copyright holders, iwa publishing).

concentration caused by different aerator configurations and operational regimes. This potentially leads to a suboptimal reactor design, wasted energy, and needless cost. Obviously, a CFD model is useful to understand the plant and troubleshoot problems after construction. But it is recommended to consider a CFD model, including biokinetics if possible, in the design phase for a new tank.

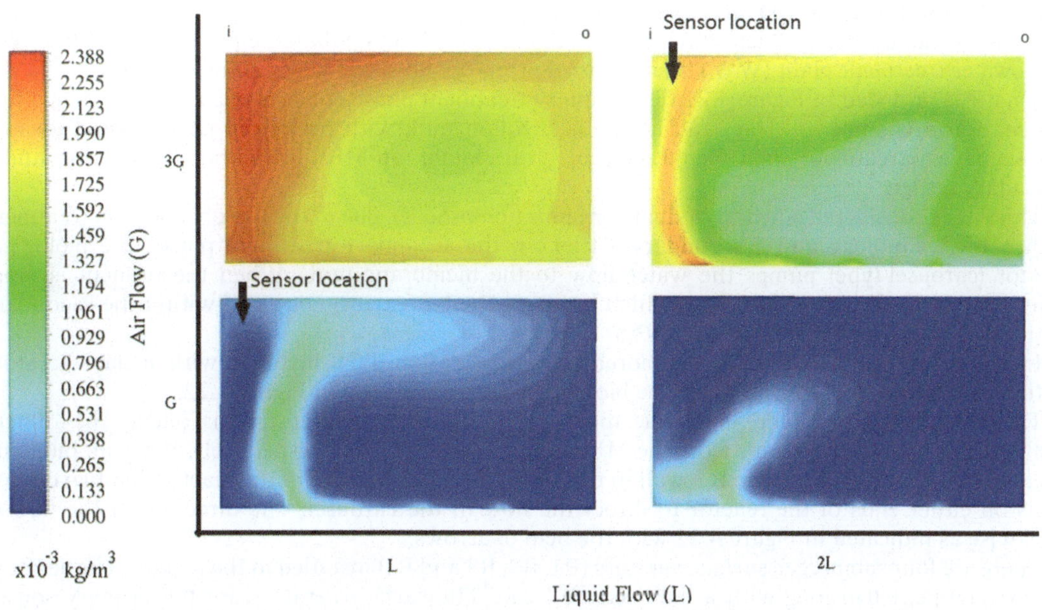

Figure 4.19 Dissolved oxygen concentrations in the middle cross section of the aerated region, at different air and liquid flow rates, for the CFD model of Eindhoven WRRF.

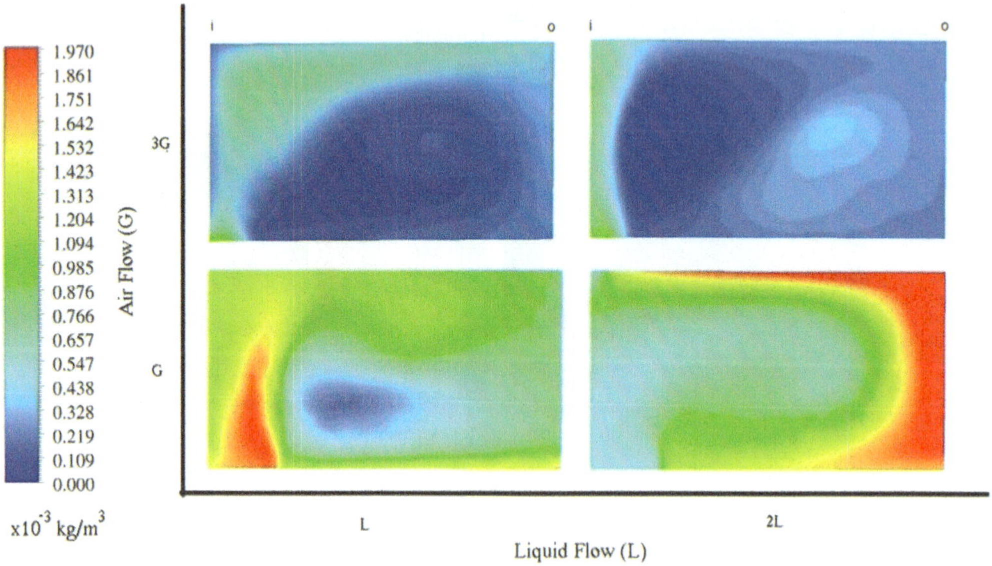

Figure 4.20 Ammonium concentrations in the middle cross section of the aerated region, at different air and liquid flow rates, for the CFD model of Eindhoven WRRF.

4.3.2 CFD modelling of the bioreactor at La Bisbal d'Empordà WWTP

Our second case study considers the wastewater treatment plant at La Bisbal d'Empordà, Spain.

4.3.2.1 Configuration of the La Bisbal WWTP

This site has an average influent flow rate of 3225 m^3/d with a population equivalent of 12 000. The wastewater treatment plant (WWTP) consists of primary and secondary treatment stages, and a site schematic is provided in Figure 4.21. The primary treatment phase consists of a coarse screen, a fine screen, a grid chamber-oil trap, and a homogenization tank with prolonged aeration and a sieve. The secondary treatment plant includes a biological reactor, an MBR (membrane bioreactor) and a secondary settler.

This site is considered a hybrid or dual configuration process, due to the use of biological treatment with MBR technology and a secondary settler. In the secondary treatment phase, the biological reactor (carousel-type) pumps the water flow to the membrane tank. When the influent is more than 3225 m^3/d, the remaining flow (which bypasses the bioreactor) is treated within the secondary settlement tank which can treat 3000 m^3/d.

In the modelling project, only the bioreactor, an oxidation ditch installed with surface aerators (rotors), was simulated. The layout of the bioreactor model is shown in Figure 4.22.

The bioreactor is 70 m long, 14 m wide and 4 m deep. It has three inlet streams: one for the influent and two for the recycle streams from the MBR and secondary settler, respectively. It has an outgoing stream going to the MBR. There is a wall in the middle along the length of the reactor and two curved walls on either ends of the reactor to direct the flow in the carousel. The direction of the flow is clockwise as indicated in Figure 4.22 with the help of arrows.

There are four submerged surface aerators (R1, R2, R3 and R4) installed in the reactor. The aerators are brush-type, 6.0 m long with a diameter of 1.2 m. The surface aerators are the primary source of oxygen and provide mechanical energy to maintain the sludge flow and avoid potential sludge settling. Therefore, the function of the surface aerators is particularly important for the performance

Figure 4.21 La Bisbal d'Empordà WRRF.

of the bioreactor. However, operating all the aerators continuously results in high operational cost. Currently, only two aerators are used during the day (when the electricity price is high) and all four are only used at night (when the electricity price is low). This strategy reduces the operational cost, but the performance of the bioreactor is compromised. Hence, it is important to evaluate the performance of the reactor under both conditions.

4.3.2.2 Measurements

Flow measurements were performed for the validation of the hydrodynamic model of the reactor, using a simple flow measuring device (MC20, PCE inst.), with velocities measured 2.0 m downstream of rotor R2 (this location was chosen due to the ease of access). All the measurements were performed during the day, that is when only two rotors are in operation.

The measurements were performed at seven sample points across the width, and at three depths (at the surface, 1.0 and 2.0 m deep). Each measurement was performed for 2–3 minutes and repeated three times.

Dissolved oxygen measurements were taken using a DO probe (497ADO, Emerson) located 2.0 m downstream from R2 (the same location as the velocity measurements) and at R4 (see Figure 4.22). These measurements were performed at three points across the width, but only at one depth due to practical limitations. These measurements were performed primarily to calibrate the oxygen flux coming into the reactor from the surface aerators. Again, the measurements were time averaged and repeated three times.

4.3.2.3 Simulation scenarios

Two scenarios: (a) two operational rotors, and (b) four operational rotors, were simulated. It was assumed that the plant was operating at average diurnal influent rate and the recycle flow from the settler was turned off.

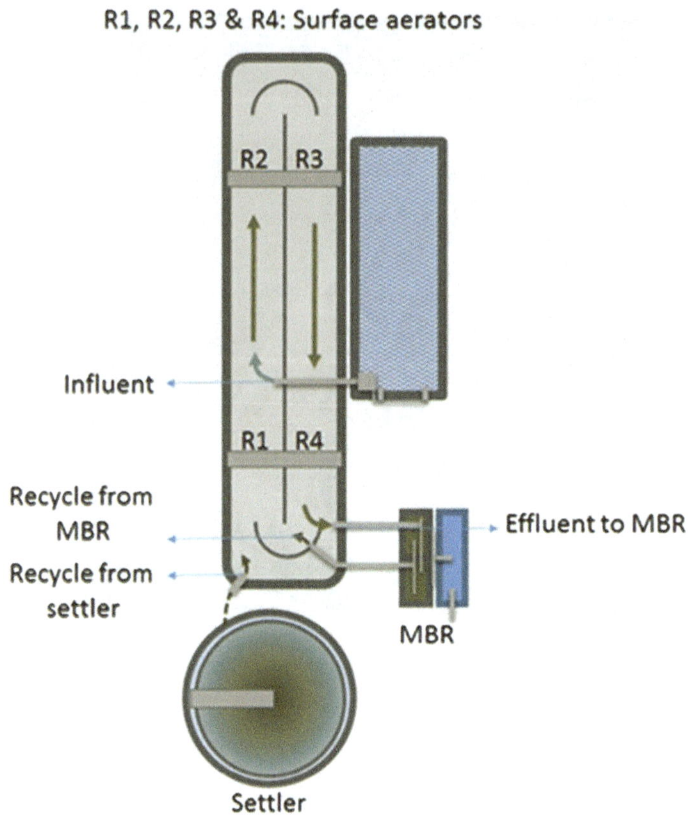

Figure 4.22 Configuration of the oxidation ditch bioreactor (La Bisbal, Spain).

4.3.2.4 Computational fluid dynamic modelling

A stepwise approach to the creation of the CFD model, as outlined in Chapter 2 and in this chapter, was followed.

4.3.2.4.1 Geometry development

In the first step, a three-dimensional geometrical model of the reactor was created. The size and location of the inlets, outlets and surface aerators were based on the dimensions of the site. This can be seen in Figure 4.23. The outlet was divided into three sections, because effluent is pumped to the MBR using three separate pumps. The surface aerators (rotors) were named individually *R1, R2, R3* and *R4* for reference. *CW1* and *CW2* were two curved walls at either end of the reactor, and they are used for reference during the discussion. Similarly, the wall in the middle is referred to as the *Middle Wall.* Additionally, arrows show the direction of flow in the reactor.

The four surface aerators were modelled as long thin cylinders with a hollow core mimicking the shaft of the rotors. The modelling of rotors is illustrated in Figure 4.24. The inner diameter is the size of the shaft and the outer diameter is the size of the blades. The individual rotor blades were not modelled, but it is assumed that the rotor acts merely as a source of extra momentum. Hence, the rotor was divided equally into four segments (across the cross section) to implement a tangential momentum source in each segment.

4.3.2.4.2 Meshing

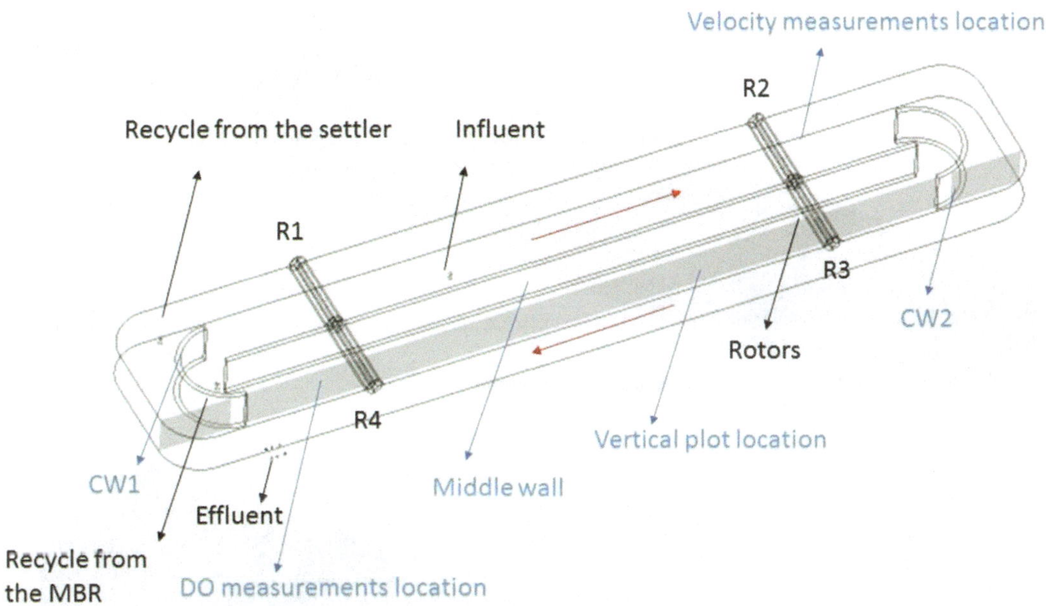

Figure 4.23 The three-dimensional model of the oxidation ditch reactor at La Bisbal, Spain (blue text introduces the terminology to be used in the results and discussion section).

Detailed meshing of the reactor was performed. Two separate meshes were generated for the two operating conditions.

In the two-rotor case, R2 and R4 are operating. The four-rotor mesh consisted of 800,000 cells, whereas the two-rotor mesh consisted of 610,000 cells. The mesh was kept fine in specific areas (such as near the rotors, and the inlet and outlet boundaries) where high gradients of pressure and velocity were expected. The resulting meshes for both cases can be seen in Figure 4.25. The only difference

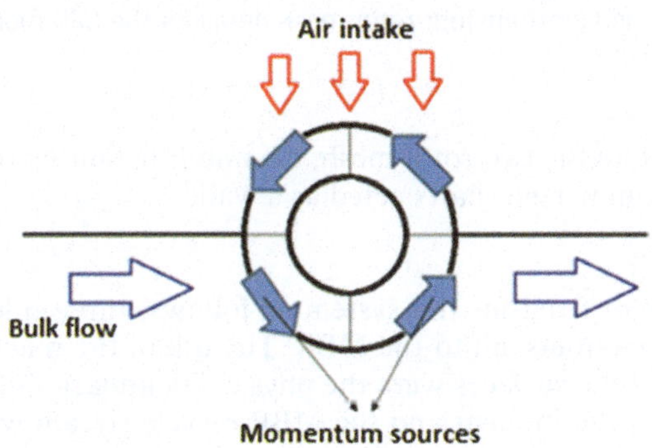

Figure 4.24 Illustration of the rotor modelling.

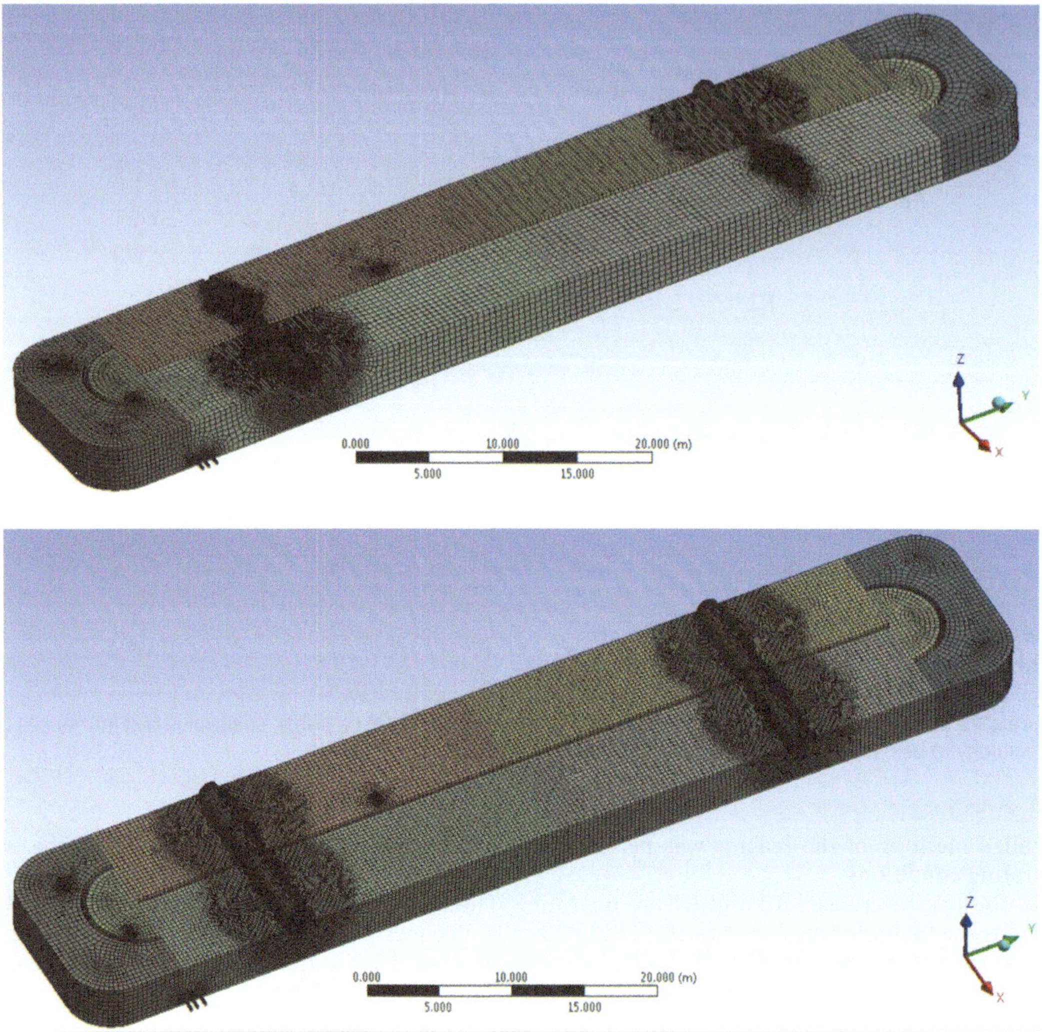

Figure 4.25 (top) The two-rotor and (bottom) four-rotor mesh detail for the CFD model of the oxidation ditch at La Bisbal, Spain.

between the meshes was that, in the two-rotor mesh, the non-functioning rotors (R1 and R3) were not included, and their partially immersed shafts acted as a wall.

4.3.2.4.3 Boundary conditions

The boundary conditions were fixed for the system as follows: three inlets (influent, MBR recycle and settler recycle) and three outlets, all to the MBR. The top of the reactor was a free surface (i.e., uncovered). The remainder of the surfaces were the physical boundaries of the reactor.

The flows coming through the influent and the MBR recycle stream were known, and they were set with '*velocity inlet*' boundary condition. The settler recycle was turned off so this was a '*wall*' boundary condition. The flow going out of the system through effluent streams was set with an '*outflow*' boundary condition. The top surface of the reactor was set with a '*symmetry*' boundary condition. The tops of all the rotors were set with a '*wall*' boundary condition to be able to define an oxygen flux through them. The remaining boundaries were defined as '*walls*'.

The inlet boundary conditions for the influent and the MBR recycle stream were calculated based on the average flow data. The average influent flow rate was 3225 m^3/d and the MBR recycle flow was 2995 m^3/d.

4.3.2.4.4 Flow and other considerations

Only steady-state simulations were performed, bearing in mind that the biokinetic model would be subsequently integrated and therefore transient modelling was likely to be impractical. The flow inside the reactor was multiphase, due to the presence of solids (biomass) in the liquid, and air being present. Solids concentration was considered, and density included as a function of the total suspended solids fraction (Equation (4.4)).

Due to the presence of surface aerators, the flow inside the reactor would be turbulent and a $k-\varepsilon$ model was used. The convergence criterion was set to be in the order of 10^{-4} for all the flow variables (i.e., continuity, momentum, turbulent kinetic energy and turbulent dissipation). The solution time for the hydrodynamic modelling of the four-rotor case was approximately 4–5 hours using four parallel processors. However, the solution time reduced to 3–4 hours for the two-rotor case (due to the lower number of cells in the mesh).

4.3.2.4.5 Biokinetic modelling

The composition of the influent and the MBR recycle stream was taken from plant data and is presented in Table 4.5 in ASM1 component form. However, the complete composition for the MBR recycle stream was unavailable. Only ammonium and nitrate concentrations were known and, therefore, the rest of the components were assumed based on similar systems.

The oxygen intake due to surface aeration was modelled by defining a boundary flux at the top surface of the rotors.

4.3.2.5 Results and discussion

4.3.2.5.1 Comparison between the velocity measurements and the CFD model

In Figure 4.26, a comparison between measured velocities and the velocities predicted by the CFD modelling is presented.

Table 4.5 The influent and MBR recycle stream composition for La Bisbal, Spain.

No.	ASM1 Components	Influent (mg/l)	MBR Recycle (mg/l)
1	S_I	30.8	30.8
2	S_S	75.01	1
3	X_I	684.9	1000
4	X_S	119.3	40
5	$X_{B,H}$	23.8	900
6	$X_{B,A}$	10.1	95
7	X_P	0	0
8	S_O	0	0
9	S_{NO}	2.7	3.6
10	S_{NH}	18.4	4.5
11	S_{ND}	0.4	0.4
12	X_{ND}	4	4

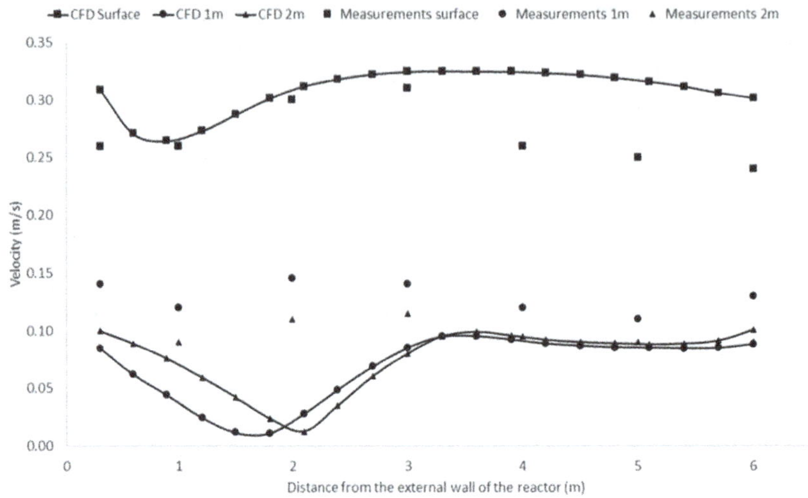

Figure 4.26 Comparison between the velocity measurements and the velocities predicted by the CFD model at La Bisbal, Spain.

The velocities predicted by the CFD near the surface were in good agreement with the measured values, however, the predicted velocities at depths of 1.0 and 2.0 m were lower than expected.

This difference could be due to the approximation method used for modelling the rotors, and thus could be improved by more detailed modelling of the rotor brush arrangement. Although the impact of the solids concentration was taken into account by modelling density as a function of total suspended solids, a more detailed modelling of solids transport may be needed to accurately model the potential sludge settling behaviour and hence improve the hydrodynamic prediction.

4.3.2.5.2 Hydrodynamic results

The hydrodynamic results for both cases (i.e., two-rotor and four-rotor) are presented separately.

4.3.2.5.2.1 The two-rotor case

The velocity contour plots for the two-rotor case in horizontal cross sections at different depths (near the top, middle and bottom) in the reactor are shown in Figure 4.27. Note, that the flow is in a clockwise direction.

The impact of the rotors on the flow in the reactor is clearly visible in Figure 4.27. Higher velocities are observed in the vicinity of the rotors, whereas low velocity regions are visible near the middle wall and in the corners of the reactor behind the curved walls (CW1 and CW2). Two swirl regions are formed, one along the middle wall and one behind the curved wall. Both are due to boundary layer detachment phenomena.

Figures 4.28 and 4.29 also highlight velocity profile changes as a function of depth. Close to the rotors, high velocity areas form near the surface, but lower velocities exist near the base of the reactor. A significant swirl region is formed, and as the fluid moves at higher velocity near the surface, a region of lower pressure is created which draws fluid up from the bottom.

4.3.2.5.2.2 The four-rotor case

Similarly, the velocity plots for the four-rotor case at different depths are shown in Figure 4.30.

The velocities in the reactor are higher in comparison with the two-rotor case. However, lower velocities can still be seen near the middle wall and in the corners of the reactor.

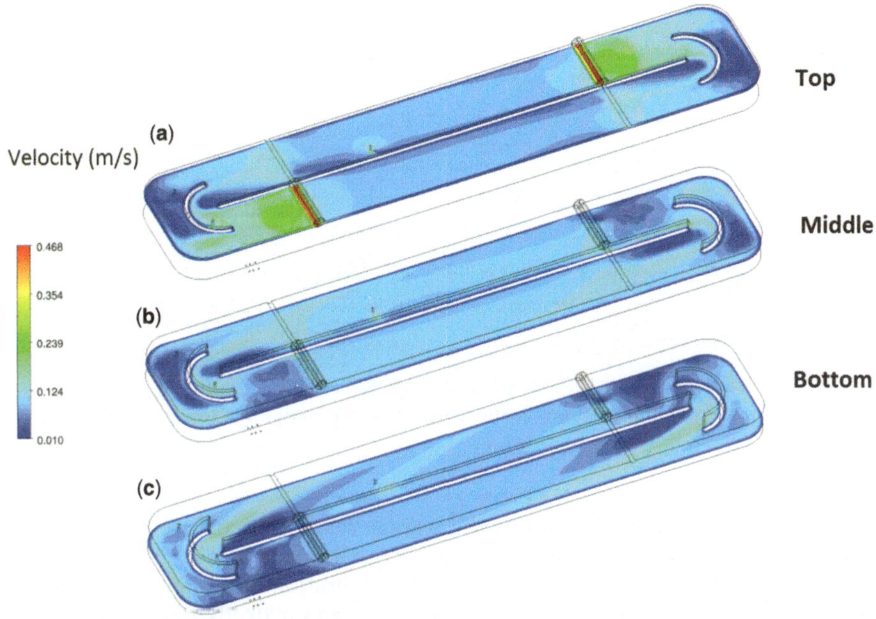

Figure 4.27 Velocity contour plots near the (a) top, (b) middle, and (c) bottom of the reactor model for the two-rotor case at La Bisbal, Spain.

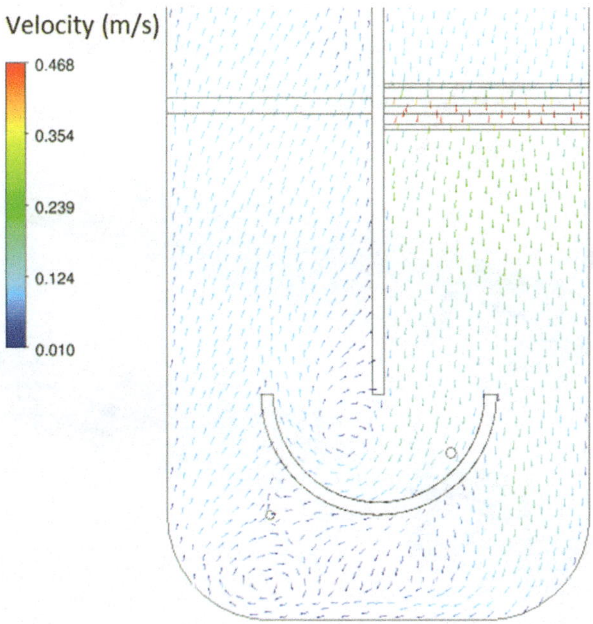

Figure 4.28 Velocity vector plots near the top surface of the reactor for the two-rotor case at La Bisbal, Spain.

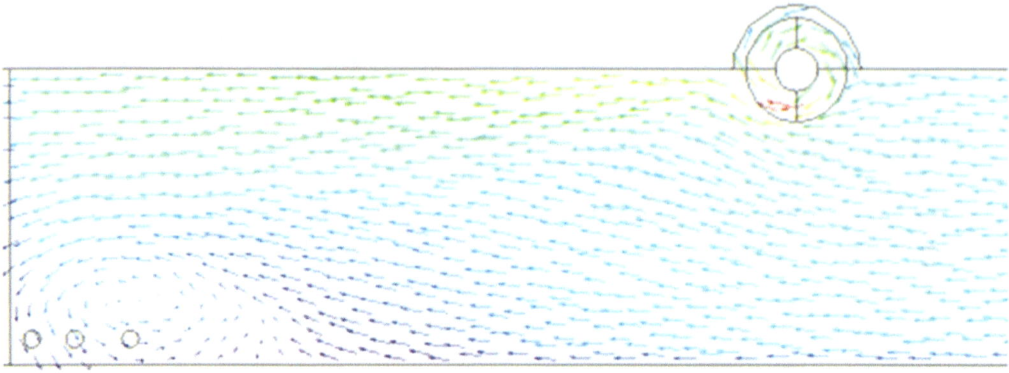

Figure 4.29 Velocity vector plot in the sidewise vertical cross section of the reactor near the R4 rotor at La Bisbal, Spain.

The impact of operating two or four rotors on the average velocities at different depths is summarized in Figure 4.31. It shows that the average velocities for the four-rotor case are almost twice as high compared with the two-rotor case. These are important operational differences that are usually not considered when modelling such systems using a tanks-in-series hydrodynamic approach.

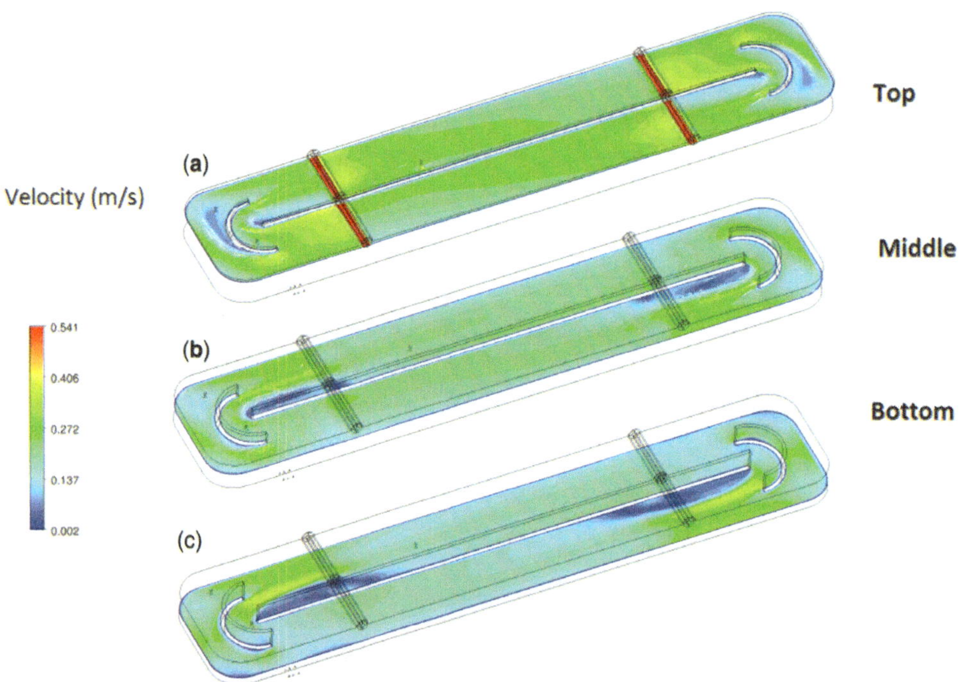

Figure 4.30 Velocity contour plots near the (a) top, (b) middle, and (c) bottom of the reactor model for the four-rotor case at La Bisbal, Spain.

Figure 4.31 Comparison of average velocities at different depths for 2-rotor and 4-rotor cases at La Bisbal, Spain.

4.3.2.5.3 CFD-biokinetic modelling results

4.3.2.5.3.1 The two-rotor case

Dissolved oxygen (DO) concentration plots for La Bisbal, at different depths of the reactor are presented in Figure 4.32.

The DO levels varied across the reactor between 0.05 and 1.08 mg/l. Higher DO levels were observed in the region next to the rotors, however, the DO level decreased significantly away from the rotors. The DO level was higher near the top surface of the reactor, but Figure 4.32 clearly demonstrates a large depletion in DO in the middle and near the bottom of the reactor.

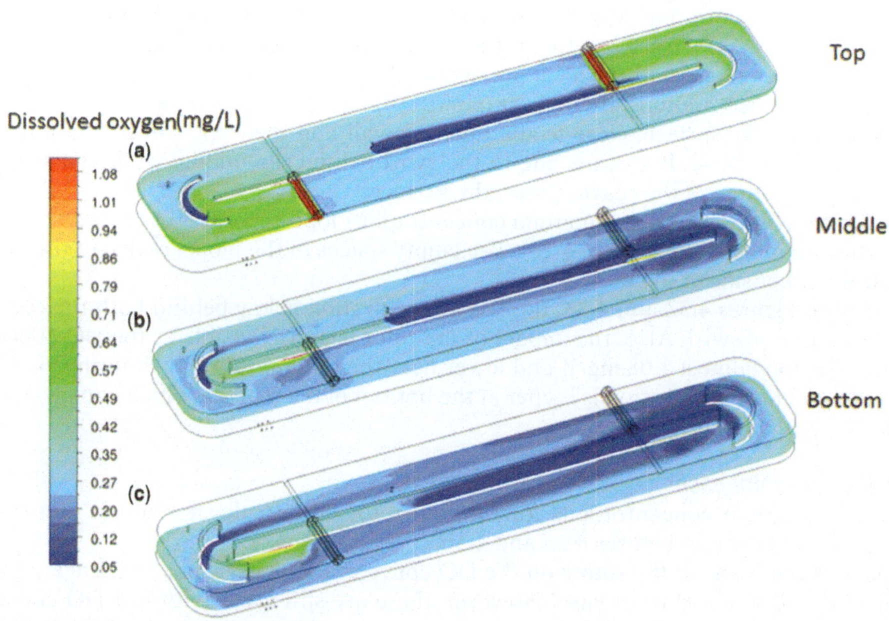

Figure 4.32 Dissolved oxygen concentration plots near the (a) top, (b) middle, and (c) bottom of the reactor model for the two-rotor case at La Bisbal, Spain.

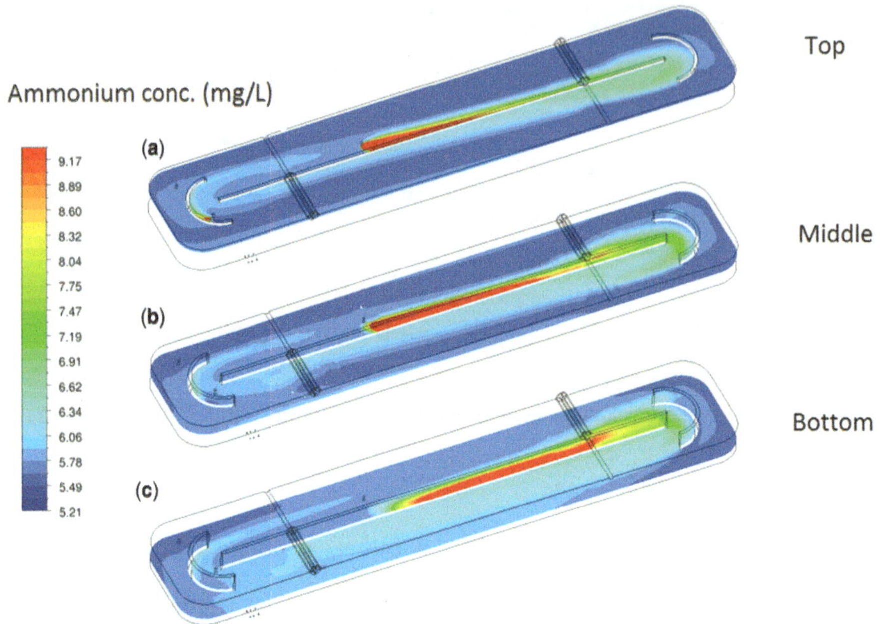

Figure 4.33 Ammonium concentration plots near the (a) top, (b) middle, and (c) bottom of the reactor model for the two-rotor case at La Bisbal, Spain.

The ammonium concentrations at different depths are presented in Figure 4.33.

The ammonium concentrations vary between 5.21 and 9.17 mg/l although the blue contours do not really show good conversion in the reactor, as they remain above 5.0 mg/l. This variation is due to the hydrodynamics of the reactor and the local (un)availability of dissolved oxygen.

The ammonium concentrations are highest (almost 9.1 mg/l) near the middle wall, as the influent flow is introduced at the surface, close to the dividing wall and then flows in the bulk direction of the carousel. After the bend, it spreads due to the swirl behind the middle wall, and spreads more uniformly near the bottom of the reactor, than the surface.

Figure 4.34 shows the DO and ammonium concentrations for a vertical cross section of the reactor (the plot location is indicated in Figure 4.23). The empty spaces in the plots are due to the presence of curved walls on either side of the tank.

As observed in Figures 4.32 and 4.33, the DO concentration is low behind both the curved walls due to the formation of swirl. Also, the impact of the rotor is clearly visible as the DO concentration is higher near the top (almost 1.08 mg/l) and lower near the bottom (0.05 mg/l) in the vicinity of the rotor. The ammonium concentration is higher at the bottom of the reactor, which can be explained by the lack of dissolved oxygen.

4.3.2.5.3.2 The four-rotor case

The DO and ammonium concentrations for the four-rotor case in the horizontal cross section at different depths are shown in Figures 4.35 and 4.36.

The impact of operating all the rotors on the DO concentrations is visible, as there are higher DO levels compared with the two-rotor case. However, there are still areas with low DO concentration due to the swirl near the bends. The DO concentrations are higher near the top surface and lower near the bottom, and the ammonium concentration distribution is similar to the two-rotor case, although the values are lower and vary between 2.16 and 5.9 mg/l.

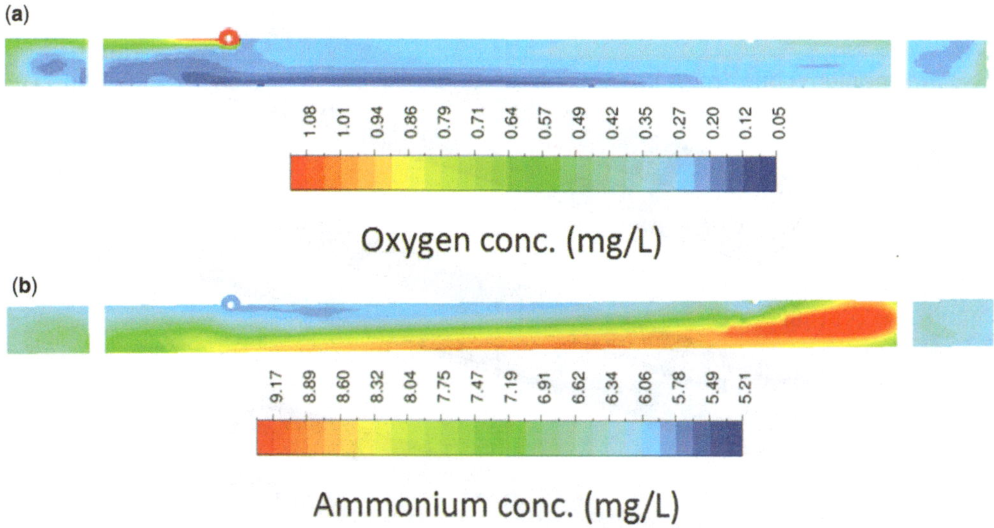

Figure 4.34 (a) Dissolved oxygen, and (b) ammonium concentrations at a vertical cross section of the reactor for the two-rotor case at La Bisbal, Spain.

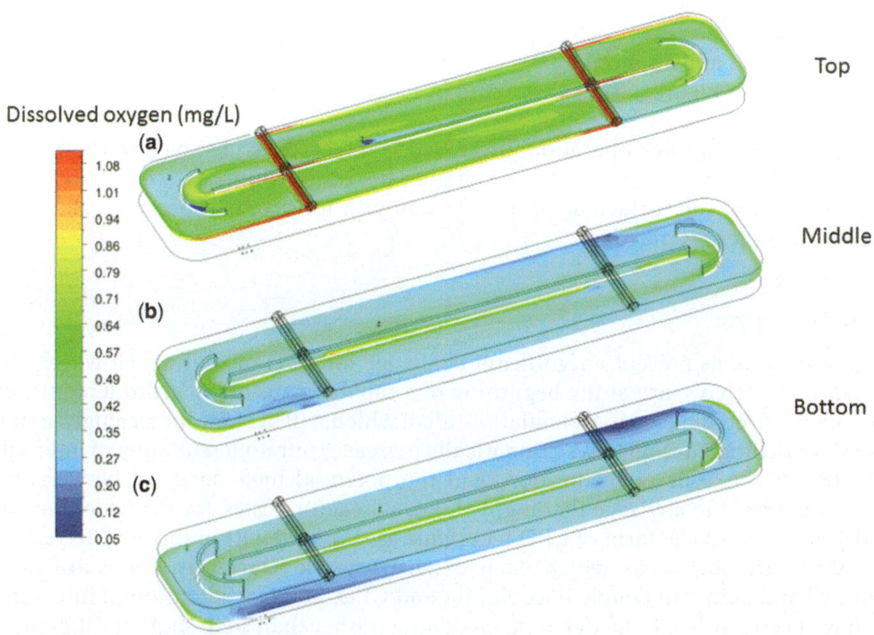

Figure 4.35 Dissolved oxygen concentration plots near the (a) top, (b) middle, and (c) bottom of the reactor model for the four-rotor case at La Bisbal, Spain.

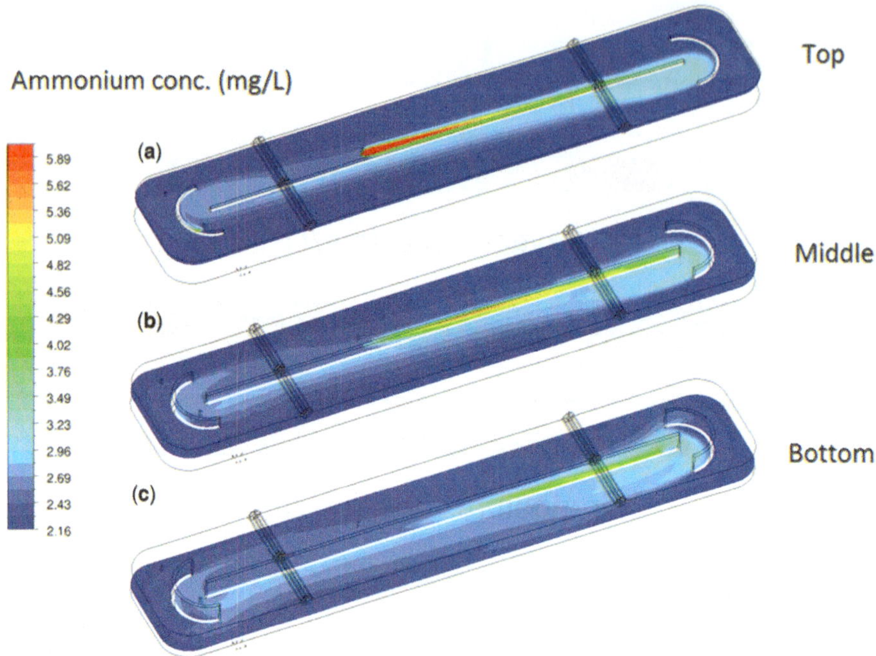

Figure 4.36 Ammonium concentration plots near the (a) top, (b) middle, and (c) bottom of the reactor model for the four-rotor case at La Bisbal, Spain.

The dissolved oxygen and ammonium concentration plots presented for a vertical cross section of the reactor are shown in Figure 4.37. As observed in Figure 4.37, the DO is highest at the surface and lowest near the bottom.

Compared with the two-rotor operation, increased DO levels are observed in the region between the rotors, and the ammonium concentration is higher near the bottom. However, this highlights the operational limitation of the surface aerators, as even with four rotors operating, the area near the base of the reactor is (nearly) anoxic.

4.4 RESEARCH NEEDS

Suspended growth systems present a significant challenge for CFD modelling. However, this chapter has shown that it is likely we are at the beginning of a rapid expansion in hydrodynamic, gas/liquid/solid and biokinetic three-dimensional simulation, all of which will become increasingly standard in the water industry. Modelling internal flows dramatically increases our understanding of internal substrate distribution, dissolved oxygen concentration, and the potential inefficiency of blowers, mixers and bioreactors. It is not hard to envisage that highly tailored reactor design for the water and wastewater industry will incorporate some form of CFD technique as standard within the next 10 years.

However, there are still areas which need extensive work. Currently, there are limits to our understanding of and ability to couple flocculation and CFD, and consideration of interactions at the micro-scale have been limited. Fully dynamic modelling (rather than snap shots at different flow rates) is still not practical. It is likely there will be a key need for broader applications of the CFD technique, with a range of non-Newtonian flows, complex rheological properties, more detailed biokinetic models, and greenhouse gas emissions based on velocity predictions at the surface. Models which

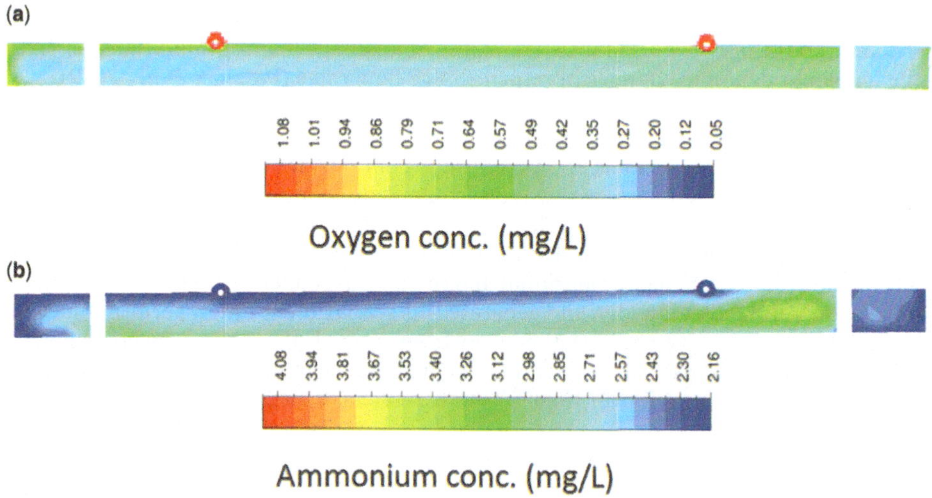

Figure 4.37 (a) Dissolved oxygen, and (b) ammonium concentrations at a vertical cross section of the reactor for the four-rotor case at La Bisbal, Spain.

cover the range of activated sludge and other processes are necessary given the diversity of the field. There is also a need to apply CFD techniques to emerging low energy units, such as granular activated sludge, anaerobic mainline treatment units (including anaerobic MBR), suspended carrier systems, and side stream anaerobic ammonium removal (which is critically dependent on local conditions).

The bubbly flow regime in an AS tank and the intensity of collisions, agglomeration, breakup and deformation also promotes a large diversity in the shape and size of bubbles (Portela, 2013; Shaikh & Al-Dahhan, 2007; Takacs, 2005). The impact of bubble size on KLa and oxygen mass transfer in aeration tanks has been recognized (Fayolle *et al.*, 2007; Vermande *et al.*, 2007), however, for the CFD modelling of aeration in conventional AS tanks, the assumption of a non-distributed scalar property (a constant bubble diameter) is still dominant (Sobremisana *et al.*, 2012). Consequently, extended multiphase models which incorporate bubble/particle size distributions will require the use of population balance models (PBMs) to describe variations in population over time. An assessment of several solution methods for PBMs, namely the discrete class size method (Hounslow *et al.*, 1988), the standard method of moments, SMM (Randolph & Larson, 1971) and the quadrature method of moments, QMOM (Marchisio *et al.*, 2003) have been highlighted, although their application to modelling of aeration systems with suspended solids is still scarce and limited to a few examples (Bridgeman *et al.*, 2009; Dhanasekharan *et al.*, 2005; Morchain *et al.*, 2014; Nopens *et al.*, 2015; Sobremisana *et al.*, 2012; Wang, 2010).

When considering an AS tank in detail, experimental studies have also demonstrated the influence of floc size on flocculation behaviour and the settleability of activated sludge (Nopens *et al.*, 2015), as well as the biokinetic reaction environment within a single floc (Sobremisana *et al.*, 2012). Therefore, an estimation of floc size distribution may provide further insight into modelling activated sludge mixing and settling (Samstag *et al.*, 2012), which could allow a better prediction of hydraulic capacity and treatment efficiency during high hydraulic loads associated with wet weather flows (Nielsen *et al.*, 2000; Sharma *et al.*, 2013).

In engineering practice, the behaviour of AS tanks is inseparably linked to the performance of a clarifier, as efficient BOD and nutrient recovery in the AS process depends on solids retention time (SRT), concentration of MLSS, and the microbiology, which is in turn affected by F/M (feed/mass) ratio, aeration conditions and the occurrence of toxic shock loadings (hard COD). There is a significant drive towards dynamic modelling of a coupled aeration tank + clarifier system, and the prediction of interconnected flows and mass transport in unsteady loading conditions, for example, to evaluate the impact of dynamic changes in return sludge flow rates (Patziger *et al.*, 2012).

When considering the complexity of CFD models, the ability to predict clarifier behaviour simultaneously with a complete solid-liquid-gas model of the AS tank, is inevitably limited by the availability of typical CFD resources. As a result, common modelling practice is to avoid the interference between multiple unit processes, and assume a constant separation efficiency for a clarifier downstream (Le Moullec *et al.*, 2010a, 2010b, 2011) predicted in agreement with standard guidelines (Copp, 2000; Laurent *et al.*, 2014). It is surely time to create a CFD model, validated with experimental data, which includes both a three-dimensional biokinetic suspended growth system and a simple settlement tank, operating dynamically, and generate this scenario with high performance computing infrastructure. This would move the discussion on from snap shots of nutrient concentration for a single tank to the dynamic impact of wet weather events, or a variation in blower output and influent concentration on the aeration tank over time.

REFERENCES

Alvarado A., Vedantam S., Goethals P. and Nopens I. (2012). A compartmental model to describe hydraulics in a full-scale waste stabilization pond. *Water Research*, **46**(2), 521–530, https://doi.org/10.1016/j.watres.2011.11.038

Amerlinck Y. (2015). Model refinements in view of wastewater treatment plant optimization: improving the balance in sub-model detail. PhD thesis, Ghent University. Faculty of Bioscience Engineering, Ghent, Belgium.

Andersson B., Andersson R., Håkansson L., Mortensen M., Sudiyo R. and Van, Wachem B. (2011). Computational Fluid Dynamics for Engineers. Cambridge University Press, Cambridge, UK.

Ardern E. and Lockett W. T. (1914). Experiments on the oxidation of sewage without the aid of filters. *Journal of the Society of Chemical Industry*, **33**(10), 523–539, https://doi.org/10.1002/jctb.5000331005

Boyle-Gotla A., Jensen P. D., Yap S. D., Pidou M., Wang Y. and Batstone D. J. (2014). Dynamic multidimensional modelling of submerged membrane bioreactor fouling. *Journal of Membrane Science*, **467**, 153–161, https://doi.org/10.1016/j.memsci.2014.05.028

Brannock M. W. D., Wang Y. and Leslie G. (2010). Evaluation of full-scale membrane bioreactor mixing performance and the effect of membrane configuration. *Journal of Membrane Science*, **350**(1–2), 101–108, https://doi.org/10.1016/j.memsci.2009.12.016

Bridgeman J., Jefferson B. and Parsons S. A. (2009). Computational fluid dynamics modelling of flocculation in water treatment: a review. *Engineering Applications of Computational Fluid Mechanics*, **3**(2), 220–241, https://doi.org/10.1080/19942060.2009.11015267

Climent J., Basiero L., Martínez-Cuenca R., Berlanga J. G., Julián-López B. and Chiva S. (2018). Biological reactor retrofitting using CFD-ASM modelling. *Chemical Engineering Journal*, **348**, 1–14, https://doi.org/10.1016/j.cej.2018.04.058

Cobb E. D. (1993). Broad-band acoustic Doppler current profiler. *Flow Measurement and Instrumentation*, **4**(1), 35–37, https://doi.org/10.1016/0955-5986(93)90008-7

Cockx A., Do-Quang Z., Audic J. M., Liné A. and Roustan M. (2001). Global and local mass transfer coefficients in waste water treatment process by computational fluid dynamics. *Chemical Engineering and Processing*, **40**(2), 187–194, https://doi.org/10.1016/S0255-2701(00)00138-0

Copp J. (2000). The COST Simulation Benchmark – Description and Simulator Manual. COST. European Cooperation in the field of Scientific and Technical Research, Brussels, Belgium.

De Clercq B. (2003). Computational fluid dynamics of settling tanks: development of experiments and rheological, settling, and scraper submodels. PhD thesis, Ghent University, Ghent, Belgium.

Dhanasekharan K. M., Sanyal J., Jain A. and Haidari A. (2005). A generalized approach to model oxygen transfer in bioreactors using population balances and computational fluid dynamics. *Chemical Engineering Science*, **60**(1), 213–218, https://doi.org/10.1016/j.ces.2004.07.118

Do-Quang Z., Cockx A., Liné A. and Roustan M. (1998). Computational fluid dynamics applied to water and wastewater treatment facility modeling. *Environmental Engineering and Policy*, **1**(3), 137–147, https://doi.org/10.1007/s100220050015

Ducoste J. J. and Clark M. M. (1998). The influence of tank size and impeller geometry on turbulent flocculation: I. Experimental. *Environmental Engineering Science*, **15**(3), 215–224, https://doi.org/10.1089/ees.1998.15.215

Fan L., Xu N., Wang Z. and Shi H. (2010). PDA Experiments and CFD simulation of a lab-scale oxidation ditch with surface aerators. *Chemical Engineering Research and Design*, **88**(1), 23–33, https://doi.org/10.1016/j.cherd.2009.07.013

Fayolle Y., Cockx A., Gillot S., Roustan M. and Héduit A. (2007). Oxygen transfer prediction in aeration tanks using CFD. *Chemical Engineering Science*, **62**(24), 7163–7171, https://doi.org/10.1016/j.ces.2007.08.082

Glover G. C., Printemps C., Essemiani K. and Meinhold J. (2006). Modelling of wastewater treatment plants – how far shall we go with sophisticated modelling tools? *Water Science and Technology*, **53**(3), 79–89, https://doi.org/10.2166/wst.2006.078

Gresch M., Armbruster M., Braun D. and Gujer W. (2011). Effects of aeration patterns on the flow field in wastewater aeration tanks. *Water Research*, **45**(2), 810–818, https://doi.org/10.1016/j.watres.2010.09.009

Henze M., Grady Jr, C. P., Gujer W., Marais G. V. R. and Matsuo T. (1987). Activated Sludge Model No. 1. IAWQ Scientific and Technical Report No. 1, London, UK.

Henze M., Gujer W., Mino T. and van Loosdrecht M. (2006). Activated Sludge Models ASM1, ASM2, ASM2d and ASM3. IWA Publishing, London, UK.

Hounslow M. J., Ryall R. L. and Marshall V. R. (1988). A discretized population balance for nucleation, growth, and aggregation. *AIChE Journal*, **34**(11), 1821–1832, https://doi.org/10.1002/aic.690341108

Huang W. and Li K. (2013). CFD Simulation of flows in stirred tank reactors through prediction of momentum source. In: Nuclear Reactor Thermal Hydraulics and Other Applications. Donna Guillen (ed.). London: IntechOpen, 2013. 10.5772/51754

Jiang C.-Y., Huang W.-D., Wang G., Wang Y.-Z. and Xie R.-H. (2010). Numerical computation of flow fields in an oxidation ditch by computational fluid dynamics model. *Huanjing Kexue yu Jishu*, **33**(8), 135–140.

Karpinska A. M. and Bridgeman J. (2016). CFD-aided modelling of activated sludge systems – a critical review. *Water Research*, **88**, 861–879, https://doi.org/10.1016/j.watres.2015.11.008

Karpinska A. M. and Bridgeman J. (2017). Towards a robust CFD model for aeration tanks for sewage treatment – a lab-scale study. *Engineering Applications of Computational Fluid Mechanics*, **11**(1), 371–395, https://doi.org/10.1080/19942060.2017.1307282

Karpinska A. M. and Bridgeman J. (2018). CFD As a tool to optimize aeration tank design and operation. *Journal of Environmental Engineering*, **144**(2), 05017008, https://doi.org/10.1061/(ASCE)EE.1943-7870.0001307

Karpinska A. M., Dias M. M., Boaventura R. A. R. and Santos R. J. (2015). Modeling of the hydrodynamics and energy expenditure of oxidation ditch aerated with hydrojets using CFD codes. *Water Quality Research Journal of Canada*, **50**(1), 83–94, https://doi.org/10.2166/wqrjc.2014.036

Kinnear D. J. and Deines K. (2012). Acoustic Doppler current profiler clarifier velocity measurement. *Proceedings of the Water Environment Federation*, **2001**(9), 500–515, https://doi.org/10.2175/193864701790861596

Laurent J., Samstag R. W., Ducoste J. M., Griborio A., Nopens I., Batstone D. J., Wicks J. D., Saunders S. and Potier O. (2014). A protocol for the use of computational fluid dynamics as a supportive tool for wastewater treatment plant modelling. *Water Science and Technology*, **70**(10), 1575–1584, https://doi.org/10.2166/wst.2014.425

Laursen J. (2006). WWTP process tank modelling: the integration of advanced hydrodynamicand microbiological models. PhD thesis, Aalborg University, Aalborg, Denmark.

Le Moullec Y., Potier O., Gentric C. and Pierre Leclerc J. (2008). Flow field and residence time distribution simulation of a cross-flow gas–liquid wastewater treatment reactor using CFD. *Chemical Engineering Science*, **63**(9), 2436–2449, https://doi.org/10.1016/j.ces.2008.01.029

Le Moullec Y., Gentric C., Potier O. and Leclerc J. P. (2010a). Comparison of systemic, compartmental and CFD modelling approaches: application to the simulation of a biological reactor of wastewater treatment. *Chemical Engineering Science*, **65**(1), 343–350, https://doi.org/10.1016/j.ces.2009.06.035

Le Moullec Y., Gentric C., Potier O. and Leclerc J. P. (2010b). CFD Simulation of the hydrodynamics and reactions in an activated sludge channel reactor of wastewater treatment. *Chemical Engineering Science*, **65**(1), 492–498, https://doi.org/10.1016/j.ces.2009.03.021

Le Moullec Y., Potier O., Gentric C. and Leclerc J. P. (2011). Activated sludge pilot plant: comparison between experimental and predicted concentration profiles using three different modelling approaches. *Water Research*, **45**(10), 3085–3097, https://doi.org/10.1016/j.watres.2011.03.019

Lei L. and Ni J. (2014). Three-dimensional three-phase model for simulation of hydrodynamics, oxygen mass transfer, carbon oxidation, nitrification and denitrification in an oxidation ditch. *Water Research*, **53**(4), 200–214, https://doi.org/10.1016/j.watres.2014.01.021

Littleton H. X., Daigger G. T. and Strom P. F. (2007a). Application of computational fluid dynamics to closed-loop bioreactors: I. Characterization and simulation of fluid-flow pattern and oxygen transfer. *Water Environment Research*, **79**(6), 600–612, https://doi.org/10.2175/106143006X136739

Littleton H. X., Daigger G. T. and Strom P. F. (2007b). Application of computational fluid dynamics to closed-loop bioreactors: II. Simulation of biological phosphorus removal using computational fluid dynamics. *Water Environment Research*, **79**(6), 600–612, https://doi.org/10.2175/106143006X136739

Lyn D. A., Stamou A. I. and Rodi W. (1992). Density currents and shear-induced flocculation in sedimentation tanks. *Journal of Hydraulic Engineering*, **118**(6), 849–867, https://doi.org/10.1061/(ASCE)0733-9429(1992)118:6(849)

Marchisio D. L., Vigil R. D. and Fox R. O. (2003). Quadrature method of moments for aggregation-breakage processes. *Journal of Colloid and Interface Science*, **258**(2), 322–334, https://doi.org/10.1016/S0021-9797(02)00054-1

Morchain J., Gabelle J. C. and Cockx A. (2014). A coupled population balance model and CFD approach for the simulation of mixing issues in lab-scale and industrial bioreactors. *AIChE Journal*, **60**(1), 27–40, https://doi.org/10.1002/aic.14238

Mullison J., Wanis P. and Symonds D. (2011). Field testing of a new ADCP. 2011 IEEE/OES/CWTM 10th Working Conference on Current, Waves and Turbulence Measurements (CWTM), Monterey, CA, USA, pp. 87–92.

Naessens W., Maere T. and Nopens I. (2012a). Critical review of membrane bioreactor models – part 1: biokinetic and filtration models. *Bioresource Technology*, **122**, 95–106, https://doi.org/10.1016/j.biortech.2012.05.070

Naessens W., Maere T., Ratkovich N., Vedantam S. and Nopens I. (2012b). Critical review of membrane bioreactor models – part 2: hydrodynamic and integrated models. *Bioresource Technology*, **122**, 107–118. Membrane Bioreactors (MBRs): State-of-Art and Future, https://doi.org/10.1016/j.biortech.2012.05.071

Nielsen M. K., Bechmann H. and Henze M. (2000). Modelling and test of aeration tank settling (ATS). *Water Science and Technology*, **41**(9), 179–184, https://doi.org/10.2166/wst.2000.0200

Nopens I. (2005). Modelling the activated sludge flocculation process: a population balance approach. PhD thesis, Ghent University, Ghent, Belgium.

Nopens I. and Wicks J. (2012). Integration of CFD and biokinetics. CFD Workshop – Presentation. 3rd IWA/WEF Wastewater Treatment Modelling Seminar – WWTmod2012, Mont-Sainte-Anne, Québec, Canada.

Nopens I., Biggs C. A., De Clercq B., Govoreanu R., Wilén B. M., Lant P. and Vanrolleghem P. A. (2002). Modelling the activated sludge flocculation process combining laser light diffraction particle sizing and population balance modelling (PBM). *Water Science and Technology*, **45**(6), 41–49, https://doi.org/10.2166/wst.2002.0092

Nopens I., Torfs E., Ducoste J., Vanrolleghem P. A. and Gernaey K. V. (2015). Population balance models: a useful complementary modelling framework for future WWTP modelling. *Water Science and Technology*, **71**(2), 159–167, https://doi.org/10.2166/wst.2014.500

Nopens I., Samstag R., Wicks J., Laurent J., Rehman U. and Potier O. (2017). To mix, or not to mix, that is the question. In: Frontiers in Wastewater Treatment and Modelling. FICWTM 2017. Lecture Notes in Civil Engineering, G. Mannina (ed.), Springer International Publishing AG, Cham Switzerland, pp. 677–683.

Parker D., Kaufman W. and Jenkins D. (1972). Floc breakup in turbulent flocculation processes. *Journal of the Sanitary Engineering Division*, **98**(1), 79–99, https://doi.org/10.1061/JSEDAI.0001389

Parker D. S., Kaufman W. J. and Jenkins D. (1970). Characteristics of Biological Flocs in Turbulent Regime, SERL Report No. 70-5. University of California, Berkeley, CA. Sanitary Engineering Research Laboratory, University of California.

Parker D. S., Kaufman W. J. and Jenkins D. (1971). Physical conditioning of activated sludge floc. *Journal (Water Pollution Control Federation)*, **43**(9), 1817–1833.

Patziger M., Kainz H., Hunze M. and Józsa J. (2012). Influence of secondary settling tank performance on suspended solids mass balance in activated sludge systems. *Water Research*, **46**(7), 2415–2424, https://doi.org/10.1016/j.watres.2012.02.007

Pereira J. P., Karpinska A. M., Dias M. M. and Santos R. J. (2009). The dissolved oxygen profile in oxidation ditches: a numerical study on its effect on nutrient removal. 2nd IWA Specialized Conference on Nutrient Management in Wastewater Treatment Processes, Krakow, Poland, pp. 97–104.

Pereira J. P., Karpinska A. M., Gomes P. J., Martins A. A., Dias M. M., Lopes J. C. B. and Santos R. J. (2012). Activated sludge models coupled to CFD simulations. In: Single and Two-Phase Flows in Chemical and Biomedical Engineering, R. Dias, R. Lima, A. Martins and T. M. Mata (eds.), Bentham Science Publishers, Dubaï, UAE, pp. 153–173.

Portela A. M. K. (2013). New design tools for activated sludge process. PhD thesis, Universidade do Porto, Porto, Portugal.

Randolph A. and Larson M. A. (1971). Theory of Particulate Processes: Analysis and Techniques of Continuous Crystallization. Academic Press, New York.

Ratkovich N., Horn W., Helmus F. P., Rosenberger S., Naessens W., Nopens I. and Bentzen T. R. (2013). Activated sludge rheology: a critical review on data collection and modelling. *Water Research*, **47**(2), 463–482, https://doi.org/10.1016/j.watres.2012.11.021

Rehman U. (2016). Next generation bioreactor models for wastewater treatment systems by means of detailed combined modelling of mixing and biokinetics. PhD thesis, Ghent University. Faculty of Bioscience Engineering, Ghent, Belgium.

Rehman U., Maere T., Vesvikar M., Amerlinck Y., Arnaldos M. and Nopens I. (2014). Hydrodynamic – biokinetic model integration applied to a full-scale WWTP. IWA World Water Congress & Exhibition, Lisbon, Portugal, p. 5.

Rehman U., Audenaert W., Amerlinck Y., Maere T., Arnaldos M. and Nopens I. (2017). How well-mixed is well mixed? Hydrodynamic-biokinetic model integration in an aerated tank of a full-scale water resource recovery facility. *Water Science and Technology*, **76**(8), 1950–1965, https://doi.org/10.2166/wst.2017.330

Samstag R. W. and Wicklein E. A. (2014). A protocol for optimization of activated sludge mixing. *Proceedings of the Water Environment Federation*, **2014**(13), 3614–3640, https://doi.org/10.2175/193864714815939453

Samstag R. W., Wicklein E. A., Reardon R. D., Leetch R. J., Parks R. M. and Groff C. D. (2012). Field and CFD analysis of Jet aeration and mixing. *Proceedings of the Water Environment Federation*, **2012**(12), 4113–4139, https://doi.org/10.2175/193864712811708301

Samstag R. W., Ducoste J. J., Griborio A., Nopens I., Batstone D. J., Wicks J. D., Saunders S., Wicklein E. A., Kenny G. and Laurent J. (2016). CFD For wastewater treatment: an overview. *Water Science and Technology*, **74**(3), 549–563, https://doi.org/10.2166/wst.2016.249

Sánchez F., Rey H., Viedma A., Nicolás-Pérez F., Kaiser A. S. and Martínez M. (2018). CFD Simulation of fluid dynamic and biokinetic processes within activated sludge reactors under intermittent aeration regime. *Water Research*, **139**, 47–57, https://doi.org/10.1016/j.watres.2018.03.067

Shaikh A. and Al-Dahhan M. H. (2007). A review on flow regime transition in bubble columns. *International Journal of Chemical Reactor Engineering*, **5**(1), 1–68, https://doi.org/10.2202/1542-6580.1368

Sharma A. K., Guildal T., Thomsen H. A. R., Mikkelsen P. S. and Jacobsen B. N. (2013). Aeration tank settling and real time control as a tool to improve the hydraulic capacity and treatment efficiency during wet weather: results from 7 years' full-scale operational data. *Water Science and Technology*, **67**(10), 2169–2176, https://doi.org/10.2166/wst.2013.129

Sobremisana A. P., Ducoste J. J. and de los Reyes F. L. (2012). Combining CFD, FLOC dynamics, and biological reaction kinetics to model carbon and nitrogen removal in an activated sludge system. *Proceedings of the Water Environment Federation*, **2011**(13), 3272–3282, https://doi.org/10.2175/193864711802721569

Stamou A. I., Theodoridis G. and Xanthopoulos K. (2009). Design of secondary settling tanks using a CFD model. *Journal of Environmental Engineering*, **135**(7), 551–561, https://doi.org/10.1061/(ASCE)0733-9372(2009)135:7(551)

Takacs G. (2005). Gas Lift Manual (PennWell Corp). PennWell Books, Tulsa, OK, USA.

Torfs E. (2015). Different settling regimes in secondary settling tanks: experimental process analysis, model development and calibration. PhD thesis, Ghent University, Ghent, Belgium.

Vermande S., Simpson K., Essemiani K., Fonade C. and Meinhold J. (2007). Impact of agitation and aeration on hydraulics and oxygen transfer in an aeration ditch: local and global measurements. *Chemical Engineering Science*, **62**(9), 2545–2555, https://doi.org/10.1016/j.ces.2005.09.016

Versteeg H. K. and Malalasekera W. (2007). An Introduction to Computational Fluid Dynamics: The Finite Volume Method, 2nd edn. Pearson Education Limited, London.

Wang T. (2010). Simulation of bubble column reactors using CFD coupled with a population balance model. *Frontiers of Chemical Science and Engineering*, **5**(2), 162–172, https://doi.org/10.1007/s11705-009-0267-5

Wicklein E., Batstone D. J., Ducoste J., Laurent J., Griborio A., Wicks J., Saunders S., Samstag R., Potier O. and Nopens I. (2016). Good modelling practice in applying computational fluid dynamics for WWTP modelling. *Water Science and Technology*, **73**(5), 969–981, https://doi.org/10.2166/wst.2015.565

Wilcox D. C. (1988). Reassessment of the scale-determining equation for advanced turbulence models. *AIAA Journal*, **26**(11), 1299–1310, https://doi.org/10.2514/3.10041

Witt P. J., Feng Y., Eick I. and Schwarz M. P. (2012). Modelling bubble flow with CFX and fluent for aluminium reduction cells. 9th Int. Conference on CFD in the Mineral and Process Industries, Melbourne, Australia.

Xie H., Yang J., Hu Y., Zhang H., Yang Y., Zhang K., Zhu X., Li Y. and Yang C. (2014). Simulation of flow field and sludge settling in a full-scale oxidation ditch by using a two-phase flow CFD model. *Chemical Engineering Science*, **109**, 296–305, https://doi.org/10.1016/j.ces.2014.02.002

Yang Y., Yang J., Zuo J., Li Y., He S., Yang X. and Zhang K. (2011). Study on two operating conditions of a full-scale oxidation ditch for optimization of energy consumption and effluent quality by using CFD model. *Water Research*, **45**(11), 3439–3452, https://doi.org/10.1016/j.watres.2011.04.007

doi: 10.2166/9781780409030_0095

Chapter 5
High-rate algal ponds

Julien Laurent[1] and Tewodros Meless Teshome[2]

[1]ICube, ENGEES, CNRS, Université de Strasbourg, Strasbourg, France; Email: julien.laurent@engees.unistra.fr
[2]Department of Hydraulic Engineering, Hawassa University, Institute of Technology, Hawassa, Ethiopia; Email: teddymett@hu.edu.et

5.1 INTRODUCTION

For a long time, microalgae have been recognized as a valuable source of human food (Priyadarshani & Rath, 2012). With the development of algal cultivation techniques, which began in the late nineteenth century, industrial algal production expanded its application range to fertilizers, animal and fish feeds, high-value bio-molecules, pharmaceuticals, cosmetics and food colourants (Andersen, 2005; Lawton *et al.*, 2017; Priyadarshani & Rath, 2012). As algal cells can accumulate a high amount of lipids and carbohydrates, they are also an ideal source for use in biofuels production (Mata *et al.*, 2010; Sirajunnisa & Surendhiran, 2016; Voloshin *et al.*, 2016).

Using microalgae for wastewater treatment presents many benefits. Being primary producers, algae use inorganic nutrients as substrates (Bellinger & Sigee, 2015; Dodd, 1986). By photosynthesis, they use light energy for their metabolism, which consumes carbon dioxide and releases soluble oxygen in the aqueous phase (Dodd, 1986). This process supports bacterial activities and increases water pH, yielding a hygienization effect (Cole, 1982; Muñoz & Guieysse, 2006). The concomitant growth of bacteria and algae also enhances flocculation between them. Thus, the algal-bacterial biomass can be harvested by simple gravitational sedimentation (Gutzeit *et al.*, 2005; Van Den Hende *et al.*, 2011, 2014).

Synergistic interactions between algae and bacteria in wastewater treatment are illustrated in Figure 5.1 (Cole, 1982; Unnithan *et al.*, 2014). With exposure to light, algae provide oxygen for bacterial oxidation processes which in turn generate carbon dioxide useful for photosynthesis. This is the photosynthetic respiration concept. The concentrations of these gases are also impacted by the gas transfer between air and liquid in which transfer rate highly depends on mixing in the reactor (Garcia-Ochoa & Gomez, 2009). Moreover, via their growth and decay, algae can release organic matter that is available as a substrate for bacteria (Bell & Mitchell, 1972; Cole, 1982). In turn, bacteria also facilitate algal growth: they degrade organic matter to provide inorganic compounds back into the bulk phase, and these are necessary for algal growth. This nutrient recycle process by bacteria is very important especially for limiting nutrients for algae such as phosphorus (Cole, 1982).

It has been shown that nitrification occurs and plays an important role in wastewater treatment application of algal-bacterial systems (Babu *et al.*, 2010; Cromar & Fallowfield, 1997; Evans *et al.*, 2005; Karya *et al.*, 2013; Mantovani *et al.*, 2020; Park & Craggs, 2010).

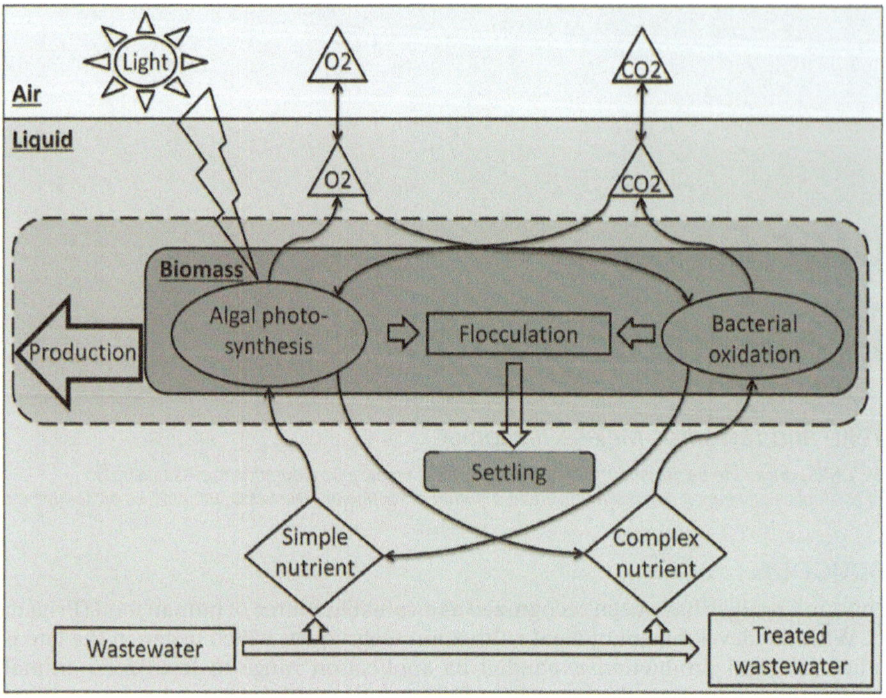

Figure 5.1 General illustration of algal-bacterial processes in wastewater under light saturation conditions (large arrows indicate main processes, small arrows indicate mass transfer between phases including air, liquid and biomass) (source: Pham, 2018).

Due to the small size of the algal cells and their low concentration in culture solution, efficient harvesting of algal biomass from water can account for 20–30% of the total production cost (Mata *et al.*, 2010; Pragya *et al.*, 2013). Therefore, algae harvesting remains one of the biggest challenges when operating the system (Christenson & Sims, 2011; Craggs *et al.*, 2015; Uduman *et al.*, 2010). One solution is to enhance algal biomass settleability by bio-flocculation (Salim *et al.*, 2011; Vandamme *et al.*, 2013). Indeed, inoculating activated sludge with algae in wastewater has been shown to improve biomass settling while maintaining good treatment efficiency (Gutzeit *et al.*, 2005; Van Den Hende *et al.*, 2011). Studies on algal-bacterial biomass indicated high gravitational settling efficiencies by flocculation between algae and bacteria (Gutzeit *et al.*, 2005; Medina & Neis, 2007; Van Den Hende *et al.*, 2014). Van Den Hende *et al.* (2014) recovered nearly 100% of algal-bacterial biomass from a pilot-scale study via the two simple harvesting steps of gravity settling and dewatering by manual filter press, requiring no chemical addition or electricity.

5.2 PROCESS DESCRIPTION

Two commercial-scale microalgae cultivation techniques, indoor photobioreactor and outdoor open ponds, are widely installed in varying sizes across different regions of the world.

5.2.1 Photobioreactors

Photobioreactors (PBRs) are closed transparent tubes or containers that are usually kept inside a controlled environment for large-scale algal production. Energy is supplied via electric lights and CO_2 is bubbled and circulated via mechanical pumps. This method enables the control and optimization of most of the influencing parameters, such as pH, temperature, nutrients and light intensity, to achieve the required end product. These systems are also less prone to contamination by bacteria. Thus, they are characterized by producing consistent and high-quality yields that can be used for food or other high-added-value products. Different designs of PBRs, (tubular, flat-plate type etc.) have been developed for different intended purposes. Highly complex PBRs are more adaptable to a wider range of operational conditions and enable the harvesting of a high amount of quality biomass. However, their construction and operational costs are very high. Though PBRs are renowned for yielding high-density and high-quality culture and provide the ability to optimize environmental factors, including light intensity, CO_2 and O_2 gas transfer rate, temperature and mixing conditions (Gupta *et al.*, 2015; Suh & Lee, 2003), there is less reported experience of PBRs using wastewater as a culturing medium (Bilad *et al.*, 2014; Judd *et al.*, 2015).

5.2.2 High-rate algal pond (HRAP) system

The high-rate algal pond (HRAP), also known as the raceway pond, was first introduced in the 1950s by W.J. Oswald in California for treating organic wastewater (Oswald & Gotaas, 1957). It consisted of meandering (folded) channels with a propeller pump for mixing and driving the flow (Oswald, 1988) (Figure 5.2). Since then, the HRAP has been applied as an algal-based wastewater treatment unit or algal cultivation facility in many places (Kumar *et al.*, 2015; Mata *et al.*, 2010; Rawat *et al.*, 2011).

Figure 5.2 Pilot-scale high-rate algal pond at Ober-Besingen wastewater treatment plant, sewerage board Lauter-Wetter (Germany). Designed and operated by THM Technical University Central Hesse, Gießen (ZEuUS Kompetenzzentrum für nachhaltiges Engineering und Umweltsysteme, competence centre for sustainable engineering and environmental systems).

It works under a similar principle to wastewater stabilization ponds (WSPs), involving the interaction of algae and bacteria, except that the liquid is under continuous circulation in the case of the HRAP. This new pond type is smaller than conventional ponds as its hydraulic retention time (HRT) is much shorter (less than one week in comparison with from three to six weeks or more). Thanks to enhancement of vertical mixing, by the propeller pump, this pond type can produce a dense algal culture supporting a treatment rate 10 times higher than a conventional WSP. Finally, the advantages of this pond type are clear including: a smaller area requirement, promotion of nutrient recovery via harvesting the biomass as well as no requirement for mechanical aeration (El Hamouri *et al.*, 2003; Oswald & Gotaas, 1957).

5.2.2.1 Geometry

The HRAP usually consists of a raceway-like channel which commonly comprises two straight ditches and two 180° curved bends at either end of the straight channel to close the loop. Hence it is essentially made of the side walls, middle separating wall or island, and bottom surface (Figures 5.2 and 5.3).

In addition to the basic physical elements, walls and bottom surface, an HRAP may or may not contain deflector baffles within the two curved channels for smoothing the flow as it changes direction, and a sump or mixing column at an appropriate location in the straight part to increase the gas/liquid contact time and efficiency of CO_2 absorption rate (Park *et al.*, 2011).

Since the initial design, over the last three decades, certain modifications have been incorporated to meet the mixing requirements and to minimize energy input for the motor (Dodd, 1986). In this system, most of the energy loss occurs at the curved 'hairpin' bends (Liu & Zhang, 2019).

In general, the geometry of the pond affects nutrient dispersion, resistance to flow near the tank walls, the volume of dead zones and the efficiency of circulation.

Experimental and numerical studies have shown the formation of dead zones in the areas where the flow separates from the boundary at sharp corners and high speed (Hadiyanto *et al.*, 2013; Sompech *et al.*, 2012). Beside significant energy loss caused by the flow separation regions, these dead zones can also induce the sedimentation of biomass and affect both microalgae and bacteria growth. The different geometries of the HRAP and configurations of the deflector baffles introduced, as well as the modification to the shape of the middle island (tested by Hadiyanto *et al.* (2013) and Sompech *et al.* (2012)) were able to minimize those undesirable effects. While the provision of deflector baffles helps to reduce flow separation, it causes an increase in energy consumption due to friction; therefore, it should be carefully considered (Mendoza *et al.*, 2013).

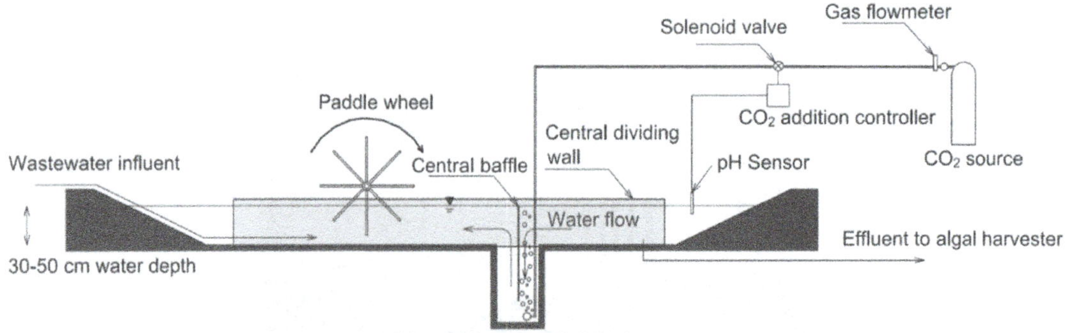

Figure 5.3 Cross-sectional side view of an HRAP with CO_2 aeration (Park *et al.*, 2011, reprinted from Bioresource Technology, Vol 108/number 1, J.B.K. Park, R.J. Craggs, A.N. Shilton, Wastewater treatment high rate algal ponds for biofuel production, Pages 35–42, Copyright 2011, with permission from Elsevier).

One of the important geometric design parameters is the length to width ratio (L/W). Computational fluid dynamics (CFD) simulation performed by Hadiyanto *et al.* (2013) revealed that a L/W ratio higher than 10 yields better performance with respect to uniform velocity distribution and gentle shear stress to algal cells. However, power consumption soared as the L/W ratio increased.

5.2.2.2 Water level

In general, the water level in an HRAP ranges from 0.1 to 1 m (Muñoz & Guieysse, 2006; Sutherland *et al.*, 2015). Like other open-pond systems, the optimal range of operating depth is from 15 to 30 cm, and the surface area for full-scale applications usually varies from 0.5 to 1 ha according to the specific need. Based on specific local conditions, the depth can be increased up to 1 m. The shallowness of the pond is deliberate to achieve a high surface-to-volume ratio for microalgae to adequately absorb sunlight (Sompech *et al.*, 2012).

Since the biomass concentration in an HRAP can be high, available light for photosynthesis could be limited by the absorbing and scattering effects of biomass and suspended matter in the upper layers preventing light from reaching the bottom layers (Grobbelaar *et al.*, 1990). Moreover, increasing water level usually leads to higher energy consumption per unit area (Hadiyanto *et al.*, 2013). Hence, the water should be kept as low as possible to maximize light penetration (Andersen, 2005). However, low water level in the reactor also allows higher temperature variation (Kumar *et al.*, 2015) and water evaporation generally has a greater impact in a reactor with lower water level due to the higher relative proportion of the lost water volume. Moreover, a low water level also increases the size of the reactor in the case of a constant HRT being required. The impacts on productivity of increasing the water level are diverse, with both positive and negative impacts having been reported (Chiaramonti *et al.*, 2013; Sutherland *et al.*, 2014).

5.2.2.3 Mixing

Mixing transports algal cells not only along the channel (axial mixing) but also vertically (vertical mixing). The latter type of mixing is important due to its transportation from dark zones of the pond to light zones and vice versa (light/dark cycle) which also decreases the impact of photoinhibition (Sutherland *et al.*, 2015). Thus, one of the most important design aspects of an HRAP is to ensure the presence of adequate vertical mixing of the culture (Kumar *et al.*, 2015). Mixing usually occurs by using a paddle wheel to provide a horizontal water velocity between 0.15 and 0.3 m/s to overcome losses due to friction and irregularities of the pond and provide a good mixing level throughout the entire reactor (Andersen, 2005). A higher mixing rate may cause high shear stress damaging algal cells, lower biomass flocculation, and an increase in energy consumption and hence operational cost (Kumar *et al.*, 2015). This can also increase the health and environmental risk of releasing bioaerosols (Sialve *et al.*, 2015). A sump of about 1.5 m depth can be added for gas mixing if necessary (Figure 5.3) (Park *et al.*, 2011).

Mixing is affected by both geometric features of the pond, including the type and size of the paddle wheel, and hydraulic conditions such as depth and flow velocity. By alternating the light-dark cycle of each cell, mixing accelerates algal growth and fuels the photosynthetic activity. Moreover, a well-mixed culture avoids heat, gas and nutrient gradients and also enhances the mass transfer rate between cell and culture medium (Grobbelaar, 1991).

However, some researchers doubt the role of mixing in the productivity of algae since other factors, such as temperature, light intensity, liquid depth, and CO_2 and nutrient distribution, have a direct influence (Ali *et al.*, 2015). Several other experimental and numerical studies, on the other hand, shows optimum mixing determines recurring light exposure of algal cells (Cheng *et al.*, 2015; Liffman *et al.*, 2013; Pruvost *et al.*, 2006), reduces settling and sedimentation of cells, and enables uniform distribution of nutrients and CO_2 (Prussi *et al.*, 2014; Richmond & Grobbelaar, 1986) in the culture while consuming minimum possible energy to circulate the flow without damaging algal cells/flocs mechanical structure (Barbosa *et al.*, 2003).

Practical experience has proved that paddle wheels are the most efficient mixing devices for HRAPs (Andersen, 2005). However, shape, alignment and number of blades, and depth and length of paddle greatly affect rotational speed and the oxygen-transfer characteristics (Ahmad & Boyd, 1988; Hreiz *et al.*, 2014). With a conventional paddle wheel having a flat bottom shape, the clearance with the free surface varies with time and so does the backflow. To keep the clearance to a minimum value to reduce backflow, at least one blade should always be inside the liquid medium. Similarly, when the submerged depth of the paddle wheel and speed of rotation increases, the standard oxygen transfer rate increases while standard aeration efficiency drops (Ahmad & Boyd, 1988). This is an important design criterion in deciding the depth of the liquid for a given diameter of paddle wheel with a particular number of blades and vice versa (Pham *et al.*, 2018). Increasing the paddle wheel diameter decreases backflow which eventually increases its efficiency (Andersen, 2005). However, as the diameter increases, construction cost and weight also increase, with implications on the overall effectiveness. Similarly, the more paddle wheel blades, the higher the paddle wheel efficiency and the lower the motor shock. However, a paddle wheel with more than eight blades is not practical from the construction point of view and does not significantly improve efficiency (Andersen, 2005). The smaller the clearance between the blade and the pond floor, the greater the efficiency of the paddle wheel. In practice, a gap of 20 mm with the side walls and the bottom surface is reasonable and allows a suitable safety margin.

5.3 CFD CONCEPTS RELEVANT TO HRAP MODELLING

Similarly to other process units described in this report, CFD can help to evaluate current HRAP designs and produce new designs in order to:

- reduce energy loss,
- optimize algal cell access to light,
- reduce sedimentation,
- improve gas/liquid transfer rates.

An HRAP constitutes a suspended growth system with a multiphase fluid. This implies similar approaches and challenges as for conventional activated sludge, which are discussed in Chapter 4.

One of the specific modelling challenges, for a system having a rotating element in general and for the HRAP in particular, is the representation of the momentum induced by the rotating paddle wheel that is usually responsible for inducing mixing and liquid velocity in such systems. Indeed, the comprehensive treatment of such components in CFD is demanding in terms of model building as well as computational resource (Amini *et al.*, 2016) since a specific numerical approach for the modelling of complex flow systems containing rotating parts is needed (Wilhelm, 2015). Hence, various methods or simplifications have been developed to simulate fluid flow problems involving a rotating device.

Inlet Velocity and Dynamic Mesh are the two common CFD methods to simulate rotating elements (Amini *et al.*, 2016). The most comprehensive one is the Dynamic Mesh method that considers the rotation of the paddle-wheel geometry and simulates the momentum transferred to the fluid through the rotating mesh (Hreiz *et al.*, 2014). In this method, linear motion or rotation of a solid body, based on the physical phenomenon, is handled by a moving or rotating mesh with or without mesh morphing within the specified boundaries. However, this approach can be computationally intensive particularly when the model is intended to be integrated with other models such as biokinetics, gas transfer, and light irradiance models.

Depending on the modelling objective, one can consider simplifying the simulation replacing the paddle wheel by introducing a momentum source term in the governing equation (Liffman *et al.*, 2013) or by assigning a constant inlet velocity on a virtual inlet boundary (Hadiyanto *et al.*, 2013).

5.3.1 Momentum source

The first option to model the effect of rotational elements such as a paddle wheel is to introduce a source term in the governing equations. This option allows the modeller to define a volume in the system domain which generates momentum, without any further complication of providing detailed geometry and finer mesh of the rotating elements. The known velocity output (with its correct x, y, and z components) and dimensions of a rotating device can be assigned to an identical shaped volume (box, sphere, cylinder or cell zones) at a similar location in the computational domain. This method can be used for the analysis of various flow problems except compressible and multiphase flows. It is also limited to implementing the average velocity that could hide actual velocity variations (Liffman *et al.*, 2013).

5.3.2 Specific boundary condition: Inlet Velocity approach

Another simplified approach to represent the rotational elements in CFD models is the Inlet Velocity method. Based on the flow direction, a virtual inlet and outlet boundaries are provided replacing the rotating device with an average velocity from this device prescribed on the inlet boundary. A small physical gap is needed between the two boundaries to realize the simulation. In the case of solving transport equations, a cyclic boundary condition is set between the inlet and outlet face to ensure the swapping of transported materials. This method assumes a steady-state condition that makes it more suitable when a steady-state physical flow field solution is expected (Hadiyanto *et al.*, 2013; Teshome *et al.*, 2019).

5.3.3 Single reference frame (SRF)

A moving reference frame, in general, is advantageous to reduce a flow problem which is unsteady in the stationary (inertial) frame to a steady type with respect to the moving frame (Ansys, 2011). For a steadily rotating frame (i.e., constant angular speed), it is possible to transform the governing fluid flow equations to the rotating frame such that steady-state simulations are possible.

Hence, fluid flow around a rotating device can be modelled using a single rotating frame of reference (SRF) that adheres to the rotating machinery when only one rotating flow domain is considered. Since a steady-state condition is assumed without any mesh motion, where the complete computational fluid region is in motion, SRF simulations are computationally less costly and are therefore predestined for elaborate multi parameter studies (Wilhelm, 2015). Nevertheless, the assumptions made in this method could compromise the result to a certain degree depending on the complexity of the problem. For HRAP, this method is not suitable as the entire computational domain should be defined as a moving reference frame.

5.3.4 Multiple reference frame (MRF)

Rotating flow applications commonly involve both steady and unsteady flow regions. An SRF cannot be applied in the computational domain as it incorporates stationary and moving zones at the same time. A multiple reference frame overcomes this drawback and enables the simulation of two or more subdomains in which either of them may be stationary or in motion (Wilhelm, 2015). A virtual arbitrary mesh interface (AMI) is introduced to connect the different regions (Dellinger *et al.*, 2018; The OpenFOAM Foundation, 2021): it has no physical meaning but is exclusively used for numerical simulations to exchange calculated flow field information at each iteration step. However, proper positioning of these interfaces in the geometry and the generation of a smooth mesh on both sides of the AMI is the challenging part of the MRF method (Wilhelm, 2015). The higher the smoothness and uniformity of the cell size on both sides of the AMI, the better the quality of the result (Wilhelm, 2015).

An MRF can be implemented in two ways: frozen rotor technique and mixing plane approach. Both methods perform a steady-state simulation. However, the first one solves the governing equation for one rotor position, normally a very quick simulation, and converges to a snapshot of the fluid flow system. It is possible to conduct the simulation at several rotor positions and average integral quantities (Wilhelm, 2015).

The mixing plane approach is an improvement on the frozen rotor technique that gives a better approximation of the circumferential average of all flow quantities in the internal interface, called the mixing plane, in a much shorter computational time compared to unsteady simulation. But local flow information, such as local vortices, are spread out over the mixing plane for which the results show only an averaged value not based on a physical time step (Wilhelm, 2015).

For a rotating (non-inertial) coordinate system on the rotor, the flow is usually steady relative to the rotating frame. Conversely, the rotor and the flow around the rotor are unsteady with respect to the fixed (inertial) reference. Under a properly prescribed interface condition, the Navier-Stokes equations are solved independently for both subdomains. The momentum equation is modified and solved for the rotating region based on the moving frame incorporating two additional force terms, Coriolis and centrifugal, which occur due to the transformation from the stationary to the moving reference as shown in Equation (5.1).

$$\frac{\partial(\rho u_r)}{\partial t} + \nabla\cdot(\rho u_r u_r) + \{2\rho\Omega\times u_r\} + \{\rho\Omega\times(\Omega\times r)\} = -\nabla p + \nabla\cdot(\mu\nabla u_r) \tag{5.1}$$

$$\frac{\partial(\rho u)}{\partial t} + \nabla\cdot(\rho u u) = -\nabla p + \nabla\cdot(\mu\nabla u) \tag{5.2}$$

$$u_r = u - \Omega\times r \tag{5.3}$$

The third and the fourth terms inside the curly brackets of Equation (5.1) are the Coriolis and centrifugal force components, respectively. For the stationary subdomain, Equation (5.2) is reduced to Equation (5.2) avoiding the relative term. In both equations p, ρ, μ and u_r stand for pressure, density, dynamic viscosity and relative velocity, respectively, Ω is the rotation vector and r is the position vector. Relative velocity is calculated as per Equation (5.3).

5.3.5 Moving mesh

Both the SRF and MRF methods described above neglect the unsteadiness of the fluid motion and provide a time-averaged solution. Unsteady flow simulation or a time-accurate solution for rotor-stator interaction can be computed by Moving Mesh modules (Ansys, 2011). Although these techniques are computationally very intensive for highly transient rotational flow, they are the most accurate option for simulating flows in multiple moving reference frames. Like the MRF case above, for the static and rotating subdomains, the governing flow equations are solved separately.

In CFD modelling, Moving Mesh can be implemented in two slightly different styles, Sliding Mesh and Dynamic Mesh, both simulating unsteady conditions. Sliding Mesh is suitable when there is a transient rotor-stator interaction in the computational domain (Ansys, 2011). It allows adjacent grids to slide relative to each other along the mesh interface surface. The Sliding Mesh method considers that cells are transferred by the motion without changing their original topology (size and shape) within a fixed boundary (i.e., the control volume remains constant). Hence, this method is more appropriate for modelling when the flow is not subjected to excessive deformation.

While the rotation or translation takes place, as long as the two AMI boundaries are based on the same geometry and finely meshed for a smooth slip, cell alignment along the arbitrary interface is not required. Since the flow is inherently unsteady, a time-dependent solution procedure is applied (Ansys, 2011) while coupling between the two regions is achieved by interpolating the flow variables on the cell faces (Wilhelm, 2015).

The Sliding Mesh technique was applied successfully for the HRAP in the studies conducted by Amini *et al.* (2016) and Hreiz *et al.* (2014).

Model construction for the Dynamic Mesh is similar to that of the Sliding Mesh technique. However, the Dynamic Mesh solver is capable of modelling fluid flows subjected to a change in shape and volume of the domain with time due to motion of one or more boundaries (Wilhelm, 2015). Hence, following the new position of the moving boundaries, the position and shape of the grid cells are updated or adjusted accordingly at each time step. The process of updating mesh topology obviously makes the calculation extremely intensive even for a small grid size. This method can be used to model boundaries moving rigidly relative to each other. A good example is the compression of fluid by a piston inside a cylinder. It can also be used for deforming boundaries such as modelling the inflation of a balloon.

5.3.6 Experimental validation

The CFD model of the HRAP often includes some simplifications of the geometry (e.g., for the paddle wheel), and the choice of turbulence model is also of significance. Model validation with experimental data is therefore required. This is often carried out by means of velocity measurements and tracer tests. The latter is an approach very well established in chemical engineering. However, its use to evaluate a CFD model and simulate contaminants transport must consider some important elements regarding both experimental and numerical procedures. This is discussed in the next section as well as in Chapters 9 and 10.

5.4 CASE STUDY: MODELLING A PILOT-SCALE HRAP

This section presents a CFD study related to a pilot-scale HRAP. The modelled HRAP is the one used in the studies performed by Pham *et al.* (2018, 2020, 2021). The method used for the paddle-wheel description is the Inlet Velocity method (Hadiyanto *et al.*, 2013; Teshome *et al.*, 2019).

5.4.1 Geometric design of HRAP

The real size of the pilot-scale HRAP is considered in the geometric design. Steady state is assumed for the flow-field and a monophasic simulation is used. Thus, the exact depths of the liquid medium (10, 15 and 20 cm) are used as height for the external wall, middle separator and flow deflectors. The same height is also used for the inlet and outlet boundaries having a width equal to the width of the channel. Actual thickness of the middle separating wall and deflectors are considered. The actual geometry is displayed in Figure 5.4.

5.4.2 Meshing the geometry

The geometry of the HRAP (Figure 5.5) was meshed using the *blockMesh* utility of the OpenFOAM package (The OpenFOAM Foundation, 2021). *blockMesh* is a utility of OpenFOAM that creates

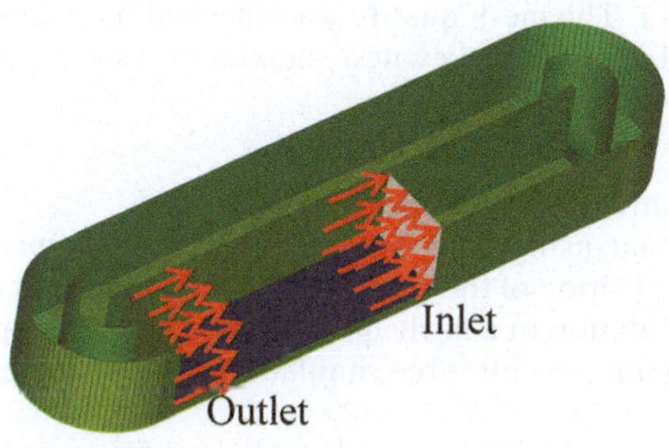

Figure 5.4 Inlet velocity schematics for the representation of the HRAP.

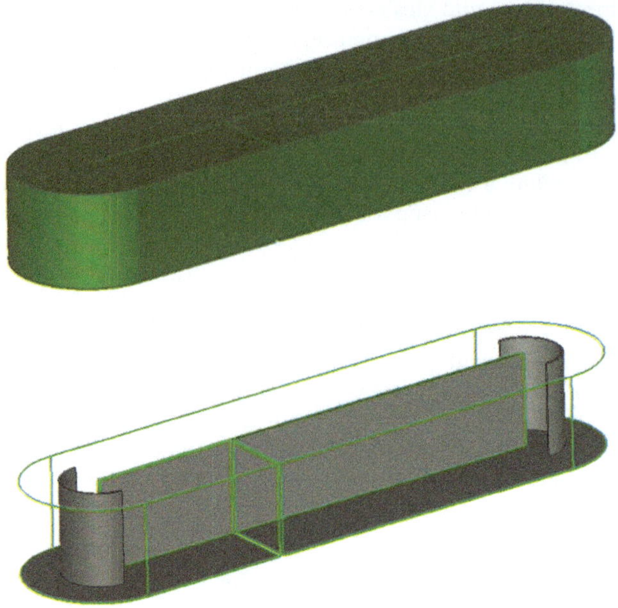

Figure 5.5 Geometry of the HRAP (top) and layout with deflector and middle wall (bottom) for the Inlet Velocity method.

parametric hexahedral meshes with grading and curved edges. Block formation and mesh generation is directed by a dictionary of code that explicitly gives the ability to control each mesh parameter. The challenge in using *blockMesh* is the absence of a graphical user interface (GUI) which is common to all OpenFOAM solvers and utilities. This slight difficulty in coding limits the use of this utility to the generation of hexahedral blocks of mesh, mainly on regular shapes. Figure 5.6 displays the mesh generated by *blockMesh* for the Inlet Velocity method.

Finer grading was used for meshing closer to the middle and external wall, and at both ends of the middle wall where the flow makes a sharp turn. Meshing started from a very fine level (482 963 cells for 10 cm, 748 710 cells for 15 cm and 1 205 236 cells for 20 cm) and was gradually made coarser until it had no influence on the simulation result and converged in a fairly reasonable time. The final number of cells in the computational domain and solution convergence time, using an intel ® Core ™ i5 HP laptop having 8 GB installed memory and 2.4 GHz CPU speed, for the three depths tested in this study are presented in Table 5.1. The mesh quality was checked in OpenFOAM ('*checkMesh*' utility): all cells were hexahedral, and maximum skewness, maximum aspect ratio, mesh non-orthogonality, and maximum cell openness were all satisfactory.

5.4.3 Solver settings and numerical simulation

CFD simulation was carried out using OpenFOAM 5.0 from the *OpenFOAM Foundation* for the *Inlet Velocity* method. Decomposition of the computational domain was not required in this case as a single core completed the calculation in a fairly quick time. The time step was set to one second with an end time of 2000 seconds. However, all three simulations for the three depths converged before the end time (Table 5.1).

To execute flow field calculations of the steady-state conditions assumed in the Inlet Velocity approach, '*simpleFoam*' solver was used. It is a steady-state solver for incompressible turbulent flow

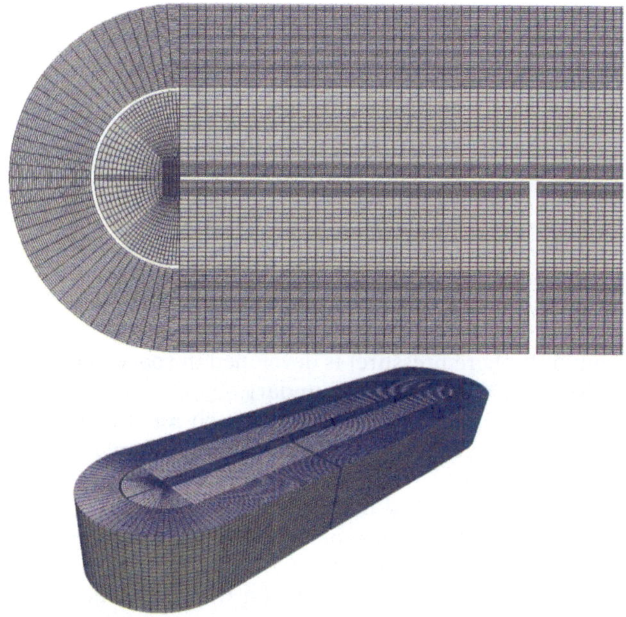

Figure 5.6 Mesh generated by *blockMesh* for the Inlet Velocity method.

Table 5.1 Simulated flow depth, number of cells, convergence time, skewness and aspect ratio for the Inlet Velocity method.

Depth (cm)	No. of Hexahedral Cells	Convergence Time (s)	Maximum Skewness	Maximum Aspect Ratio
20	417 100	1668	0.912	9.488
15	312 825	1506	0.972	9.489
10	208 550	1438	0.975	9.489

using the SIMPLE (semi-implicit method for pressure linked equations) algorithm. This algorithm is a commonly used numerical procedure to solve Navier-Stokes equations in many CFD applications (The OpenFOAM Foundation, 2021) (see also Chapter 2). An incompressible Newtonian fluid under isothermal condition is assumed. The Standard k-e turbulence model is used.

Kinematic pressure and velocity are the two key initial inputs for the solver to proceed with the calculation. Initial and boundary conditions for both pressure and velocity were set based on the physics of the HRAP (Table 5.2). The velocity value on the 'inlet' boundary, which is the one imposed by the paddle wheel, can be set by two alternative methods:

- replacing the moving zone boundary of the paddle wheel with the linear velocity calculated from the angular velocity of the rotating paddle wheel by $v=r^{*}\omega$, where v is the linear velocity (m/s) along the flow direction, r is the radius or the distance from the axis of the paddle wheel to the centre of the flow depth (m), and ω is the angular velocity of the rotating paddle wheel (rad/s).
- imposing a vertical profile that has to be previously measured experimentally for example by acoustic Doppler velocimetry (ADV).

Table 5.2 Boundary conditions for the inlet velocity approach.

	Inlet	Outlet	Free Surface	Walls
Pressure	*zeroGradient*	*zeroGradient*	*symmetry*	*zeroGradient*
Velocity	*fixedValue*	*fixedValue*	*symmetry*	*fixedValue*
nut	*calculated*	*calculated*	*symmetry*	*nutkWallFunction*
k	*fixedValue*	*fixedValue*	*symmetry*	*kqRWallFunction*
epsilon	*fixedValue*	*fixedValue*	*symmetry*	*epsilonWallFunction*

Details of the boundary conditions are given in Table 5.2.

- *zeroGradient*: the quantity (here pressure) is developed in space and its gradient is equal to zero in the orthogonal directions to the patch/boundary.
- *fixedValue*: this condition imposes a value at the patch for the given quantity (here velocity). For walls, the value is set to 0 velocity (no-slip condition), whereas for inlet and outlet it is fixed according to the desired velocity.
- *Symmetry*: here, this condition is used to mimic the free surface when using the Inlet Velocity method in single-phase simulation. It is analogous to a slip boundary condition.
- *Wall functions*: the boundary conditions for k, ε and turbulent viscosity (μ_t) are the standard ones supplied by OpenFOAM. For more details on their implementation, the reader can refer to Liu (2016).

Boundary conditions for the paddle wheel and walls are no-slip for velocity. Specification of boundary conditions for turbulence requires additional calculations. Hydraulic diameter (D_H) and Reynolds number (Re) are calculated as:

$$D_H = 4\frac{A}{P} \tag{5.4}$$

$$Re = \frac{D_H \times V \times \rho}{\mu} \tag{5.5}$$

where A and P are the wetted area and wetted perimeter, respectively, V is the average flow velocity, ρ and μ are the density and dynamic viscosity of the liquid, respectively.

Assuming a fully developed pipe flow, turbulent length scale (l) and turbulent intensity (I) are also calculated as below.

$$l = 0.038 \times D_H \tag{5.6}$$

$$I = \frac{0.16}{Re_{D_H}^{1/8}} \tag{5.7}$$

Re_{D_H} is the Reynolds number calculated based on the hydraulic diameter. The calculated values D_H, Re, l and I are 0.262 m, 81 295.61, 0.00996 and 0.0389, respectively. Based on these values, the turbulent parameters for the k-epsilon model are estimated as shown below.

$$k = 1.5(V \times I)^2 \tag{5.8}$$

$$\varepsilon = \frac{C_\mu K^{1.5}}{l} \tag{5.9}$$

where k is the turbulent kinetic energy, ε is the rate of dissipation of turbulent kinetic energy and C_μ is a model constant which in the standard version of the k-epsilon model has a value of 0.09.

5.4.4 Virtual tracer experiment

For simulating scalar transport of the pulse tracer injection, an equivalent spherical volume of numerical tracer element is introduced at the centre of the flow depth exactly where the sodium chloride solution was injected in the actual tracer experiment (Pham *et al.*, 2018). To ensure swapping of the transported tracer material from the outlet face back to the inlet face, a tracer (scalar) specific *Cyclic* boundary condition is applied. This boundary condition continually transfers all flow information, including transported materials from outlet boundary to the inlet boundary, until the simulation is ended.

Tracer transport is represented by the passive scalar transport equation (see Chapters 2, 4 and 10). Using the finally converged solutions of flow field quantities from the '*simpleFoam*' solver, the '*scalarTransportFoam*' solver in OpenFOAM is implemented to solve this transport equation separately. However, the original solver did not consider the turbulence effect in dispersion/mixing to solve the equation for the scalar transport of tracer material. For this reason, the equation is modified by adding a ratio of turbulent viscosity (μ_t) over turbulent Schmidt number (*Sc*) to the molecular diffusion coefficient. As depicted above, the turbulent viscosity is calculated from the known flow parameters. The default Schmidt number value (0.7) was used as an initial guess and two iterations were made by reducing the value down to 0.35.

$$\frac{\partial}{\partial t}(\rho C) + \nabla.\left(\rho \vec{U} C\right) = \nabla.\left(\left(\rho D_m + \frac{\mu_t}{Sc_t}\right)\nabla C\right) \tag{5.10}$$

where C represents the concentration of tracer, U the water velocity, ρ the water density, D_m the molecular diffusion coefficient (1.9×10^{-9} m^2/s), μ_t the eddy viscosity, and Sc_t the turbulent Schmidt number.

5.4.5 Results: geometrical design modifications

The following results aim to show the potential of CFD as a design tool for shape optimization of an HRAP. As discussed in the previous sections, several studies have already shown that factors like paddle wheel design, L/W ratio, middle wall geometry, presence of deflectors, and so on. can improve the hydrodynamic efficiency of the HRAP (Hadiyanto *et al.*, 2013; Hreiz *et al.*, 2014; Liffman *et al.*, 2013). As an example, comparison between the pilot-scale HRAP with and without deflectors at the bends is presented here (for a 20 cm water depth).

5.4.5.1 Pressure and energy consumption

Figure 5.7 shows the distribution of relative pressure. The pressure at the outlet boundary (in front of the paddle wheel) is set to zero and the relative pressure here is divided by the density of water (scale has been changed to improve readability). The abrupt variation of pressure at the bottom of each image is the location of the inlet velocity boundary. Without deflectors, the pressure drops before and after the paddle wheel are higher. The pressure drops and power consumption of the two designs were calculated according to the formula of Hadiyanto *et al.* (2013) and are listed in Table 5.3:

$$\Delta P = \left(p_{\text{in}} - p_{\text{out}}\right)\rho_{\text{water}} \tag{5.11}$$

The power consumption, W_{cons}, is computed via the formula of Liffman *et al.* (2013);

$$W_{\text{cons}} = \Delta P.Q = 0.01206\Delta P \tag{5.12}$$

where Q is the flowrate through the section of the pond.

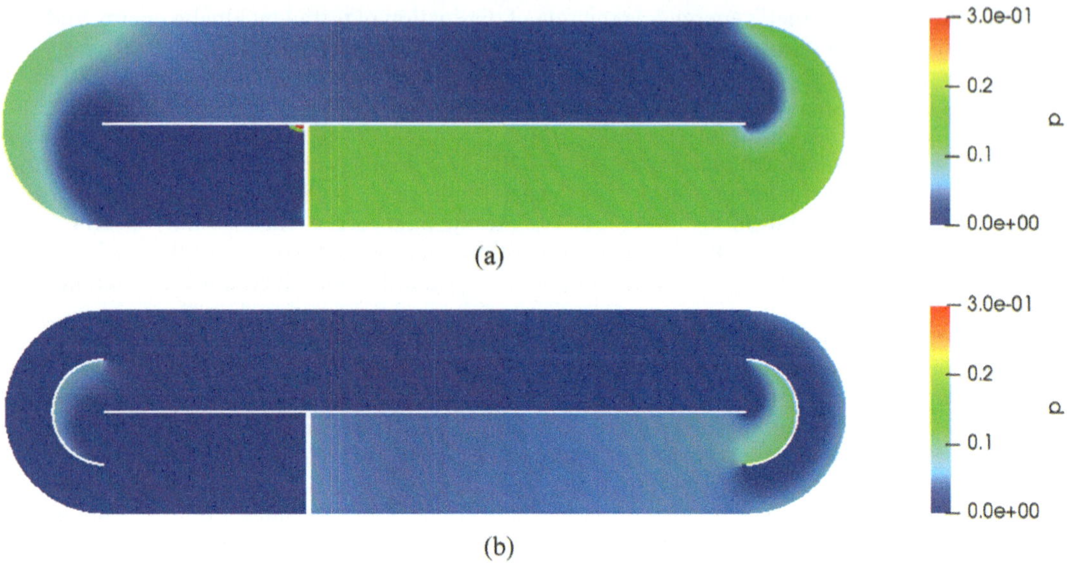

Figure 5.7 Relative kinematic pressure (p/ρ in m^2/s^2) distribution (a) without deflectors (b) with deflectors.

Table 5.3 Pressure loss of one cycle and the required power of paddle wheel.

	ΔP	W_{cons}
With deflectors	55 Pa	0.66 W
Without deflectors	121 Pa	1.46 W

The power consumption of the configuration with deflectors is reduced by 55%, which is consistent with the results from Liffman *et al.* (2013).

5.4.5.2 Velocity field

Figure 5.8 displays the velocity contours at mid-depth for the two alternative designs. Without deflectors, a significant part of the HRAP is characterized by low velocities below 0.1 m/s (dark blue parts). This proportion of 'dead-zone' volumes is significantly reduced by the addition of deflectors. Strong high-speed flow is observed after the bends where flow separation occurs due to the central wall. With deflectors, this phenomenon is reduced, and the flow is more uniform.

In addition to energy consumption, the vertical mixing is another important aspect of HRAP design that can provide recurring access to light for algal cells. Thus, the z component of the velocity field is plotted at several locations of the HRAP in Figure 5.9. One surprising result is that the configuration with deflectors provides less vertical mixing than the simpler design. If providing a uniform velocity field in the horizontal direction is fundamental to avoid dead-zones and minimize energy consumption, solutions to increase vertical mixing can also be investigated: optimizing paddle wheel design, changing the bottom geometry, adding baffles like in Cheng *et al.* (2015) and so on.

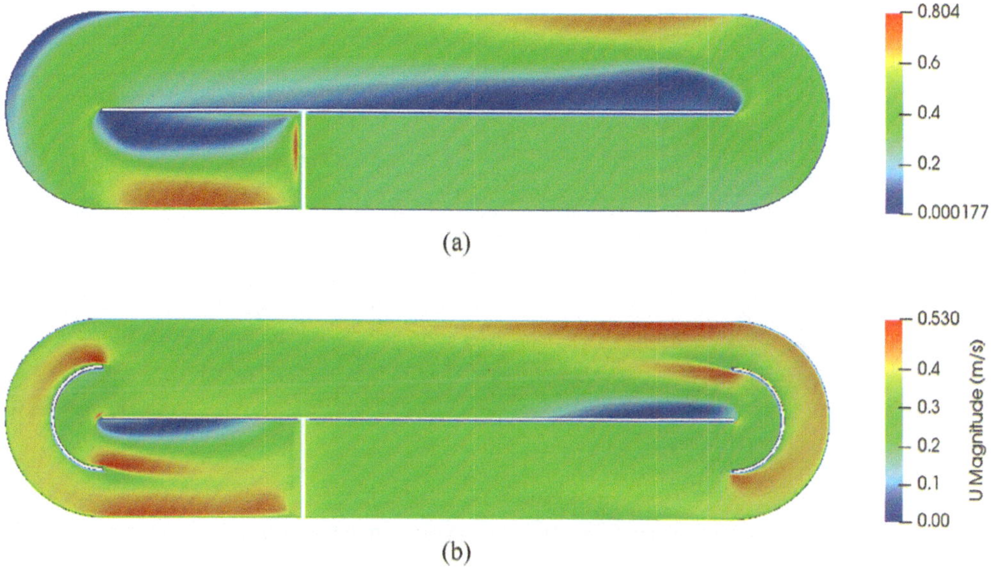

Figure 5.8 Velocity contours at mid-depth (a) without deflectors (b) with deflectors.

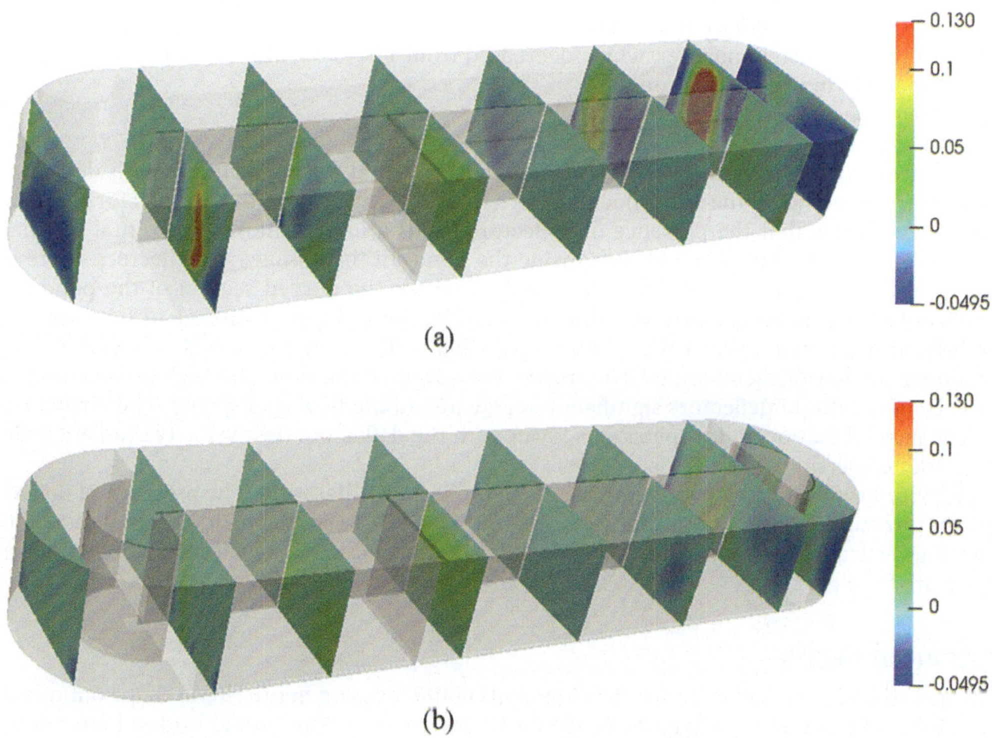

Figure 5.9 z velocity contours (m/s) for different vertical slices (a) without deflectors (b) with deflectors.

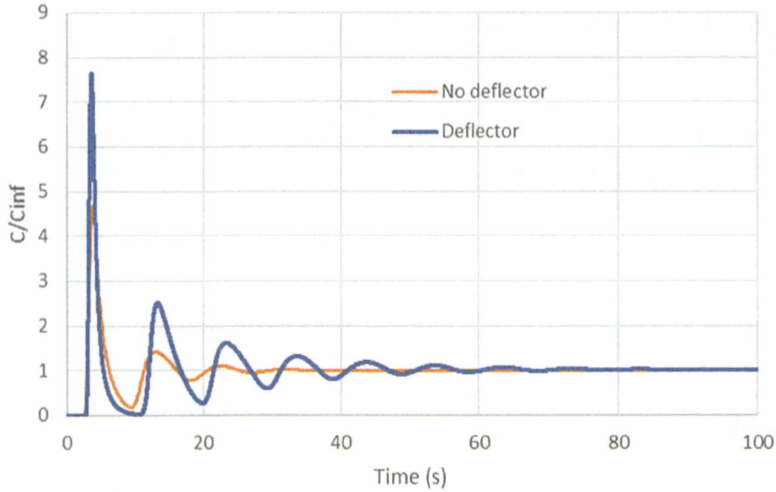

Figure 5.10 Virtual tracer experiment results for HRAP with and without deflectors at the bends ($Sc_t = 0.7$).

5.4.5.3 *Virtual tracer tests*

Based on previous simulations, virtual tracer experiments have been performed (Figure 5.10). With no deflectors, the mixing time is greatly reduced (from 60 to 30 s): non-uniformity of the flow field is higher in these conditions in both horizontal and vertical directions, yielding this faster homogenization. The estimation of the Bodenstein number, which decreased from 115 to 34, also confirms the higher level of spatial dispersion in these conditions.

5.4.5.4 *Conclusions*

Two HRAP geometrical designs have been tested and simulated using CFD: one with deflectors at the bends corresponding to the experimental HRAP, the other without these deflectors. The flow-field analysis confirms that the presence of deflectors yields a more uniform horizontal circulation velocity, leading to less dead-zone volumes inside the pond. In the absence of deflectors, there is an abrupt change in pressure and velocity after the flow passes the curved section of the pond. Since the flow is not smoothly guided into the straight channel, there is a high chance of flow separation or an extended dead zone behind the middle separating wall, leading to a higher speed of flow in the remaining section of the channel. The higher the speed of the flow, the higher the energy loss would be. Hence, without deflectors significant energy loss of the flow is observed. The virtual tracer experiment showed that overall dispersion is lower with the deflectors (less velocity gradient yielding a lower spatial dispersion).

The pressure drop analysis gave similar results to those in the literature. The presence of deflectors enabled the pressure drop to be minimized, leading to a 55% reduction in energy consumption. However, the vertical velocity component is reduced in these conditions, which can decrease recurring light exposure for algal cells.

5.5 RESEARCH NEEDS

From an operational perspective, future developments could focus on multicriteria shape optimization (size, number and position of deflectors, central wall design, etc.). This would lead to better designs, saving space and energy with a potentially higher productivity.

From a more fundamental point of view, several aspects can be considered. For example, in recent years, several biokinetic models representing microalgae/bacteria interactions have been published (Casagli *et al.*, 2021; Solimeno *et al.*, 2017; Wágner *et al.*, 2016). Similarly to activated sludge models (ASMs) (Henze *et al.*, 2006), these models consider simplified hydrodynamics; the coupling of them with CFD would be particularly relevant. However, understanding the impact of light intensity and attenuation on the dynamics of microalgae (Wágner *et al.*, 2018) is a key issue. Several models can be used to describe light dependence of algal growth ranging from low complexity such as a Haldane-type function (Casagli *et al.*, 2021) to more local or mechanistic approaches (Han, 2002; Undurraga *et al.*, 2016). When using the latter, there is a need to track the trajectories of individual cells as seen in the recent study by Pozzobon and Perré (2020) for a small-scale PBR.

REFERENCES

Ahmad T. and Boyd C. E. (1988). Design and performance of paddle wheel aerators. *Aquacultural Engineering*, **7**(1), 39–62, https://doi.org/10.1016/0144-8609(88)90037-4

Ali H., Cheema T. A., Yoon H.-S., Do Y. and Park C. W. (2015). Numerical prediction of algae cell mixing feature in raceway ponds using particle tracing methods. *Biotechnology and Bioengineering*, **112**(2), 297–307, https://doi.org/10.1002/bit.25443

Amini H., Hashemisohi A., Wang L., Shahbazi A., Bikdash M., KC D. and Yuan W. (2016). Numerical and experimental investigation of hydrodynamics and light transfer in open raceway ponds at various algal cell concentrations and medium depths. *Chemical Engineering Science*, **156**, 11–23, https://doi.org/10.1016/j.ces.2016.09.003

Andersen R. A. (2005). Algal Culturing Techniques. Elsevier, Academic Press, Amsterdam.

ANSYS FLUENT Theory Guide. (2011). ANSYS Incorporated. Canonsburg, PA, USA.

Babu M. A., Hes E. M. A., van der Steen N. P., Hooijmans C. M. and Gijzen H. J. (2010). Nitrification rates of algal-bacterial biofilms in wastewater stabilization ponds under light and dark conditions. *Ecological Engineering*, **36**(12), 1741–1746, https://doi.org/10.1016/j.ecoleng.2010.07.023

Barbosa M. J., Albrecht M. and Wijffels R. H. (2003). Hydrodynamic stress and lethal events in sparged microalgae cultures. *Biotechnology and Bioengineering*, **83**(1), 112–120, https://doi.org/10.1002/bit.10657

Bell W. and Mitchell R. (1972). Chemotactic and growth responses of marine bacteria to algal extracellular products. *The Biological Bulletin*, **143**(2), 265–277, https://doi.org/10.2307/1540052

Bellinger E. G. and Sigee D. C. (2015). Freshwater Algae, Freshwater Algae: Identification, Enumeration and Use as Bioindicators, 2nd edn. John Wiley & Sons, Inc., Hoboken, NJ, USA.

Bilad M. R., Arafat H. A. and Vankelecom I. F. J. (2014). Membrane technology in microalgae cultivation and harvesting: a review. *Biotechnology Advances*, **32**(7), 1283–1300, https://doi.org/10.1016/j.biotechadv.2014.07.008

Casagli F., Zuccaro G., Bernard O., Steyer J. P. and Ficara E. (2021). ALBA: a comprehensive growth model to optimize algae-bacteria wastewater treatment in raceway ponds. *Water Research*, **190**, 116734, https://doi.org/10.1016/j.watres.2020.116734

Cheng J., Yang Z., Ye Q., Zhou J. and Cen K. (2015). Enhanced flashing light effect with up-down chute baffles to improve microalgal growth in a raceway pond. *Bioresource Technology*, **190**, 29–35, https://doi.org/10.1016/j.biortech.2015.04.050

Chiaramonti D., Prussi M., Casini D., Tredici M. R., Rodolfi L., Bassi N., Zittelli G. C. and Bondioli P. (2013). Review of energy balance in raceway ponds for microalgae cultivation: Re-thinking a traditional system is possible. *Applied Energy*, **102**, 101–111, https://doi.org/10.1016/j.apenergy.2012.07.040

Christenson L. and Sims R. (2011). Production and harvesting of microalgae for wastewater treatment, biofuels, and bioproducts. *Biotechnology Advances*, **29**(6), 686–702, https://doi.org/10.1016/j.biotechadv.2011.05.015

Cole J. J. (1982). Interactions between bacteria and algae in aquatic ecosystems. *Annual Review of Ecology and Systematics*, **13**(1), 291–314, https://doi.org/10.1146/annurev.es.13.110182.001451

Craggs R., Park J., Sutherland D. and Heubeck S. (2015). Economic construction and operation of hectare-scale wastewater treatment enhanced pond systems. *Journal of Applied Phycology*, **27**(5), 1913–1922, https://doi.org/10.1007/s10811-015-0658-6

Cromar N. J. and Fallowfield H. J. (1997). Effect of nutrient loading and retention time on performance of high rate algal ponds. *Journal of Applied Phycology*, **9**(4), 301–309, https://doi.org/10.1023/A:1007917610508

Dellinger G., Garambois P. A., Dellinger N., Dufresne M., Terfous A., Vazquez J. and Ghenaim A. (2018) Computational fluid dynamics modeling for the design of archimedes screw generator. *Renewable Energy*, **118**, 847–857, https://doi.org/10.1016/j.renene.2017.10.093

Dodd J. C. (1986). Elements of pond design and construction. In: Handbook of Microalgal Mass Cultures, A. Richmond (ed.), CRC Press, Inc., Boca Raton, Florida. pp. 265–284.

El Hamouri B., Rami A. and Vasel J. L. (2003). The reasons behind the performance superiority of a high rate algal pond over three facultative ponds in series. *Water Science and Technology*, **48**(2), 269–276, https://doi.org/10.2166/wst.2003.0130

Evans R. A., Cromar N. J. and Fallowfield H. J. (2005). Performance of a pilot-scale high rate algal pond system treating abattoir wastewater in rural south Australia: nitrification and denitrification. *Water Science and Technology*, **51**(12), 117–124, https://doi.org/10.2166/wst.2005.0443

Garcia-Ochoa F. and Gomez E. (2009). Bioreactor scale-up and oxygen transfer rate in microbial processes: an overview. *Biotechnology Advances*, **27**(2), 153–176, https://doi.org/10.1016/j.biotechadv.2008.10.006

Grobbelaar J. U. (1991). The influence of light/dark cycles in mixed algal cultures on their productivity. *Bioresource Technology*, **38**(2–3), 189–194, https://doi.org/10.1016/0960-8524(91)90153-B

Grobbelaar J. U., Soeder C. J. and Stengel E. (1990). Modeling algal productivity in large outdoor cultures and waste treatment systems. *Biomass*, **21**(4), 297–314, https://doi.org/10.1016/0144-4565(90)90079-Y

Gupta P. L., Lee S.-M. and Choi H.-J. (2015). A mini review: photobioreactors for large scale algal cultivation. *World Journal of Microbiology and Biotechnology*, **31**(9), 1409–1417, https://doi.org/10.1007/s11274-015-1892-4

Gutzeit G., Lorch D., Weber A., Engels M. and Neis U. (2005). Bioflocculent algal–bacterial biomass improves low-cost wastewater treatment. *Water Science and Technology*, **52**(12), 9–18, https://doi.org/10.2166/wst.2005.0415

Hadiyanto H., Elmore S., Van Gerven T. and Stankiewicz A. (2013). Hydrodynamic evaluations in high rate algae pond (HRAP) design. *Chemical Engineering Journal*, **217**, 231–239, https://doi.org/10.1016/j.cej.2012.12.015

Han B. P. (2002). A mechanistic model of algal photoinhibition induced by photodamage to photosystem-II. *Journal of Theoretical Biology*, **214**(4), 519–527, https://doi.org/10.1006/jtbi.2001.2468

Henze M., Gujer W., Mino T. and van Loosdrecht M. (2006). Activated Sludge Models ASM1, ASM2, ASM2d and ASM3. IWA Publishing, London, UK.

Hreiz R., Sialve B., Morchain J. Ô., Escudié R., Steyer J. P. and Guiraud P. (2014). Experimental and numerical investigation of hydrodynamics in raceway reactors used for algaculture. *Chemical Engineering Journal*, **250**, 230–239, https://doi.org/10.1016/j.cej.2014.03.027

Judd S., van den Broeke L. J. P., Shurair M., Kuti Y. and Znad H. (2015). Algal remediation of CO2 and nutrient discharges: a review. *Water Research*, **87**, 356–366, https://doi.org/10.1016/j.watres.2015.08.021

Karya N. G. A. I., van der Steen N. P. and Lens P. N. L. (2013). Photo-oxygenation to support nitrification in an algal–bacterial consortium treating artificial wastewater. *Bioresource Technology*, **134**, 244–250, https://doi.org/10.1016/j.biortech.2013.02.005

Kumar K., Mishra S. K., Shrivastav A., Park M. S. and Yang J. W. (2015). Recent trends in the mass cultivation of algae in raceway ponds. *Renewable and Sustainable Energy Reviews*, **51**, 875–885, https://doi.org/10.1016/j.rser.2015.06.033

Lawton R. J., Cole A. J., Roberts D. A., Paul N. A. and de Nys R. (2017). The industrial ecology of freshwater macroalgae for biomass applications. *Algal Research*, **24**, 486–491, https://doi.org/10.1016/j.algal.2016.08.019

Liffman K., Paterson D. A., Liovic P. and Bandopadhayay P. (2013). Comparing the energy efficiency of different high rate algal raceway pond designs using computational fluid dynamics. *Chemical Engineering Research and Design*, **91**(2), 221–226, https://doi.org/10.1016/j.cherd.2012.08.007

Liu F. (2016). A thorough description of how wall functions are implemented in OpenFOAM. In: Proceedings of CFD with OpenSource Software, Edited by H. Nilsson, http://dx.doi.org/10.17196/OS_CFD#YEAR_2016

Liu X. and Zhang J. (2019). Computational Fluid Dynamics: Applications in Water, Wastewater, and Stormwater Treatment. American Society of Civil Engineers, Reston, VA.

Mantovani M., Marazzi F., Fornaroli R., Bellucci M., Ficara E. and Mezzanotte V. (2020). Outdoor pilot-scale raceway as a microalgae-bacteria sidestream treatment in a WWTP. *Science of the Total Environment*, **710**, 135583, https://doi.org/10.1016/j.scitotenv.2019.135583

Mata T. M., Martins A. A. and Caetano N. S. (2010). Microalgae for biodiesel production and other applications: a review. *Renewable and Sustainable Energy Reviews*, **14**(1), 217–232, https://doi.org/10.1016/j.rser.2009.07.020

Medina M. and Neis U. (2007). Symbiotic algal bacterial wastewater treatment: effect of food to microorganism ratio and hydraulic retention time on the process performance. *Water Science and Technology*, **55**(11), 165–171, https://doi.org/10.2166/wst.2007.351

Mendoza J. L., Granados M. R., de Godos I., Acién F. G., Molina E., Banks C. and Heaven S. (2013). Fluid-dynamic characterization of real-scale raceway reactors for microalgae production. *Biomass and Bioenergy*, **54**, 267–275, https://doi.org/10.1016/j.biombioe.2013.03.017

Muñoz R. and Guieysse B. (2006). Algal-bacterial processes for the treatment of hazardous contaminants: a review. *Water Research*, **40**(15), 2799–2815, https://doi.org/10.1016/j.watres.2006.06.011

Oswald W. J. (1988). Large-scale algal culture systems (engineering aspects). In Microalgal Biotechnology, M. Borowitska (ed.), Cambridge University Press, Cambridge. pp. 357–394.

Oswald W. J. and Gotaas H. B. (1957). Photosynthesis in sewage treatment. *Transactions of the American Society of Civil Engineers*, **122**(1), 73–97, https://doi.org/10.1061/TACEAT.0007483

Park J. B. K. and Craggs R. J. (2010). Wastewater treatment and algal production in high rate algal ponds with carbon dioxide addition. *Water Science and Technology*, **61**(3), 633–639, https://doi.org/10.2166/wst.2010.951

Park J. B. K., Craggs R. J. and Shilton A. N. (2011). Wastewater treatment high rate algal ponds for biofuel production. *Bioresource Technology*, **102**(1), 35–42, https://doi.org/10.1016/j.biortech.2010.06.158

Pham L. A. (2018). Algal-bacterial consortium in high rate algal pond: evaluation of performances, wastewater nutrient recovery and experimental and numerical models based design. PhD thesis, Université de Strasbourg.

Pham L. A., Laurent J., Bois P. and Wanko A. (2018). Impacts of operational conditions on oxygen transfer rate, mixing characteristics and residence time distribution in a pilot scale high rate algal pond. *Water Science and Technology*, **78**(8), 1782–1791, https://doi.org/10.2166/wst.2018.461

Pham L. A., Laurent J., Bois P. and Wanko A. (2020). A coupled RTD and mixed-order kinetic model to predict high rate algal pond wastewater treatment under different operational conditions: performance assessment and sizing application. *Biochemical Engineering Journal*, **162**, 107709, https://doi.org/10.1016/j.bej.2020.107709

Pham L. A., Laurent J., Bois P., Teshome T. M. and Wanko A. (2021). Operating a semi-continuous raceway pond allows to link pH and oxygen dynamics to the interaction between microalgae and bacteria. *Desalination and Water Treatment*, **211**, 105–116, https://doi.org/10.5004/dwt.2021.26506

Pozzobon V. and Perré P. (2020). Multiscale numerical workflow describing microalgae motion and light pattern incidence towards population growth in a photobioreactor. *Journal of Theoretical Biology*, **498**, 110293, https://doi.org/10.1016/j.jtbi.2020.110293

Pragya N., Pandey K. K. and Sahoo P. K. (2013). A review on harvesting, oil extraction and biofuels production technologies from microalgae. *Renewable and Sustainable Energy Reviews*, **24**, 159–171, https://doi.org/10.1016/j.rser.2013.03.034

Priyadarshani I. and Rath B. (2012). Commercial and industrial applications of micro algae – a review. *Journal of Algal Biomass Utilization*, **3**(4), 89–100.

Prussi M., Buffi M., Casini D., Chiaramonti D., Martelli F., Carnevale M., Tredici M. R. and Rodolfi L. (2014). Experimental and numerical investigations of mixing in raceway ponds for algae cultivation. *Biomass and Bioenergy*, **67**, 390–400, https://doi.org/10.1016/j.biombioe.2014.05.024

Pruvost J., Pottier L. and Legrand J. (2006). Numerical investigation of hydrodynamic and mixing conditions in a torus photobioreactor. *Chemical Engineering Science*, **61**(14), 4476–4489, https://doi.org/10.1016/j.ces.2006.02.027

Rawat I., Ranjith Kumar R., Mutanda T. and Bux F. (2011). Dual role of microalgae: phycoremediation of domestic wastewater and biomass production for sustainable biofuels production. *Applied Energy*, **88**(10), 3411–3424, https://doi.org/10.1016/j.apenergy.2010.11.025

Richmond A. and Grobbelaar J. U. (1986). Factors affecting the output rate of spirulina platensis with reference to mass cultivation. *Biomass*, **10**(4), 253–264, https://doi.org/10.1016/0144-4565(86)90002-8

Salim S., Bosma R., Vermuë M. H. and Wijffels R. H. (2011). Harvesting of microalgae by bio-flocculation. *Journal of Applied Phycology*, **23**(5), 849–855, https://doi.org/10.1007/s10811-010-9591-x

Sialve B., Gales A., Hamelin J., Wery N. and Steyer J.-P. (2015). Bioaerosol emissions from open microalgal processes and their potential environmental impacts: what can be learned from natural and anthropogenic aquatic environments? *Current Opinion in Biotechnology*, **33**, 279–286. Environmental biotechnology • Energy biotechnology, https://doi.org/10.1016/j.copbio.2015.03.011

Sirajunnisa A. R. and Surendhiran D. (2016). Algae – a quintessential and positive resource of bioethanol production: a comprehensive review. *Renewable and Sustainable Energy Reviews*, **66**, 248–267, https://doi.org/10.1016/j.rser.2016.07.024

Solimeno A., Parker L., Lundquist T. and García J. (2017). Integral microalgae-bacteria model (BIO_ALGAE): application to wastewater high rate algal ponds. *Science of the Total Environment*, **601–602**, 646–657, https://doi.org/10.1016/j.scitotenv.2017.05.215

Sompech K., Chisti Y. and Srinophakun T. (2012). Design of raceway ponds for producing microalgae. *Biofuels*, **3**(4), 387–397, https://doi.org/10.4155/bfs.12.39

Suh I. S. and Lee C.-G. (2003). Photobioreactor engineering: design and performance. *Biotechnology and Bioprocess Engineering*, **8**(6), 313–321, https://doi.org/10.1007/BF02949274

Sutherland D. L., Turnbull M. H. and Craggs R. J. (2014). Increased pond depth improves algal productivity and nutrient removal in wastewater treatment high rate algal ponds. *Water Research*, **53**, 271–281, https://doi.org/10.1016/j.watres.2014.01.025

Sutherland D. L., Howard-Williams C., Turnbull M. H., Broady P. A. and Craggs R. J. (2015). Enhancing microalgal photosynthesis and productivity in wastewater treatment high rate algal ponds for biofuel production. *Bioresource Technology*, **184**, 222–229. Advances in biofuels and chemicals from algae, https://doi.org/10.1016/j.biortech.2014.10.074

Teshome T. M., Pham L. A. and Laurent J. (2019). Inlet velocity vs dynamic mesh methods for virtual tracer experiments in high rate Algal Pond. Watermatex 2019, Copenhagen, Denmark.

The OpenFOAM Foundation. (2021). OpenFOAM v9 User Guide. https://cfd.direct/openfoam/user-guide (accessed 12.20.21)

Uduman N., Qi Y., Danquah M. K., Forde G. M. and Hoadley A. (2010). Dewatering of microalgal cultures: a major bottleneck to algae-based fuels. *Journal of Renewable and Sustainable Energy*, **2**(1), 012701, https://doi.org/10.1063/1.3294480

Undurraga D., Poirrier P. and Chamy R. (2016). Microalgae growth kinetic model based on the PSII quantum yield and its utilization in the operational curves construction. *Algal Research*, **17**, 330–340, https://doi.org/10.1016/j.algal.2016.05.020

Unnithan V. V., Unc A. and Smith G. B. (2014). Mini-review: a priori considerations for bacteria–algae interactions in algal biofuel systems receiving municipal wastewaters. *Algal Research*, **4**(1), 35–40, https://doi.org/10.1016/j.algal.2013.11.009

Van Den Hende S., Vervaeren H., Desmet S. and Boon N. (2011). Bioflocculation of microalgae and bacteria combined with flue gas to improve sewage treatment. *New Biotechnology*, **29**(1), 23–31, https://doi.org/10.1016/j.nbt.2011.04.009

Van Den Hende S., Beelen V., Bore G., Boon N. and Vervaeren H. (2014). Up-scaling aquaculture wastewater treatment by microalgal bacterial flocs: from lab reactors to an outdoor raceway pond. *Bioresource Technology*, **159**, 342–354, https://doi.org/10.1016/j.biortech.2014.02.113

Vandamme D., Foubert I. and Muylaert K. (2013). Flocculation as a low-cost method for harvesting microalgae for bulk biomass production. *Trends in Biotechnology*, **31**(4), 233–239, https://doi.org/10.1016/j.tibtech.2012.12.005

Voloshin R. A., Rodionova M. V., Zharmukhamedov S. K., Nejat Veziroglu T. and Allakhverdiev S. I. (2016). Review: biofuel production from plant and algal biomass. *International Journal of Hydrogen Energy*, **41**(39), 17257–17273, https://doi.org/10.1016/j.ijhydene.2016.07.084

Wágner D. S., Valverde-Pérez B., Sæbø M., Bregua de la Sotilla M., Van Wagenen J., Smets B. F. and Plósz B. G. (2016). Towards a consensus-based biokinetic model for green microalgae – the ASM-A. *Water Research*, **103**, 485–499, https://doi.org/10.1016/j.watres.2016.07.026

Wágner D. S., Valverde-Pérez B. and Plósz B. G. (2018). Light attenuation in photobioreactors and algal pigmentation under different growth conditions – model identification and complexity assessment. *Algal Research*, **35**, 488–499, https://doi.org/10.1016/j.algal.2018.08.019

Wilhelm D. (2015). Rotating flow simulations with open-FOAM. *International Journal of Aeronautical Science & Aerospace Research*, **1**, 1–1.

doi: 10.2166/9781780409030_0115

Chapter 6
Sedimentation

Alonso G. Griborio[1], Maria Elena Valle-Medina[2], Ed Wicklein[3] and Julien Laurent[2]

[1]Hazen and Sawyer, Hollywood, FL, USA; Email: agriborio@hazenandsawyer.com
[2]ICube, ENGEES, CNRS, Université de Strasbourg, Strasbourg, France; Email: mariaelena.valle-medina@engees.unistra.fr; julien.laurent@engees.unistra.fr
[3]Carollo Engineers, Seattle, WA, USA; Email: ewicklein@carollo.com

6.1 INTRODUCTION

Primary sedimentation (Figure 6.1) is one of the key processes for total suspended solids (TSS) and biochemical oxygen demand (BOD) removal in wastewater resources recovery facilities (WRRFs) and for reducing -the load to the downstream biological processes (Ahuja & Griborio, 2017). Primary sedimentation exhibits flocculation, and discrete, hindered, and compression settling of near neutral buoyancy colloidal mineral and organic particles. Process performance is highly influenced by flow, including those induced by multiphase solid-liquid interactions in sludge blankets. These factors make it a clear target for the application of CFD. As such, primary sedimentation was one of the first unit processes to receive attention from CFD researchers.

Secondary sedimentation (Figure 6.1) is traditionally the last major process step in secondary treatment, whether that occurs by fixed film (trickling filter or suspended media) or suspended growth (activated sludge) processes. The first function of secondary sedimentation is to separate solids from the secondary treatment process prior to tertiary treatment such as disinfection and/or effluent discharge, thereby ensuring as low effluent suspended solid concentration as possible. A second function is to concentrate sludge for return to biological reactors and/or for wastage to a solids handling unit process. In recent years membrane separation has proven to be a significant competitor to gravity sedimentation for secondary sludge, but an alternative that, while providing higher levels of capture in a much smaller footprint, has the disadvantage of increased energy consumption. A third function would be storage of sludge during periods of diurnal flow variations.

Secondary sedimentation would appear to have similar challenges to primary sedimentation, but is actually governed by some different mechanisms, including near neutral solids density, discrete settling, prominent hindered and compressive settling behavior and higher underflow rates. Modeling goals are also directed by the multiple functions of final effluent clarification, activated sludge thickening, sludge storage and flocculation.

Editor(s): Julien Laurent, Randal Samstag, Jim Wicks and Ingmar Nopens

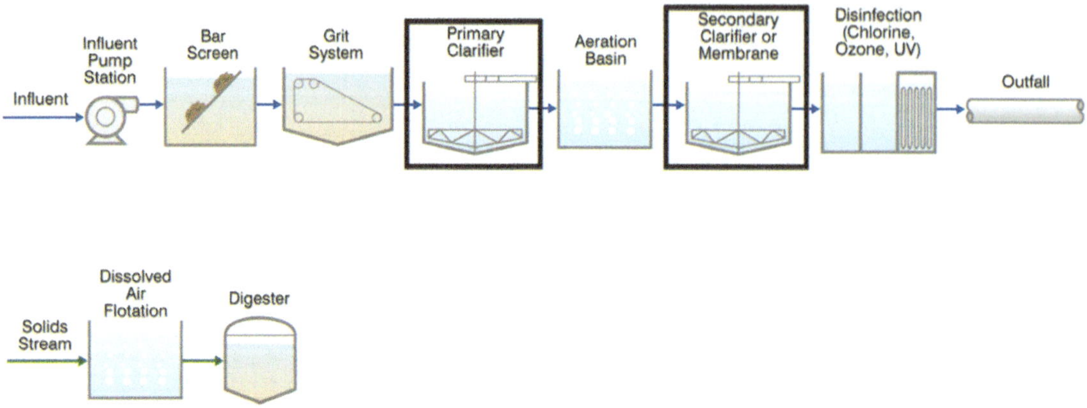

Figure 6.1 Placement of the primary and secondary clarifiers in a wastewater resources recovery facility.

Secondary sedimentation has received considerable scrutiny in the literature. It has been the subject of a comprehensive Scientific and Technical Report published by the International Association on Water Quality (Ekama *et al.*, 1997) and a comprehensive group evaluation as a part of a Water Environment Federation Manual of Practice (Water Environment Federation, 2005).

6.2 HISTORICAL BACKGROUND

One of the earliest documented studies on the theory of sedimentation goes back to the late 19th century with the work by Seddon (1889) who provides observations and preliminary theory on the use of sedimentation to improve water quality. Hazen (1904) provided the first detailed consideration of the theory of sedimentation. At that time, Hazen noted that little was published on the theory of sedimentation but significant advancements had been made in actually building and operating sedimentation basins. The clarifier modeling field can trace its beginning to the work by Hazen, who introduced the surface overflow rate (SOR) concept, a design concept which has been widely used and still is a major criterion in settling tank design (Ahuja & Griborio, 2017).

The SOR is usually expressed in cubic meters per square meter of surface area per day ($m^3.m^{-2}.d^{-1}$) or in gallons per square feet per day ($gpd.sf^{-1}$ in English units).

Hazen recognized the importance of the hydraulic regime of the tank mentioning that a 'well-baffled basin will do as much work as a much larger basin without baffles.' He recognized the importance of inlet control: 'The kinetic energy of the water as it enters is still capable of producing vortex motion after long periods, but which can be much reduced by controlling the water at the entrance so that it has only a very low velocity.' He also recognized the impact of wind and changes in temperature on buoyancy-induced flow. Hazen's theory states that the hydraulic retention time should be equal to the time needed for a particle to settle from the top to the bottom of the tank; in this way all the particles with settling velocities higher than that of the design particle will be removed.

Hazen's theory has many simplifications that make its application unrealistic, particularly for secondary sedimentation tanks (SSTs), for example:

- Hazen assumes a uniform horizontal velocity field where turbulence is not considered; in reality, the flow field in an SST is turbulent and heterogeneous in nature, and the high solids loading and low hydraulic loading lead to density-dominated flows.

- Hazen assumes that the settling rate of the particle is constant and is independent of the flow; this can be true for discrete particles settling, but in an SST the flocculent, hindered and compression settling usually dominate the process, and these are influenced by the concentration of suspended solids and the biological nature of the flocs.

Thomas R. Camp was a leader during the middle decades of the 20th century in consolidating the basic theory of Hazen (1904) and extending it with application to grit chamber and sedimentation tank design (Camp, 1973). Camp's first significant paper (Camp, 1936) applied Stokes' law to the settling of particles, proposed a test procedure to establish the distribution of settling velocities in raw water or sewage that is still useful today, and derived an expression for 'ideal' settling:

$$r = \frac{v}{v_o} = \frac{Av}{Q} \tag{6.1}$$

where, r = the removal as a ratio of particles settling at any velocity, v = less than the overflow rate, v_o, A = the surface area of the tank, Q = the rate of discharge.

Camp recognized that this applied only to an ideal tank, without considering the influence of turbulence, multidimensional, asymmetrical flow, and vortex formation. In his paper from 1945 (Camp, 1945) he presents the equation developed by Dobbins (1944) which is a partial differential transport equation for sedimentation in a two-dimensional domain:

$$V' \frac{\partial c}{\partial x} = \epsilon_y \frac{\partial^2 c}{\partial y^2} + \left(v + \frac{\partial \epsilon_y}{\partial y} \right) \frac{\partial c}{\partial y} + \epsilon_x \frac{\partial^2 c}{\partial x^2} \tag{6.2}$$

where, $V' = \partial x / \partial t$ = the velocity at height, y, c = the concentration of solid particles, x and y = the horizontal and vertical spatial coordinates, ϵ_y = dispersion coefficient in the vertical direction, ϵ_x = dispersion coefficient in the horizontal direction, v = the particle settling velocity.

Results from this equation have been used as verification for a finite difference solution of the full equation by Jobson and Sayre (1970). Since the early work from Hazen, Camp and Dobbins, a number of other analytical approximations have been made to simplify multidimensional clarifier geometries into a single dimensional finite difference problem, mainly by imposing radial homogeneity or symmetry. Takamatsu *et al.* (1974) applied radial symmetry and a wall boundary condition, together with discretization in the vertical dimension to identify the optimal depth of a clarifier. The boundary condition for the bottom (sludge) boundary identified in this work has been widely used.

As described in Griborio (2004), and still applicable today, the solids flux theory is the most commonly applied method for the design of secondary sedimentation tanks, while primary sedimentation tanks (PSTs) are commonly designed and operated solely based on SOR and hydraulic retention time (HRT) (Patziger *et al.*, 2016; Rodriguez *et al.*, 2021). According to Ekama *et al.* (1997), the solids flux method is applied in two phases:

- zone settling and thickening considerations are applied, leading to the determination of the surface area and depth;
- the internal features are included in the tank, which should optimize the clarification efficiency.

The second phase is typically performed following some semi-empirical rules and basically relying on the engineer's experience (Griborio & McCorquodale, 2005). As discussed in Ekama *et al.* (1997) the solids flux theory can significantly overpredict or underpredict the clarifier maximum allowable solids loading rate (SLR) leading to over or under prediction of the required clarifier area. In general, a 20% reduction in the predicted SLR is recommended to prevent undersizing the clarifier area when using the solids flux method. However, there is evidence that this factor may still lead to undersized units and in many cases to over design of the clarifier infrastructure, resulting in unnecessary expenditure.

The uncertainty related to the solids flux theory is partially due to its inherited simplifying assumptions, such as: the effluent suspended solids and tank hydrodynamics are ignored; the sludge withdraw is assumed to be one dimensional with no short circuiting; clarifier storage is ignored and the flow and solids accumulation are assumed to be at steady state; and the Vesilind (1968) equation (Equation (6.4)) is assumed to apply at the critical flux boundary and compression or two-phase effects are assumed to be negligible.

Wahlberg *et al.* (2003) presented a method to estimate primary clarifier efficiency based on historical data. They presented the following equation for the estimation of TSS:

$$E_{TSS} = E_{TSS_{MAX}}(1 - e^{-\lambda/SOR}) \tag{6.3}$$

where, E_{TSS} = total suspended solids removal efficiency (%), $E_{TSS_{MAX}} = 1 - (TSS_{non}/TSS_{PI})$, TSS_{non} = non-settleable total suspended solids, mg.L^{-1}, TSS_{PI} = influent total suspended solids, mg.L^{-1}, SOR = surface overflow rate, gallons per day per square foot, λ = a settling parameter with dimensions of gallons per day per square foot.

A similar expression was developed for prediction of removal of chemical oxygen demand. The approach proposed by Wahlberg is also presented in the Water Environment Federation Manual of Practice on Clarifier Design (Water Environment Federation, 2005).

Many authors have criticized the use of simplistic models, such as the models presented by Camp, and application of the solids flux theory (Wahlberg *et al.*, 2003), for the evaluation and design of sedimentation tanks (Goula *et al.*, 2008; Griborio, 2004; Kinnear, 2004; Patziger *et al.*, 2016; Rodriguez *et al.*, 2021 – just to name a few). Goula *et al.* (2008) indicated that sedimentation tanks appear to be a simple phase-separating device, but in fact an intricate balance of forces is present.

Sedimentation tanks are characterized by interesting hydrodynamic phenomena, such as circulation zones, bottom currents, inflow dynamic behavior, flow viscosity, turbulence, density currents (Razmi *et al.*, 2009; Tamayol *et al.*, 2010) and thermal density currents induced by variable influent temperatures as well as surface heat exchange (Griborio *et al.*, 2014). Therefore, short circuiting and dead areas are created, which reduce the effective sedimentation volume of the tanks, meaning solids may be resuspended, affecting the performance of these tanks currents (Razmi *et al.*, 2009; Tamayol *et al.*, 2010). Since design standards and some literature-based resources do not take into account important site-specific factors nor the internal features of the clarifiers, designing with such simplifications can lead to over or under designs (Ahuja & Griborio, 2017).

CFD is an ideal tool for the evaluation, design and optimization of primary and secondary sedimentation tanks. CFD can accurately represent the clarifier internal configuration and include the major processes and factors affecting clarifier performance such as (Griborio, 2004): (A) hydrodynamics, (B) settling, (C) turbulence, (D) sludge rheology, (E) flocculation, and (F) thermal effects. Compared to traditional evaluation methods, techniques based on CFD are of high-precision (Gao & Stenstrom, 2020a; Karpinska & Bridgeman, 2016), and allow visualization of the internal conditions in the clarifier such as position of the sludge blanket and flow pattern. CFD is a powerful tool for the evaluation of clarifier capacity and for determining clarifier performance under different sets of internal modifications. This is particularly important during clarifier design or rehabilitation work, in order to determine the set of internal features or modifications that results in optimum performance. CFD models can be applied in one-, two- and three-dimensions, but 2D and 3D models are required for a real definition and incorporation of the hydrodynamics and other processes and factors described above. Ahuja and Griborio (2017) presented a comparison of the different modeling tools available, including application of the solids flux theory and 1D, 2D and 3D CFD models. This comparison is presented in Table 6.1.

Table 6.1 Comparison of clarifier modeling tools.

Level	Strength	Application	Weakness
Limiting Solids Flux Theory (Example: State Point Analysis)	Simple	Zone settling (secondary clarifiers preliminary design and operational)	Ignores hydrodynamics, flocculation and internal configuration. Cannot predict effluent quality.
1D CFD	Computational speed	All types of settling including two-phase flows (applicable to primary and secondary clarifiers)	Ignores hydrodynamics, flocculation and internal configuration.
2D CFD	Computational speed compared to 3D. Includes all major processes, factors and geometric considerations affecting clarifier performance and capacity.	All clarifiers where there is a dominant flow direction; dynamic simulations. Models such as 2Dc (McCorquodale *et al.*, 2005) are available for design engineers.	Ignores lateral non-uniformity in solids and momentum.
3D CFD	Completeness of governing equations; high spatial resolution.	All clarifiers. Steady state simulations where a dominant flow direction cannot be assumed. Typically, 3D models are built using available platforms such as ANSYS Fluent, OpenFOAM, CFX and Flow–3D.	Long execution times; high level of expertise required.

Source: Adapted from Ahuja and Griborio (2017).

Primary and secondary sedimentation were among the first unit processes to receive attention from CFD researchers. Summary descriptions of some of the most important research follow.

6.3 SLUDGE BULK FLUID MODELING OF CLARIFIERS

The sludge settling process can be considered as a multiphase phenomenon in both PSTs and SSTs where water is a continuous phase and the sludge is the dispersed phase. However, full implementation of a multiphase approach is still computationally very demanding, and several simplifications have been made to balance accuracy with computational time. As described by Gao and Stenstrom (2020b), the accurate numerical simulation of the sludge settling process relies on the accurate assessment of the sludge transport phenomena and the character of interaction between the phases. Thus, to model sludge settling different approaches can be applied (see also Chapter 2):

Euler-Lagrange model. This approach is mainly applied when the dispersed phase has no significant impact on the bulk fluid, for example within grit removal processes. Each one of the particles within the fluid is tracked through a flow domain. Thus, the momentum equation for the particle describes the motion of the particle in the continuous medium. One fundamental consideration in this model is that the dispersed phase must have a low volume fraction (maximum 10–12%). This approach has been successfully employed to estimate the trapping efficiency of settlers by Isenmann (2016) and Isenmann *et al.* (2021).

Euler and transport model (scalar approach). The fluid mechanics governing equations are only solved for the continuous phase (i.e. water). The sludge mass motion is described by a scalar transport equation that uses the computed flow field for the advective term. This approach requires the coupling of solids transport to momentum and continuity equations. This is performed using specific relations for viscosity and density coupled between the phases through the relations presented in Section 6.4.3 (Samstag *et al.*, 2012; Wicklein & Samstag, 2009). This is maybe the most commonly used approach to model the sludge transport (De Clercq, 2003; Griborio, 2004; Lakehal *et al.*, 1999; Samstag *et al.*, 2016; Xanthos *et al.*, 2011; Zhou & McCorquodale, 1992a).

Euler-Euler model. The sludge is considered as a second continuous medium, having similar conservation equations to the continuous phase. Therefore, two sets of Navier-Stokes equations will be solved (Brennan, 2001): one set for the sludge particles and the other set for the water. They are solved by introducing the volume fraction and the mechanisms for the exchange of momentum and mass between the phases which make it more complex and computationally demanding (Karpinska & Bridgeman, 2016).

Mixture Model. The model is based on the assumption that only one continuity and one momentum equation are solved for the mixture. The distribution of the dispersed phase is modeled with an advection-dispersion equation that introduces the concept of drift velocity (Brennan, 2001). It is assumed that in the horizontal direction, both phases act as just one. To employ such a model, one must consider that both phases are isothermal, non-reactive, incompressible and non-phase-changing. The model uses the volume fraction of the dispersed phase: implicitly the density of the activated sludge is included, since the volume fraction is expressed as the ratio of the sludge concentration divided by the sludge density (i.e. the floc density as defined in Nopens (2005)). This approach has been applied successfully in Valle-Medina (2019), Valle Medina and Laurent (2020) and Rodriguez *et al.* (2021). Using this approach, different particle sizes can be incorporated by adding separate transport equations for each class, and the particle interactions and external forces are incorporated using separate equations.

A detailed review of the different approaches used to simulate particle motion in primary and secondary settling tanks can be found in Gao and Stenstrom (2020b).

6.3.1 Primary settling

A number of analytical approximations have been taken to simplify multidimensional clarifier geometries into a single dimensional finite difference problem, mainly by imposing radial homogeneity or symmetry. Takamatsu *et al.* (1974) applied radial symmetry and a wall boundary condition, together with discretization in the vertical dimension to identify the optimal depth of a clarifier. The boundary condition for the bottom (sludge) boundary identified in this work has been widely used. The first two-dimensional clarifier model was presented by Larsen (1977). Larsen's model was developed for a rectangular secondary clarifier and was based on the equations of motion, continuity and an exponential equation relating settling velocity to concentration. Larsen introduced the concept of stream function and vorticity, and the generation of vorticity by internal density gradients and shear along solid boundaries. Diffusivity was assumed equal to eddy viscosity, which was computed on the basis of the Prandtl mixing length theory (see Chapter 2).

One of the first multidimensional numerical models of primary sedimentation was introduced by Imam and McCorquodale (1983). The model used a finite difference discretization of the governing equations to simulate the settling of several classes of solids in the flow domain. Results were compared to several different theoretical estimates by Camp (1945) and showed a more realistic lower rate of removal. The work assumed neutral density and therefore would not be expected to be effective in adequately predicting the influence of underflow or sludge blanket solids concentrations.

Abdel-Gawad and McCorquodale (1984a) presented a strip integral numerical model to simulate the performance of primary clarifiers, with emphasis on the prediction of the velocity field and concentration distribution. The hydrodynamic sub-model was calibrated using a measured velocity distribution in a physical model. Solids were modeled as classes of discrete particles with settling velocities based on batch tests. The model was calibrated by measurements in full-scale rectangular tanks. The predicted concentration profiles and removal efficiency were in good agreement with the measured values. The sludge blanket and underflow, however, were not modeled.

A 2D, neutral density, turbulent CFD model of rectangular primary sedimentation tanks was developed by Stamou *et al.* (1989) who compared predicted solids profile and residence time distribution (RTD) curves to field measurements. Turbulent flow was modeled using a $k-\varepsilon$ approach (Rodi, 2000) (see Chapter 2). They found that variation in the turbulent Schmidt number from 0.5 to 1.0 resulted in little difference in the flow through curve (FTC) (for detailed discussion about turbulent Schmidt number, see also Chapter 10). Wang and Maxey (1993) explored the problem of discrete settling in a flow field of homogeneous turbulence from the perspective of theoretical fluid mechanics. This is sophisticated work with considerable attention to numerical and experimental issues. They produced results for the increase in mean Stokes drag settling speed in uniform turbulence of zero mean velocity (batch conditions).

A generalized settling model (GSM) that includes a model for hindered settling in addition to discrete settling was presented in Mazzolani *et al.* (1998). This was specifically intended to improve upon discrete settling models (DSMs) and monodisperse settling models (MSMs) that do not recognize that typical sedimentation tanks for water and wastewater 'typically treat highly heterodisperse suspensions.' They discuss how their model is different to the single exponential model (Vesilind, 1968) and the double exponential model of Takács *et al.* (1991) (Equations (6.4) and (6.5)). They positively compare results of a 2D numerical simulation of their GSM to discrete and basic hindered models.

Griborio *et al.* (2014) provide a report on a comprehensive evaluation of two primary sedimentation tanks evaluated with 2D and 3D CFD models. The projects included field testing and evaluation of alternative geometries and recommendations for improvement. Notably, the work with the 2D model included full solids transport with density coupling, a five-component settling model including a compression model, and flocculation using the 2Dc/3Dc model Griborio (2004). The 3D model did not include these features but was used to estimate velocity profiles in the inlet and outlet.

Patziger and Kiss (2015) provide a 2D PST model which includes a settling function for primary sludge comprising flocculation, hindered settling, compression and rheological features. Model calibration and verification took place based on full-scale measurements of velocity and concentration patterns in a wide range of hydrodynamic and TSS load conditions, with SORs ranging from 5 m.h^{-1} to 10 m.h^{-1}. Beyond a good prediction of effluent suspended solids concentration this approach resulted in an accurate calculation of the entire flow and concentration pattern as well as the sludge bed movements.

Liu and García (2011) developed a 3D model for the evaluation of the hydrodynamics and internal configuration of PSTs. A two-fluid model using a Eulerian technique was used to capture the behavior of the solid-water mixture. Hindered settling was modeled using the Takacs double exponential equation (Takács *et al.*, 1991) (Equation (6.5)) and the sludge was modeled as a Bingham plastic with parameters estimated from experiments (see Section 6.4.3.2 in this chapter and the rheology section in Chapter 8). They measured particle size distribution and used it as a model input, and investigated the PST performance under different configurations with emphasis on the tank and the feedwell depths. Due to heavy computational effort, only a limited number of particle size classes were considered, which is one of the limitations of the application of Eulerian models to simulation of settling tanks.

Rodriguez *et al.* (2021) presented the development and calibration of a PST 3D model. Their model uses a multiphase mixture model including a 'granular option' to represent interparticle collisions. They implemented three particle sizes (10, 50 and 80 μm) based on particle size distribution measurements and calibrated their model by comparing measured effluent quality and internal sludge profiles to the model prediction. Rodriguez *et al.* included the simulation of a rotating moving scraper using a sliding mesh methodology.

Table 6.2 presents a summary of important contributions to the literature and important characteristics of their CFD approaches.

Table 6.2 CFD characteristics in research on primary clarifiers.

Reference	Solution Technique	Dimensions	Turbulence Model	Multiphase Approach	Mesh/CFD Implementation	Density Couple
Imam and McCorquodale (1983)	Vorticity/Stream Function	2D	Constant eddy viscosity	Solids transport	73 by 39/FORTRAN	Neutral
Abdel-Gawad and McCorquodale (1984a, 1984b)	Strip Integral Method/Finite Element	2D	Mixing length	Solids transport	Coarse grid/ FORTRAN	Neutral
Mazzolani *et al.* (1998)	SIMPLEC	2D	$k-\varepsilon$	Solids transport	Not specified	Yes?
Razmi *et al.* (2009)	SIMPLEC	2D	$k-\varepsilon$	Air/Water	680 by 53/Fluent	Neutral
Stamou *et al.* (1989)	SIMPLEC	2D	$k-\varepsilon$	None	60 by 30/FORTRAN	Neutral
Wang and Maxey (1993)	Pseudospectral	3D	Random Fourier modes	Stokes settling	323 to 963/Not specified	Neutral
Liu and García (2011)	Finite Volume	3D	$k-\varepsilon$	Two-fluid model/ Eulerian technique	1.2 million cells/ OpenFOAM	Yes
Griborio *et al.* (2014)	Finite Volume Vorticity/Stream Function	2D	Mixing length	Solids transport/settling function developed for primary sludge	Variable, up to 200 by 150/FORTRAN	Yes
Patziger and Kiss (2015)	Finite Volume	2D	RNG $k-\varepsilon$	Settling function developed for primary sludge	Ansys Fluent	Yes
Rodriguez *et al.* (2021)	Finite Volume/ SIMPLE	3D	$k-\varepsilon$	Mixture model with three particle sizes	500 000 grids/Ansys Fluent	Yes

6.3.2 Secondary settling

McCorquodale and his students and associates at the University of Windsor and the University of New Orleans have been leaders in the application of CFD to sedimentation, using customized computer codes based either on the vorticity/stream function method of Roache (1972) or the methods of Patankar (1980). Zhou and McCorquodale (1992c) investigated 2D flow in a radial flow (circular) clarifier. They performed simulations using their CFD code based on Patankar (1980) and incorporating turbulent flow ($k-\varepsilon$ turbulence model), solids transport, and settling by a double exponential function (Takács *et al.*, 1991) (Equation (6.5)). The model also included coupling of the concentration profile with the fluid flow through fluid density (see Section 6.4.3.1). They compared the resulting velocity profile in their model for the cases with and without solids input and underflow pumping. Their results demonstrated the 'density waterfall' that occurs in radial flow clarifiers as a result of solids settling immediately after introduction at the tank inlet. Their comparisons in CFD and physical model tests showed that with no solids present, the velocity pattern in the tank is completely different from when solids are included. They investigated sedimentation in a tank with three different feedwell diameters. Results showed that performance was best for the tank with the smallest feedwell. Results from the larger feedwell were very similar to results from simulations of the case with no feedwell at all. More general applications of this CFD technique were presented by these authors for rectangular clarifiers in Zhou and McCorquodale (1992a) and for circular clarifiers in Zhou and McCorquodale (1992b, 1992c). Other research by McCorquodale and his associates has been crucial to development of the state of the art of clarifier modeling, including McCorquodale *et al.* (1991) and McCorquodale and Zhou (1993). Griborio (2004) covers development and validation of a five-component settling model, flocculation models, and details for quasi-3D CFD for activated sludge clarifiers. This CFD model uses a vorticity/stream function approach to eliminate pressure from the fluid equations and a simple mixing length turbulence model. This work also contains an extensive reference list.

Flocculation (see Chapter 4) is also linked to clarifier performance. The impact of flocculation processes on secondary clarifier performance via simulation was extensively assessed within McCorquodale's group and is further detailed in Section 6.4.2 as well as in Griborio (2004).

Armbruster *et al.* (2001) present results from use of the mathematical model presented in Lakehal *et al.* (1999). Results were compared to experiments by Krebs *et al.* (1995). Concentration profiles from simulation of dynamic loading over 24 hours of flow based on data from a treatment plant in Germany were presented. De Clercq (2003) developed a 2D CFD model in a commercial package incorporating solids transport, density couple, and flocculation. The model was applied to a radial flow (circular) activated sludge clarifier. A Herschel-Bulkley rheological model was incorporated by user-defined functions (see Section 6.4.3.2 in this chapter and also the rheology section in Chapter 8).

A 2D finite element analysis of flows in secondary settling tanks was presented in Kleine and Reddy (2005). They incorporated solids settling using a double exponential function (Takács *et al.*, 1991) (Equation (6.5)) and included the influence of density on the momentum equations. The $k-\varepsilon$ equation was used to simulate turbulence and compared to a constant turbulent viscosity approach. The model was compared to field tests for radial flow circular activated sludge settling tanks and to results from the commercial sedimentation program, SettlerCAD. In comparison to field results the best fit of CFD results was with a turbulent Schmidt number greater than 1 (for detailed discussion about turbulent Schmidt number, see also Chapter 10).

Patziger *et al.* (2008) developed an SST model that included an RNG $k-\varepsilon$ model for turbulence, a settling function consisting of four parts, and a viscosity model which consists of three functions depending on the range of the local concentration: one for the low concentration ranges (Newton), one for the medium concentration ranges (modified linear) and one for the high concentration ranges (Hershel Bulkley). In order to be able to investigate return sludge flow strategies they coupled the model with a simplified reactor cascade model to include the effects of retention time, mixing of water and sludge streams, and suspended solids concentration entering the SST in each time step.

Burt (2010) developed 2D CFD models calibrated by field and laboratory tests. The CFD models were constructed in a commercial software using an algebraic slip model with user-defined functions for definition of density, settling velocity and rheology. An extended drift flux (EDF) model was compared to a multiple drift flux (MDF) model which tracked 10 different size groups.

Gong *et al.* (2011), Ramalingam *et al.* (2009, 2012) and Xanthos *et al.* (2013) presented the development of 3D models and their application for the evaluation of rectangular settling tanks. These models developed by the same group in City College New York, included incorporation of a flocculation sub-model following the same model incorporated by Griborio (2004). Their model included discrete settling velocities for small, medium and large flocs and an exponential-type equation (Vesilind, 1968) (Equation (6.4)) to represent both zone and compression settling regions.

Gao and Stenstrom (2018) compared the performance of three different types of turbulence models, Standard $k-\varepsilon$, Realizable $k-\varepsilon$, and RNG $k-\varepsilon$, on the evaluation of circular SSTs using a 2D model. They found that the effluent TSS was not sensitive to changes in the turbulence model but some differences were observed in the prediction of the sludge blanket height (SBH). Gao and Stenstrom did not conclude which of the three turbulence models resulted in more accurate simulation but indicated that the Realizable $k-\varepsilon$ and RNG $k-\varepsilon$ predicted higher capacity in one of the case studies. They used the Takacs double exponential equation (Takács *et al.*, 1991) to simulate sludge settling (Equation (6.5)) and a Bokil model (Bokil, 1972; Bokil & Bewtra, 1975) to represent sludge rheology.

Gao and Stenstrom (2020a, 2020b) used 3D CFD modeling to study the effect of surface wind on the performance of circular and rectangular clarifiers. In these models a constant horizontal velocity was assumed on the free surface along the wind direction. The models used the Standard$k-\varepsilon$ model for turbulence closure including a buoyancy damping term in the turbulence kinetic energy equation to describe the turbulent and density-stratified flow. Similarly to their 2D model application, they used the Takacs equation (Equation (6.5)) for settling and the Bokil model for sludge rheology and a single transport equation to simulate the advection, diffusion and deposition of solids in the settling tank.

Table 6.3 presents a summary of important contributions to the literature and their CFD approaches.

6.4 PROCESS DESCRIPTION AND FEATURES TO BE INCLUDED IN CFD MODEL

Understanding the similarities and differences between primary and secondary tanks is key to the appropriate field characterization for model set-up, calibration and successful application to primary and secondary settling tanks. The model set-ups are similar in secondary and primary tanks; however, the order of magnitude of some of the processes and factors may differ greatly. The following sections describe the most important sub-models/processes required for modeling primary and secondary sedimentation.

This and the following section indicate minimum standards for a reliable CFD and some of the more advanced implementations and the general steps associated with good modeling practices.

6.4.1 Solids settleability

Evaluation of raw sewage sedimentation by CFD requires determination of the settling characteristics of the sewage. The settling characteristics are very distinct for primary and secondary settling tank influents. The removal process in PSTs is dominated by discrete settling and/or floating particles with a very high fraction of non-settleable particles. The non-settleable fraction limits the achievable ESS and BOD removals. In PSTs, this fraction can exceed 40% of the total influent concentration while in SSTs this fraction is very small (about 1%). The determination of this fraction is critical to modeling the removal efficiency of PSTs. The removal of floatables (scum) is also an important part of achieving good overall performance in primary tanks. Although both PSTs and SSTs are designed to remove TSS, the removal efficiency of PSTs is in the order of 50–70% while SSTs efficiencies are approximately 99% (Griborio *et al.*, 2014). There is very little zone settling in primaries, but the compression parameters

Table 6.3 CFD characteristics in research on secondary sedimentation.

Reference	Solution Technique	Dimensions	Turbulence Model	Multiphase Approach	Mesh/CFD Implementation	Density Couple
Larsen (1977)	Vorticity/Stream Function	2D	Prandtl Mixing Length	Active scalar for solids concentration	Fortran	Yes
Krebs (1991)	Not specified	2D Cartesian	Constant turbulent viscosity	Active scalar for solids volume fraction	PHOENICS	Yes
Zhou and McCorquodale (1992a, 1992b)	SIMPLE	2D Cartesian	$k-\varepsilon$	Active scalar for solids concentration	12 by 24, 12 by 42, and 24 by 42 rectangular mesh	Yes
Samstag *et al.* (1992)	Vorticity/Stream Function	2D	Monk Anderson eddy viscosity	Active scalar	Fortran	Yes
Brouckaert and Buckley (1999)	Not specified	2D Axisymmetric	Not specified	None	Grid not specified/PHOENICS	No
Lakehal *et al.* (1999)	Finite Volume SIMPLEC	2D	$k-\varepsilon$ with plastic viscosity in shear stress terms	Solids transport Takacs settling function	202 by 98 non-orthogonal grid	Yes
Armbruster *et al.* (2001)	Finite Volume	2D Axisymmetric	$k-\varepsilon$	Takacs settling model	60 by 150 to 150 by 250 curvilinear grids	Yes
Kleine and Reddy (2005)	Finite Element (Galerkin weighted residual)	2D Axisymmetric	$k-\varepsilon$ and constant turbulent viscosity	Active scalar for solids	85 by 58 rectangular mesh	Yes
De Clercq (2003)	Fluent	2D Axisymmetric	$k-\varepsilon$	Active scalar for solids and temperature	Gambit triangular mesh/Fluent	Yes
Griborio and McCorquodale (2006); Griborio (2004)	Vorticity/Stream Function	2D	Mixing length	Solids transport/ settling function developed for primary sludge	Variable, up to 200 by 150/ FORTRAN	Yes
Patziger *et al.* (2008)	Fluent	2D Axisymmetric	RNG $k-\varepsilon$	Active scalar for solids	Fluent ~ 2 00 000 cells	Yes?
Wicklein and Samstag (2009)	Finite Volume SIMPLE and Fluent	2D	$k-\varepsilon$	Solids transport versus drift flux	40 by 20 rectangular and hexahedral fine mesh grids	Yes
Burt (2010)	Drift Flux	2D Axisymmetric	$k-\omega$	Solids transport	CFX – 10 000 cells	Yes
Samstag *et al.* (1992)	SIMPLE and Fluent	2D Cartesian and 3D	$k-\varepsilon$	Active scalar for solids	Coarse rectangular mesh in Fortran and Gambit polyhedral mesh with Fluent CFD	Yes
Xanthos *et al.* (2013)	Finite Volume	3D	$k-\varepsilon$		2 million cells/Fluent	Yes
Wicklein *et al.* (2018)	Finite Volume SIMPLE and Fluent	3D	$k-\varepsilon$	Active scalar for solids	Fluent	Yes
Gao and Stenstrom (2018)	Finite Volume SIMPLE	2D	Standard $k-\varepsilon$ Realizable $k-\varepsilon$ RNG $k-\varepsilon$	Solids transport	428 000 cells and 218 700 cells/Fluent	Yes
Gao and Stenstrom (2020a)	Finite Volume SIMPLE	3D	Standard $k-\varepsilon$	Solids transport	4.8 million cells/Fluent	Yes
Valle Medina and Laurent (2020)	Finite Volume	2D Axisymmetric	Buoyant $k-\varepsilon$	Mixture Model, solved for the center of the mixture mass	Axisymmetric of 7790 cells/ OpenFOAM	Related to the volume fraction

are very important. All types of settling (discrete, flocculant, hindered and compression settling) are important in secondary settling tanks.

Different approaches have been used to simulate the settling behavior in PSTs. In general, Patziger and Kiss (2015) indicated that Stokes' law has been primarily used for describing settling of particles in PSTs but that settling functions describing settling velocity as a function of sludge concentration provide a more accurate approach particularly related to the compression properties of primary sludge. Liu and García (2011) and Patziger and Kiss (2015) used the Takacs equation (Takács *et al.*, 1991) (Equation (6.5)) or a similar approach, while Griborio *et al.* (2014), McCorquodale *et al.* (2007) and Rodriguez *et al.* (2021) used different particles sizes and equations to simulate discrete, flocculant, hindered and compression settling.

Almost all research into the performance of sedimentation models incorporates a model for batch settling which is assumed to apply in the larger-scale 3D world. The Vesilind function relies on an exponential empirical relation of hindered settling velocity with local solids concentration (Vesilind, 1968):

$$V_s = V_0 e^{-r_H X} \tag{6.4}$$

where V_s is the settling velocity at sludge concentration X, V_0 is the theoretical maximum settling velocity of the particles, and r_H is a settling parameter, characteristic of the hindered settling regime.

Takács *et al.* (1991) introduced a correction factor in Vesilind's function to account for the settling of smaller slow-settling particles always present at diluted concentrations in settling tanks. They also included a concentration of non-settleable particles as the minimum attainable concentration in the clarifier. Their generalized equation is:

$$V_s = V_0 \left(e^{-r_H (X - X_{\min})} - e^{-r_p (X - X_{\min})} \right) \tag{6.5}$$

where $X_{\min}$ is the minimum attainable settling concentration (non-settleable portion of the influent concentration), and r_p is a settling parameter, characteristic of the low solids concentration.

The Takacs function has been widely used to simulate the settling properties in SST CFD models with many case studies in the literature, including recent applications such as Gao and Stenstrom (2018, 2019, 2020b). However, its use has been presented as a shortcoming in clarifier modeling (Ekama *et al.*, 1997; Lakehal *et al.*, 1999; Mazzolani *et al.*, 1998). The limitations of this model have been documented abundantly and numerous research advances have been made to overcome these limitations (Plósz *et al.*, 2012; Torfs *et al.*, 2017a, 2017b). To overcome some of the limitations of the Takacs equation, Griborio (2004) presented the development and application of a five-component settling model that has been subsequently used in several applications (Gong *et al.*, 2011; Griborio & McCorquodale, 2016a; Griborio and McCorquodale, 2006, 2014; McCorquodale *et al.*, 2007; Ramalingam *et al.*, 2012, 2009; Xanthos *et al.*, 2013). This model includes non-settleable, floatable particles, discrete settling velocities for small, medium and large flocs, flocculent settling and exponential type equations (Vesilind equation) to represent both zone and compression settling regions. Table 6.4 presents the regions and predominant settling regimes based on the total suspended solids concentration along with the corresponding settling velocity sub-models.

Here *FSS* is the flocculated suspended solids threshold, X is the suspended solids concentration, V_s is the settling velocity, V_{si} is the discrete settling velocity of the i floc class, f_i is the i settling fraction for discrete settling, V_0 and r_H are settling parameters for the hindered settling model, and V_c and k_c are settling parameters for the compression settling model.

Valle Medina and Laurent (2020) adapted the one-dimensional partial differential equation of Bürger-Diehl (Bürger *et al.*, 2013) (Equation (6.6)) in the simulation of a 2D circular SST. This equation has the advantage of enabling switching among different constitutive relations for the hindered settling

Table 6.4 Settling regions based on the TSS concentration – Griborio (2004) five-regions settling model.

Total Suspended Solids Concentration (X)	Settling Regime	Settling Model	Equation
$X \leq FSS$	Non-settleable	$V_s = 0$	$V_s = 0$
$FSS < X \leq$ Discrete Threshold (Xd)	Discrete settling	Individual floc settling velocity	$V_s = V_{si}$ for $i = 1, 2, \ldots, n$ $X = \sum_{i=1}^{n} X_i + \mathrm{FSS}$ $X_i = f_i(X - \mathrm{FSS})$
Discrete Threshold (Xd) $< X \leq$ Hindered Threshold (Xh)	Flocculent settling	Transition zone	$V_s = \varsigma V_o e^{-k_1 X} + (1 - \varsigma) V_{sd}^{*}$ $\varsigma = \frac{(X - Xd)}{(Xh - Xd)}$ $V_{sd}^{*} = \sum_{i=1}^{3} f_i V_{si}$
Hindered Threshold (Xh) $< X \leq$ Compression Threshold (Xc)	Hindered settling	Exponential model	$V_s = V_o e^{-r_H X}$
X > Compression Threshold (Xc)	Compression settling	Exponential model	$V_s = V_c e^{-k_c X}$

(Diehl, 2015; Takács *et al.*, 1991; Vesilind, 1968), compression (effective solids stress of De Clercq *et al.* (2008)), and dispersion terms, without affecting the solvability of the model.

$$
\begin{aligned}
\frac{\partial X}{\partial t} = & -\frac{\partial}{\partial z}(X v_{\mathrm{hs}}(X)) \\
& -\frac{\partial}{\partial z}(X v_{\mathrm{conv}}(z,t)) \\
& +\frac{\partial}{\partial z} + \left(d_{\mathrm{comp}}(X)\frac{\partial X}{\partial z}\right) \\
& +\frac{\partial}{\partial z} + \left(d_{\mathrm{disp}}(z, Q_{\mathrm{in}}(t))\frac{\partial X}{\partial z}\right) \\
& +\frac{Q_{\mathrm{in}}(t) X_{\mathrm{in}}(t)}{A}\delta(z)
\end{aligned}
\tag{6.6}
$$

The first and second terms on the right side of the equation stand for particle displacement by hindered settling and advection (through the bulk velocity), respectively. The third term models the sludge compression. The fourth term describes the dispersion of the mixture at the inlet (turbulence effect). The last term models the feed mechanism at a given height of the system. The equation to describe the settling behavior in a 2D or 3D approach becomes (Valle Medina & Laurent, 2020):

$$
\begin{aligned}
\frac{\partial \alpha_d}{\partial t} = & -\nabla \cdot (\alpha_d \overrightarrow{v_m}) \\
& -\nabla \cdot \left(\frac{\alpha_d \rho_c}{\rho_m}\overrightarrow{v_{dj}}\right) \\
& +\nabla \cdot (d_{\mathrm{comp}} \nabla \alpha_d) \\
& +\nabla \cdot (\Gamma \nabla \alpha_d)
\end{aligned}
\tag{6.7}
$$

Equation (6.7) includes the description of the hindered settling regime (second term on the right side) and the compression regime (third term of the right side). It must be remarked that such an approach is based on the mixture approach, this is the reason why the equation is in terms of the volume fraction (α_d). Usage of this model greatly improved the prediction of the sludge blanket height and settling velocities of the particles within the sludge blanket height (Valle-Medina, 2019).

6.4.2 Flocculation

Flocculation can be important in both primary and secondary clarifier performance. The impact of flocculation processes on secondary clarifier performance via simulation was extensively assessed within McCorquodale's group. Griborio (2004) covers development and validation of a five-component settling model (see Table 6.4), flocculation models, and details for quasi-3D CFD for activated sludge clarifiers. This CFD model uses a vorticity/stream function approach to eliminate pressure from the fluid equations and a simple mixing length turbulence model. This work also contains an extensive reference list. The same model was used by Gong *et al.* (2011) in analysis of sedimentation in a rectangular sedimentation tank using a 3D turbulent commercial CFD model. Griborio and McCorquodale (2006) used this model to investigate the influence of center-well geometry on flocculation and clarifier performance. They found that the major benefit of optimum center-well geometry is in improving hydrodynamic performance and that flocculation had less influence.

Griborio *et al.* (2014) stated that flocculation occurs in both primary and secondary settling tanks. Typically, the flocculation is greater in SSTs because of the higher concentrations and the presence of exocellular polymers in activated sludges. Hence the changes in the flocculation state of the activated sludge caused by the make-up sludge flocs, greatly impact the sludge compression (Torfs *et al.*, 2015). Nevertheless, flocculation does occur in primary clarifiers and should be included in the model set-up.

While the effect of flocculation on the performance of SSTs and the causes which promote flocculation in SSTs have been widely acknowledged and studied; its actual application in SST modeling is still scarce and limited to very few examples (Karpinska & Bridgeman, 2016). Griborio (2004) and McCorquodale *et al.* (2005) used a differential equation described by Parker *et al.* (1970, 1971) to describe the overall kinetics of flocculation in turbulent mixing. The same model, represented by the following equation, was later used by Gong *et al.* (2011), Ramalingam *et al.* (2009) and Xanthos *et al.* (2011,; 2013) in 3D CFD modeling applications:

$$\frac{dn}{dt} = K_B \cdot X \cdot G^m - K_A \cdot X \cdot n \cdot G \qquad (6.8)$$

where X is the mixed liquor suspended solids (MLSS) concentration (g.L^{-1}), G is the root-mean-square velocity gradient (s^{-1}), K_A is a floc aggregation coefficient (L.g^{-1}), K_B is a floc breakup rate coefficient (number.s^{m-1}.g^{-1}), m is the floc breakup rate exponent (dimensionless), and n is the primary particle number concentration (number.L^{-1}).

6.4.3 Fluid properties: density and rheology

The assumption of neutral density and viscosity has been common in CFD simulations of raw sedimentation tanks. This is clearly not appropriate in sludge blankets where density gradients and viscosity effects are certainly present. Equations to represent the impact of solids on density and viscosity should be included. Also, the impact of temperature on these properties must be considered when thermal effects are being considered.

6.4.3.1 Density

The hydrodynamics of primary and secondary settling tanks is heavily influenced by density gradients and viscosity effects, and therefore the assumption of a neutral density is a gross simplification. An equation of state must be included in the CFD model to accurately represent the impact of the density current and density stratification in the clarifier.

McCorquodale *et al.* (2007) presented the following equation to estimate the density of the mixture, ρ, as a relation to the water reference density and the suspended solids concentration:

$$\rho = \rho_{\text{ref}} + \left(1 - \frac{1}{Ss}\right)X \tag{6.9}$$

or for multiple particle classes:

$$\rho = \rho_{\text{ref}} + \left(\Sigma X_i - \Sigma\left[\frac{X_i}{S_{si}}\right]\right) \tag{6.10}$$

where X_i is the suspended solids concentration for class i in mass per unit volume of mixture, S_{si} is the specific gravity of the dry solids for class i, and ρ_{ref} is the water reference density which is a function of water temperature (T) and water dissolved solids content (TDS).

Liu and García (2011) used the following equation for a Eulerian technique:

$$\rho = \alpha\rho_d + (1 - \alpha)\rho_{dw} \tag{6.11}$$

where α is the total volume fraction of the particles, and ρ_d and ρ_w are the density of particles and water, respectively.

Locatelli (2015) found that the content of inorganic matter in activated sludge greatly impacts the settling velocity even at the same MLSS concentration. If the percentage of the non-volatile suspended solids (NVSS) increases, then the settling velocity of the activated sludge is higher. Sludge samples taken directly from a WRRF showed that a sludge with initial MLSS concentration of 4.6 kg.m^{-3} has a settling velocity four times higher than a sludge with initial MLSS of 4.2 kg.m^{-3}. In this case, the first sludge had a 39% NVSS content and the second sludge had a 28% NVSS content (Valle-Medina, 2019). This illustrates the importance of sludge composition, and among other properties, density, on the performance of the SST.

6.4.3.2 Sludge rheology

The flow of thickened sludge at the bottom of settling tanks largely depends on the flow characteristics of the material itself. In this respect, the shear stress defines the amount of momentum transport between parallel liquid layers. It is a function of both the shear rate and the viscosity (De Clercq, 2003). This is an area that needs much more attention in settling tanks, and perhaps more importantly in primary units. Concentrations in the sludge blankets of PSTs are typically three to five times higher than those in secondary tanks; these high concentrations increase the effective viscosity of the sludge in PSTs. The sludge removal strategies are also different, for example continuous high rate withdrawal is common in secondary settling tanks, while intermittent or low rate withdrawal is usually used in primary tanks. The sludge rheology and density coupled with the methods of sludge withdrawal are important in determining whether or not rat-holing (short circuiting) occurs.

Several models have been proposed to simulate the rheology of non-Newtonian sludges and there does not appear to be a consensus on which model is best for sewage, for example the Ostwald (pseudoplastic model), Bingham (plastic model), and Herschel-Bulkley equations (yield-pseudoplastic model) (see also Chapter 8 for discussion on non-Newtonian behavior). A study performed by Khalili Garakani *et al.* (2011) made a comparison among seven different rheology models. They found that a Carreau model provides the best viscosity prediction for an MLSS range of 2.41–30 g.L^{-1}. However at low MLSS (less than 10 g.L^{-1}) the Bingham model can give an accurate description of sludge viscosity.

Other approaches have been presented by Bokil and Bewtra (1975) and De Clercq (2003). De Clercq (2003) developed a 2D CFD model using a commercial package incorporating solids transport, density couple, flocculation and sludge rheology. The model was applied to a radial flow (circular) activated sludge clarifier. A Herschel-Bulkley rheological model was incorporated by user-defined

Table 6.5 Rheological models used for sedimentation CFD applications.

Type of Model/ Reference	Mathematical Expression	Comments
Bingham plastic model (Armbruster *et al.* 2001; Dahl *et al.* 1994; Lakehal *et al.* 1999)	$\tau = \tau_0 + \eta_p \left(\frac{dv}{dy} \right)$	where τ_o is the yield stress, η_p is the plastic viscosity. τ_o and η_p are both functions of the suspended solids concentrations. τ_o, the initial resistance of the sludge to deformation, must tend to zero as the concentration approaches zero
Herschel-Bulkley model (De Clercq, 2003)	$\tau = \left[\frac{\tau_o}{\gamma^*}(1 - e^{-m\gamma^*}) + \eta_p \gamma^{*(n-1)} \right] \left(\frac{dv}{dy} \right)$	where γ^* is the magnitude of the strain rate, m is the stress growth exponent, and n is the flow behavior index
Bingham plastic model (Liu and García, 2011)	$\tau = \left(\frac{\tau_y}{\lvert\sigma\rvert^{\frac{1}{2}}} + \mu_p + \mu + \mu_t \right) \sigma$ $\sigma = \sigma \quad \lvert\tau\rvert^{\frac{1}{2}} \geq \tau_y$ $\sigma = 0 \quad \lvert\tau\rvert^{\frac{1}{2}} < \tau_y$	where τ is the shear stress tensor, σ is the shear rate tensor, $\lvert\tau\rvert$ and $\lvert\sigma\rvert$ are the second invariant of the stress tensor and strain tensor. τ_y is the yield stress and μ_p is the plastic viscosity

functions with successful results; however, when comparing their model to the Bingham model and to the Bokil model, they found that the Bokil model resulted in the best prediction of the sludge blanket height. Based on these results and the simplicity of its application, Griborio (2004) and McCorquodale *et al.* (2007) implemented the Bokil model and calibrated it to field data. This model was subsequently used by Gao and Stenstrom (2018, 2020a, 2020c) in 2D and 3D applications.

The relationship proposed by Bokil (1972) is represented by the following equation (McCorquodale *et al.*, 2007):

$$\begin{aligned} \nu &= 1 \times 10^{-6}\, e^{1.386X} && \text{for } X \leq 1\,\text{g.L}^{-1} \\ \nu &= 2.9 \times 10^{-6}\, e^{0.322X} && \text{for } X > 1\,\text{g.L}^{-1} \end{aligned} \tag{6.12}$$

in which X is SS in g.L^{-1} and ν is the mixture kinematic viscosity in m^2.s^{-1}.

Table 6.5 presents other rheological models used in CFD applications.

Additional information about the application of rheological models can be found in Chapter 8 on Anaerobic Digestion. This chapter also presents important numerical considerations when implementing a rheological model in CFD.

6.4.4 Hydraulic regime

The hydraulic regime of raw sewage sedimentation tanks has been documented (Imam *et al.*, 1983; Imam & McCorquodale, 1983), to be characterized by a bottom current towards the tank outlet with occurrence of a reverse flow eddy towards the tank inlet at the top of the tank profile. A profile of experimental results compared to Imam's 2D CFD model is presented in Figure 6.2. The density currents in primary clarifiers are not as strong as in secondary clarifiers. Typically, SSTs have strong bottom density currents induced by the high MLSS in the secondary influent while PSTs have weak solids-induced density currents. Nevertheless, Stamou *et al.* (1989) reported that primary tanks with concentrations as low as 100 mg.L^{-1} can exhibit density currents. Thermal density currents induced by variable influent temperatures as well as surface heat exchange are relatively more important in PSTs than SSTs and may need to be included in the PST model for an accurate representation of the tank hydrodynamic behavior.

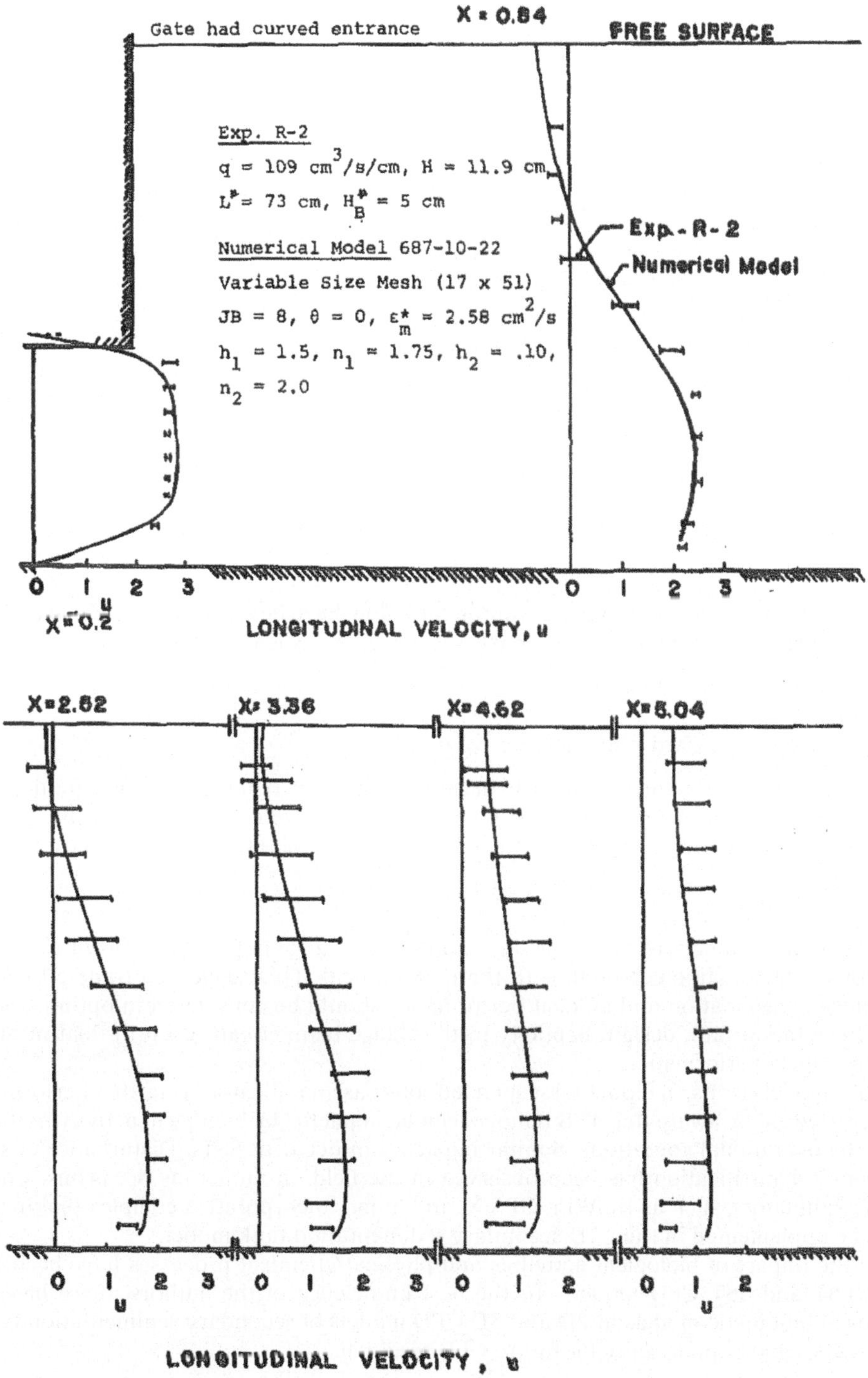

Figure 6.2 Velocity profiles for a rectangular sedimentation tank from Imam and McCorquodale (1983).

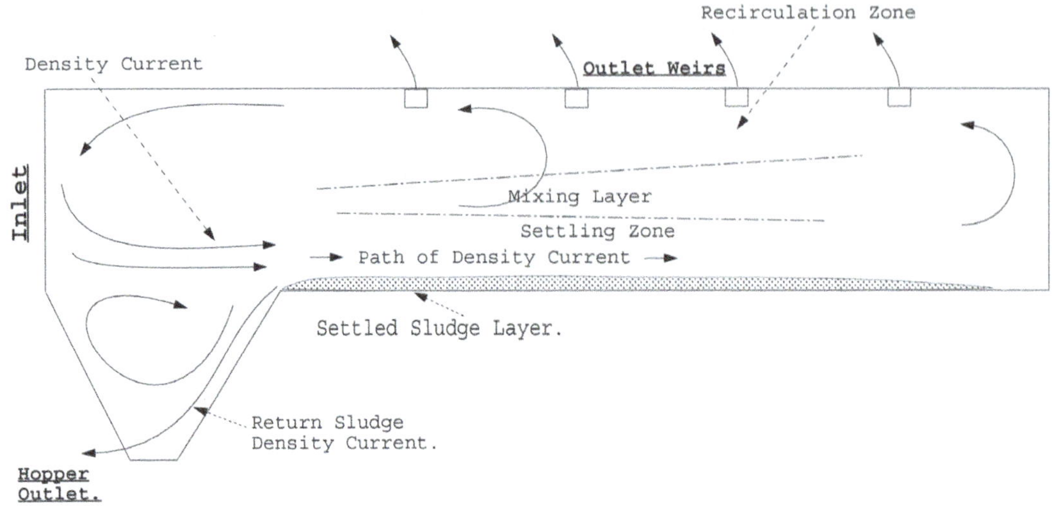

Figure 6.3 Principal flow processes in a rectangular clarifier (Brennan, 2001).

The main features of secondary sedimentation tank flow have been recognized since early work by Larsen (1977), using a custom-built flow measuring device, revealed these features:

- density driven bottom current;
- reverse flow eddy;
- sludge movement towards the inlet.

Figures 6.3 and 6.4 present the main hydraulic features in rectangular and circular clarifiers, respectively. Figures 6.5 and 6.6 present a velocity magnitude profile and a solids profile, respectively, of a radial flow circular tank developed using a 3D drift flux OpenFOAM model (Marques, 2015).

6.4.5 Significance of biological activity

Within PSTs, biological activity in the sludge blanket and sludge hoppers can result in a significant discharge of soluble organic compounds to the tank effluent. This suggests a trade-off between the goals of sludge consolidation and effluent performance should be considered in optimization of raw sewage sedimentation tank design. Septicity in the sludge blanket can result in floating sludge and significantly impact performance.

Most PSTs models focus on modeling suspended solids as an indicator of clarifier performance, and the modeler needs to be aware that TSS removal can be impacted by biological activity in the tank, as well as by the operational conditions. Similar impacts can occur in SSTs. Disturbance of secondary clarification by denitrification has been observed in the field on numerous occasions. Commercial whole-plant simulators, such as BioWin, GPS-X, and Sumo, incorporate a complex biokinetic model which can be implemented in their 1D secondary sedimentation tank model.

To date, the impact of biological activities and physical/chemical processes have been neglected in current PST and SST CFD models. To the best knowledge of the authors, there have been no applications of biokinetic models in 2D and 3D CFD models of secondary sedimentation tanks. This area represents a good opportunity for future CFD research.

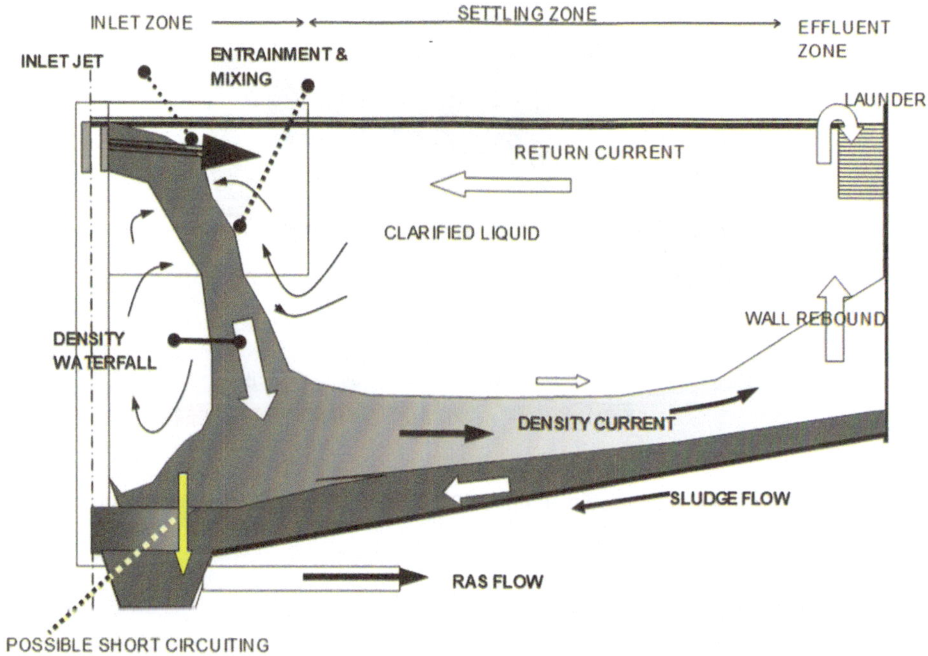

Figure 6.4 Principal flow processes in a circular clarifiers (after Griborio, 2004). RAS: return activated sludge.

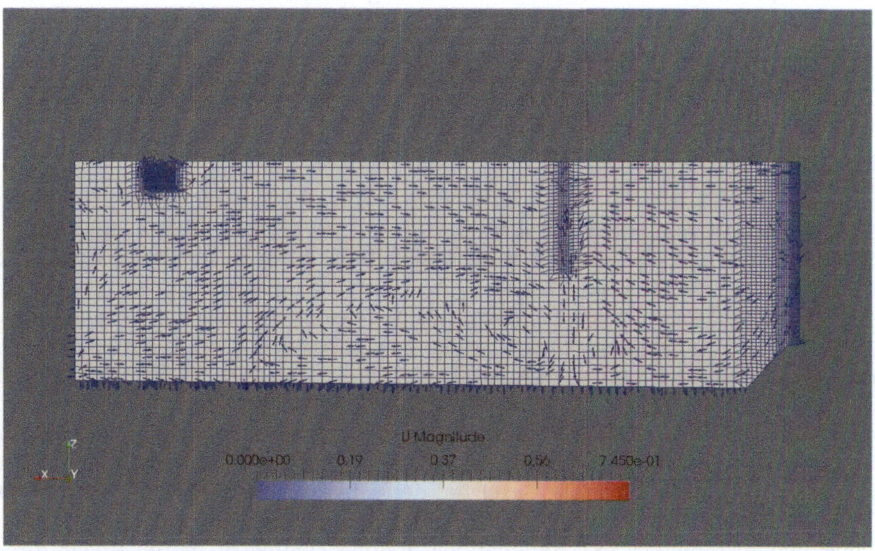

Figure 6.5 Velocity vectors from a simulation of a radial flow clarifier (from Marques, 2015).

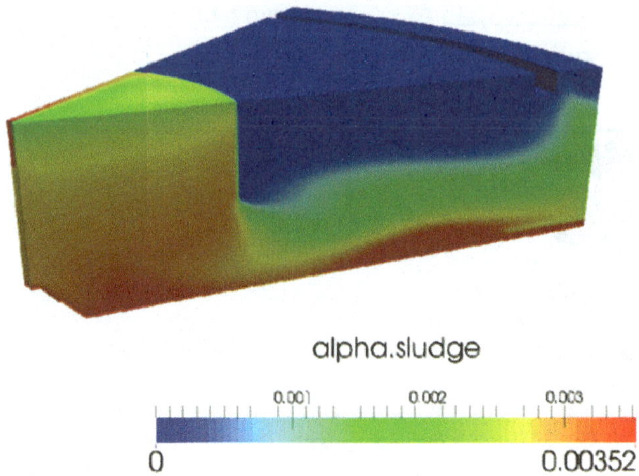

Figure 6.6 Solids profile for a secondary sedimentation CFD simulation developed in open source software (from Marques, 2015).

6.5 CFD MODELING APPROACH

6.5.1 Modeling simplifications

As for other CFD applications, modeling assumptions and simplifications are critical components of the CFD model development and its successful application. The modeling assumptions must be selected according to the modeling objectives and the specific clarifier to be modeled. Table 6.1 presented a comparison of the different modeling tools available, including application of the solids flux theory and 1D, 2D and 3D CFD models. According to Ahuja and Griborio (2017), each model has its application depending on the modeling objectives. Once it has been decided that a CFD model is required, other important considerations include (see also Good Modeling Practice guidelines in Chapter 2):

- definition of the spatial dimension (2D vs 3D);
- dynamic (time dependent modeling) vs steady state modeling;
- single phase versus multiphase modeling.

Clarifier modeling was the pioneer application for unit processes in wastewater treatment plant applications and is perhaps the most mature application. Most cases studies presented in the literature are 2D applications assuming that the dominant hydrodynamics processes of circular SSTs take place in the radial direction, and therefore can be reasonably simplified to a 2D axisymmetric approach (Chu *et al.*, 2015; Esping *et al.*, 2015; Gao & Stenstrom, 2018; Griborio, 2004; Griborio & McCorquodale, 2016a; Lakehal *et al.*, 1999; Patziger *et al.*, 2016, 2012; Patziger & Kiss, 2015; Siczka *et al.*, 2015; Tamayol *et al.*, 2010). In circular 2D applications the model uses the full radial and vertical momentum equations. Griborio and McCorquodale (2016a) and McCorquodale *et al.* (2005) kept the angular momentum equation but still used the axisymmetric assumption in the tangential direction to eliminate all of the terms with partial derivatives in the angular direction but preserve the swirl component of the flow.

Two-dimensional rectangular SST applications assume symmetry in the linear direction and the momentum and transport equations are laterally averaged to yield a 2D solution (Griborio & McCorquodale, 2016a). Many successful applications are also presented in the literature for rectangular settling tanks.

The main advantage of 2D applications is the computational time and the ability to perform long-term dynamic simulations that can predict clarifier performance under transient loading conditions. This is particularly useful in consulting engineering applications for the evaluation of wet weather strategies where different sludge and flow management inventory strategies are considered. Hence, we expect 2D modeling to be the consulting engineering tool of choice for secondary clarifier simulations, and for design and optimization of existing tanks. However, given its inherited assumption, it is important to understand the potential limitations of 2D applications. The prediction accuracy of 2D models decreases as the geometry and flow become asymmetrical (Morse *et al.*, 2016).

Two-dimensional models are limited in their ability to predict swirl flow pattern generated by inlet vanes (such as those created by energy dissipating inlets (EDIs)), in both circular and rectangular SSTs, and in resolving the corner effects of effluent launders for rectangular SSTs. Also, simulation of swirl effects induced by rotating scrapers cannot accurately be incorporated (Gong *et al.*, 2011; Xanthos *et al.*, 2013). To the limitation in inlet and swirl effects, Gong *et al.* (2011) added that return activated sludge (RAS) withdrawal and turbulence are non-symmetrical and intrinsically 3D flow phenomena in nature. Given the 3D nature of the clarifier geometry, hydrodynamics and turbulence, and the advancement in computer speed, we expect an increase in the development and application of 3D models. However, currently, 3D applications are still limited to short-term, steady-state simulations.

6.5.2 Shape and geometry

Primary and secondary sedimentation tanks have been designed in almost every conceivable shape but the most common configurations are circular and rectangular tanks. Early designs favored the use of rectangular tanks, which are still relevant today, particularly for treatment plants with specific space constraints. Designers in the last four decades, appear to prefer the design of circular tanks, particularly for SSTs. In fact, an Environmental Protection Agency (EPA) report from the 1980s (Tekippe, 1987) reported that 19 out of 20 consulting engineering firms surveyed, preferred the use of circular tanks over rectangular units. In spite of this overwhelming preference, the report found 'little data has been generated to objectively compare the performance of circular and rectangular basins.' There is little evidence to support a conclusion that one gross tank shape is manifestly superior to another, all other geometric factors being optimal. McCorquodale *et al.* (2006) indicated that both circular and rectangular tanks have been shown to be capable of excellent performance and both have experienced problems. They mentioned that the hydrodynamics of each configuration under similar hydraulic loading differ in fundamental aspects that determine what types of geometric modifications are likely to improve the clarifier performance.

The choice of clarifier configuration depends on site-specific space constraints, operator familiarity, maintenance requirements and the judgement of the engineer (Ahuja & Griborio, 2017). This is an area where additional calibrated CFD could be very helpful in determining the optimum shape for gravity primary or secondary clarifiers.

As discussed by (McCorquodale *et al.*, 2006), the hydrodynamics of circular and rectangular clarifier configurations, under similar hydraulic loads, differ in some fundamental aspects:

- the Reynolds number in the circular center-fed tank decreases significantly from the inlet to the launder, while in the rectangular tank it remains more or less constant;
- the inlet zone in circular tanks is closer to being two-dimensional compared to the rectangular inlet;
- the sludge removal blade velocities are variable in the circular tank and constant in the rectangular tank;
- the aspect ratio of a typical rectangular tank (L/H) is much larger than the corresponding (R/H) in a circular tank;
- the floor slope in a circular tank with scrapers is generally larger than a rectangular tank.

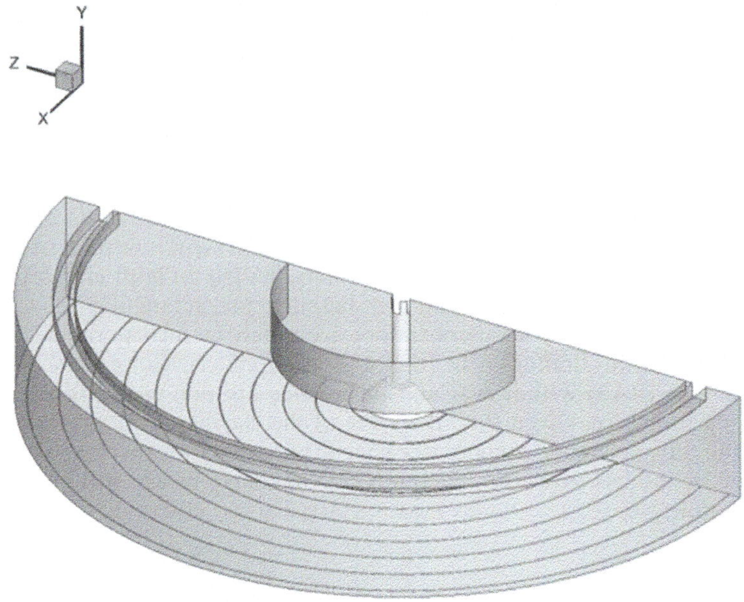

Figure 6.7 Geometric model of a typical radial flow sedimentation tank (from Wicklein *et al.*, 2018).

These factors will determine what type of geometric modifications will improve the clarifier performance and which will not be effective for each configuration.

As indicated by Wicklein *et al.* (2016) the accuracy of a CFD simulation depends on the spatial discretization of the domain which must include the main physical features that will affect the flow. Therefore, accurate geometric information is required to build an accurate CFD model. It is necessary to describe the internal clarifier configuration including: inlet pipe and influent ports; energy dissipating inlet; feedwell or center well; location and dimensions of the effluent weir and launder; and location and dimension of the sludge hopper and other internal baffles, such as density current baffles.

Figure 6.7 presents a geometric model of a typical radial flow, circular tank design with an inlet feedwell and an in-board effluent launder. Sludge withdrawal from a tank of this type would be by rotating suction arms or by rotating scraper mechanism. Figure 6.8 presents a 2D representation of a CFD grid for modeling of a rectangular tank. In this tank, feed is from the left side of the modeled tank image through a perforated baffle wall, effluent withdrawal is from a series of cross tank weirs, and sludge is withdrawn from a hopper part of the way downstream from the inlet. Figure 6.9 presents a geometric model of a square, radial flow tank with effluent baffles around the perimeter of the tank and sludge withdrawal by a center scraper mechanism.

6.5.2.1 Inlet, outlet configurations and baffling

Hazen (1904) already noted the importance of inlet arrangements on tank performance. The widely-used 10 States Standards (Great Lakes-Upper Mississippi River Board of State Sanitary Engineers, 1973) starts its discussion of settling tanks with a section on inlets saying, 'Inlets should be designed to dissipate the inlets velocity, to distribute the flow equally and to prevent short-circuiting.'

Over the last two decades, CFD models have been used to explore the flow pattern of PSTs and SSTs and to optimize their design, and generally suggest that inlet design should focus on dissipating the kinetic energy or velocity head of the sewage, avoiding short circuiting, alleviating density current effects and minimizing blanket disturbances (Rodriguez *et al.*, 2021). Xanthos

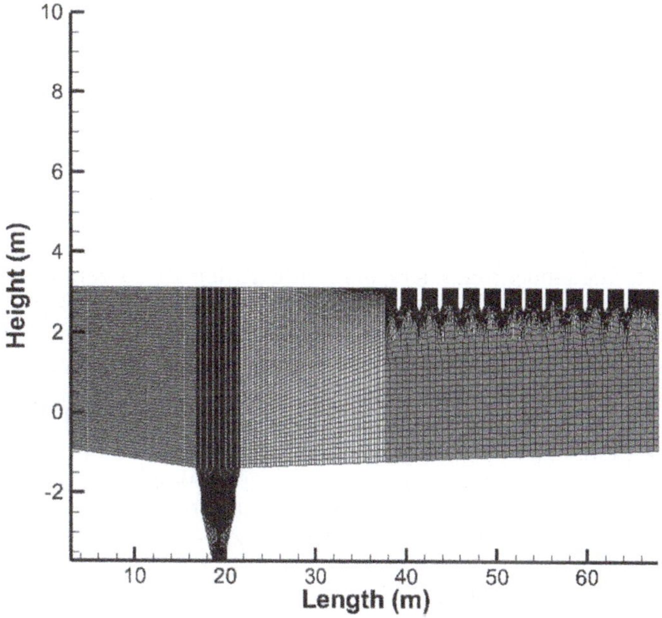

Figure 6.8 CFD grid for a rectangular tank model.

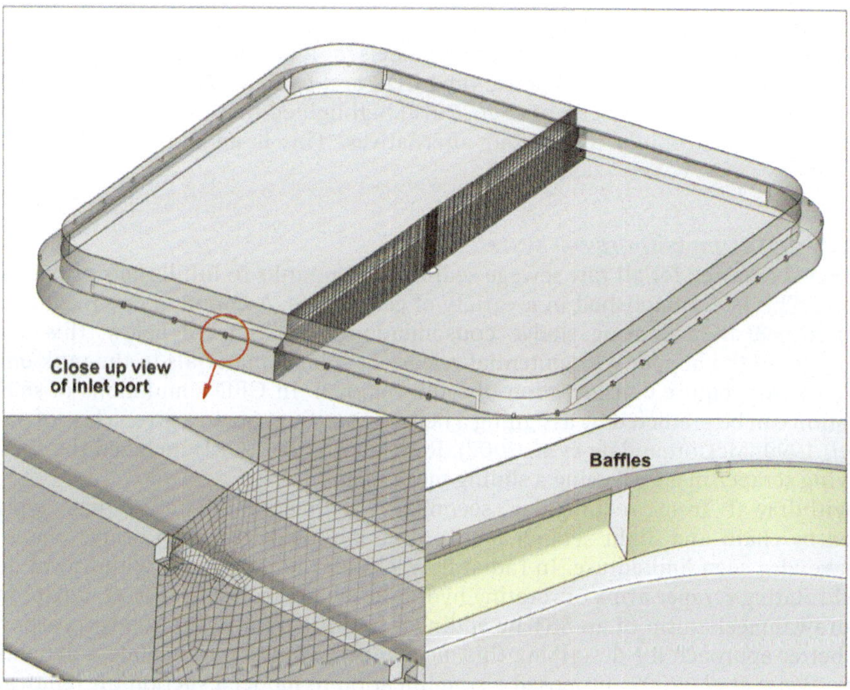

Figure 6.9 Geometric model of a square, radial flow sedimentation tank.

et al. (2011) recognized that inlet and outlet configurations are important for optimization of performance of sedimentation tanks. Inlet configurations can range from simple openings in a tank wall or vertical center inlet pipe to highly baffled inlets designed for maximum energy dissipation and/or enhancement of flocculation. Inlets should be designed to dissipate the influent flow energy and allow for the even distribution of flow in all directions in the tank and should promote the aggregation of remaining dispersed particles while minimizing floc breakup. A properly designed inlet configuration should prevent rat-holing, minimize re-entrainment, and prevent disturbance of the sludge blanket.

Baffling is used in both primary and secondary clarifiers to dissipate energy, promote flocculation, prevent scum from entering effluent launders, and mitigate density currents and other currents that could impair clarifier performance. Energy-dissipating inlets, center wells (feedwells), peripheral/Stamford baffles, and scum baffles are typically included in circular clarifiers. Various types of baffles or diffuser assemblies are used to dissipate energy where flow enters rectangular clarifiers. These can include flat plate target baffles mounted 2–3 feet from the wall in front of the inlet ports or special inlet diffuser assemblies with two or more lateral outlet ports at each inlet location. Additional baffles can be installed at different locations of a rectangular clarifier to improve tank hydrodynamics. CFD modeling has been used in many applications to study the merits of a baffling configuration as well as the optimal position and size of the baffles (Esping *et al.*, 2012; Griborio and McCorquodale, 2006, 2011; Liu & García, 2011; Razmi *et al.*, 2009; Rodriguez *et al.*, 2021; Tamayol *et al.*, 2008, 2010).

Xanthos *et al.* (2011) used a 3D CFD model to evaluate the effects of different in-tank baffle configurations on the hydrodynamic behavior of a rectangular Gould II type SST. They indicated that proper placement of the internal baffles can significantly improve the SST hydraulics and prevents issues such as short circuiting and reversed flow. Properly designed baffles can promote a plug-flow pattern, reduce the impact of the density currents, and provide additional zones for flocculation thus enhancing the overall efficiency of the SST and promoting a high-quality effluent.

Figure 6.10 presents a comparison of geometric models for four different types of inlet arrangement: target baffle, tangential inlet EDI, concentric tube EDI, and baffled EDI. Calibrated 3D modeling of these four inlet configurations (Wicklein *et al.*, 2018) highlighted that the commonly-used tangential inlet performed the most poorly of the four alternatives. This is an example of the usefulness of 3D CFD.

6.5.2.2 Sludge and scum withdrawal systems

Sludge removal is required for all raw sewage sedimentation tanks to fulfill their primary function of solids removal. This is accomplished in a variety of geometries. A common one includes a relatively deep hopper aimed at facilitating sludge consolidation. As discussed below, this can result in biological activity in the hopper with potential release of soluble materials to the tank effluent. Most regulatory standards require consideration of scum removal. In CFD simulations of sedimentation, scum production can be predicted by assigning a negative settling rate to a given class of particles (e.g., Lakehal *et al.* 1999; McCorquodale *et al.* 2007). Rodriguez *et al.* (2021) included the simulation of a rotating moving scraper in a PST using a sliding mesh methodology.

Sludge withdrawal from rectangular secondary sedimentation tanks has typically been accomplished by chain and flight scraper mechanisms although traveling bridge plates or suction manifolds have also seen limited use. In radial flow, circular tanks, sludge withdrawal designs have incorporated rotating scraper arms or rotating hydraulic suction tubes. Das *et al.* (2016) modeled the sludge withdrawal mechanism of an SST by adding a rotating term in the solids transport equation. However, a better approach for describing this mechanism would be to include a dynamic mesh for the geometry. Weiss *et al.* (2007) proposed a simplification to model a suction-lift removal system by using negative source terms on all field equations.

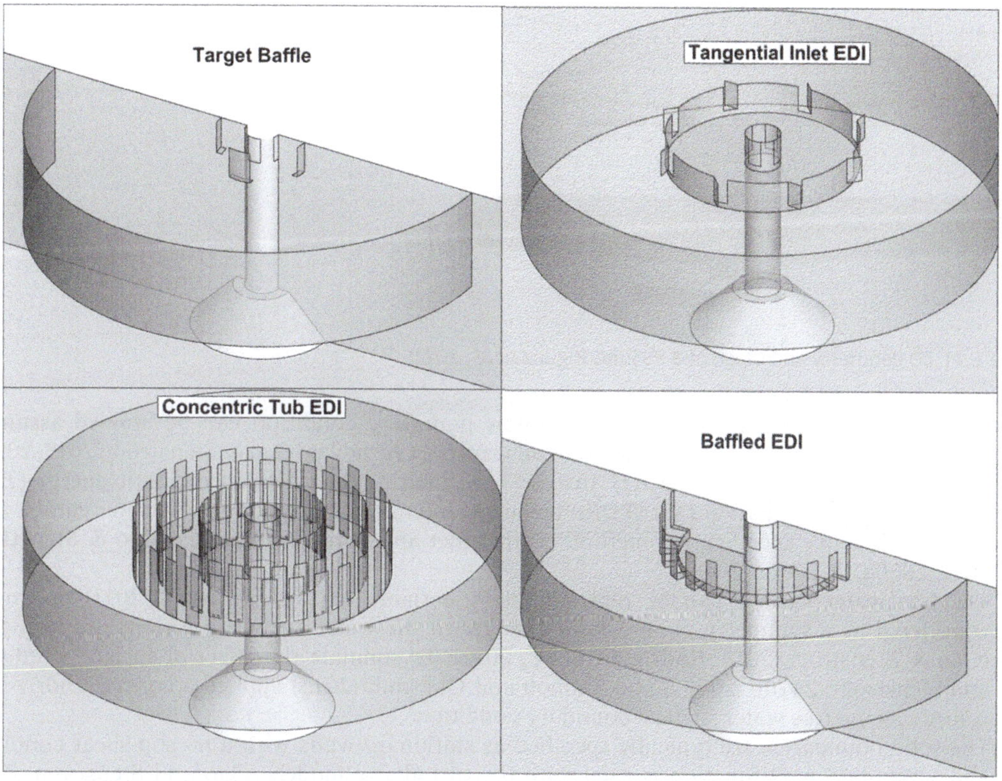

Figure 6.10 Alternative inlet configurations for a circular, radial flow sedimentation tank (from Wicklein *et al.*, 2018).

6.5.3 Mesh and boundary conditions

Wicklein *et al.* (2016) provided good modeling practices related to mesh generation and evaluation of mesh quality (see also Chapter 2). As a general guideline, the mesh should be refined in local areas where it is likely to have steep velocity gradients or when the internal sludge may exhibit fluctuating behavior. For primary and secondary clarifiers, mesh refinement may likely be required in the inlet zone and inside the energy-dissipating inlet and feedwell; also in the area around or next to the sludge hopper and the clarifier outlet.

A mesh independence study is recommended for clarifiers, but this must be accomplished based on a comparison to real data. Section 6.5.5 describes general recommended methods for data collection and for model calibration and validation of settling tank CFD models. An overview of validation for CFD models is also provided in Chapter 9.

As for any CFD application, appropriate setting of the boundary conditions is important for an accurate and realistic representation of the real phenomena. Typical boundary conditions applied to PST and SST CFD model applications include the following:

- Velocity inlet boundary conditions are typically applied to the inlet and sludge withdraw boundaries (Gao & Stenstrom, 2018; Liu & García, 2011; Rodriguez *et al.*, 2021). At the inlet, the velocity is specified according to the influent flow and the area of the inlet pipe. At the outlet, the velocity depends on the outlet pipe and the primary sludge or the return sludge flow.

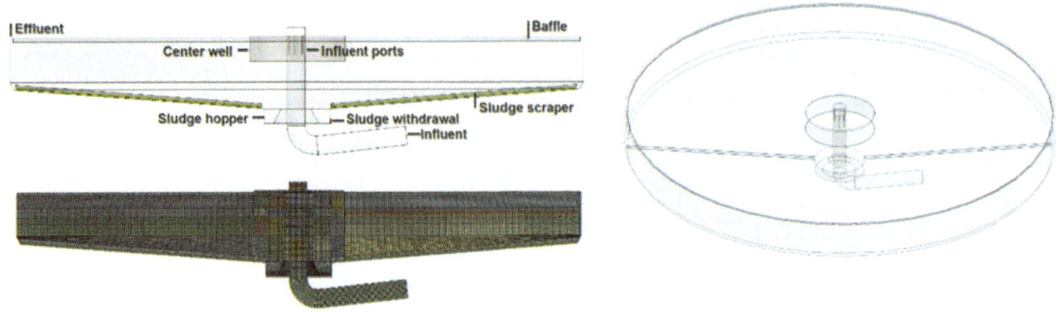

Figure 6.11 3D geometry and mesh of a PST (Rodriguez *et al.*, 2021).

- For the surface outlet, a pressure or outflow boundary condition can be applied assuming hydrostatic pressure and zero gradient for all transport variables (Gao & Stenstrom, 2018, 2020a, 2020b, 2020c; Liu & García, 2011). Instead of estimating the turbulence kinetic energy (TKE) and turbulence dissipation rate (TDR), turbulent intensity and hydraulic diameter can be used as the turbulence specification method on the inlet and outlet boundaries (Gao & Stenstrom, 2018, 2020a, 2020b, 2020c).
- The free water surface can be represented by a rigid lid (Liu & García, 2011) or can be considered as a symmetry plane where changes in the water surface positions are negligible (Gao & Stenstrom, 2018; Rodriguez *et al.*, 2021). To simulate the effect of surface wind Gao and Stenstrom (2020b, 2020c) and Stamou and Gkesouli (2015) applied a constant horizontal velocity as the free water surface boundary condition.
- The solid boundaries are typically specified as stationary walls with a no slip-shear condition and a standard wall function is used with the turbulence models (Gao and Stenstrom, 2018, 2020a, 2020b, 2020c; Rodriguez *et al.* 2021). To simulate a moving scraper, Rodriguez *et al.* (2021) use a 'moving wall' with the 'no slip' option selected for the floor boundary.

While available techniques to represent a moving field in CFD applications are fully described in the literature, the rotation of scrapers and clarifier mechanisms in general have been largely neglected in CFD applications. Implementation of these requires a special boundary condition (Rodriguez *et al.*, 2021). Available techniques include the multiple reference frame (MRF) approach and the sliding mesh (SM) approach which are further described in Chapter 8 on digestion and Chapter 5 on high-rate algal ponds. In the work of De Clercq (2003), the scraper itself is not incorporated in the 2D CFD mesh: its radial velocity component is imposed on the liquid parcel where it resides at a given time. Rodriguez *et al.* (2021) used the SM approach to include the simulation of conventional plow-type scrapers in a primary clarifier. Figure 6.11 presents the geometry of the PST modeled, and the mesh developed, by Rodriguez *et al.* (2021).

6.5.4 Turbulence modeling

Basically, three different approaches to modeling turbulence have been used in clarifier models:

- a constant eddy viscosity;
- relation of the eddy viscosity to the local mean velocity gradient G and the mixing length l_m;
- relation of the eddy viscosity to the turbulence kinetic energy (k) and the turbulence dissipation rate (ε).

The constant eddy viscosity was used in early applications and does not account for the heterogeneous nature of the eddy viscosity in settling tanks. The mixing length was first introduced in the pioneering work from Larsen and since has been use in many successful applications (Abdel-Gawad & McCorquodale, 1984a, 1984b, 1985a, 1985b; Chu *et al.*, 2015; Griborio, 2004; McCorquodale *et al.*, 2007). In Griborio's work the mixing length model was configured in cylindrical coordinates in terms of the mean velocity gradient, G, in the following form:

$$\nu_t = G l_m^2 \tag{6.13}$$

where:

$$G = \sqrt{\left(\frac{\partial u}{\partial y}\right)^2 + \left(\frac{\partial v}{\partial r}\right)^2} \tag{6.14}$$

And l_m was solved from a Poisson equation:

$$\frac{\partial^2 l_m}{\partial r^2} - \frac{1}{r}\frac{\partial l_m}{\partial r} + \frac{\partial^2 l_m}{\partial y^2} = K_{l_m} \tag{6.15}$$

where u and v represent velocities, respectively, in the radial and vertical directions, r and y, and K_{l_m} is a calibration constant.

Larsen (1977) already compared the Prandtl mixing length model to the $k-\varepsilon$ model. He found little difference in modeling results for rectangular sedimentation tanks and preferred to use the simpler Prandtl model. The shortcomings and virtues of the Prandtl and $k-\varepsilon$ models for use in secondary clarifiers were likewise discussed in Griborio (2004) who also elected to use this simpler model to simulate secondary sedimentation.

Griborio (2004) argued that the application of the mixing length model relies on a deep understanding of the clarifier flow field and the results are dependent on model calibration. Gao and Stenstrom (2018) questioned the accuracy of the mixing length model and recommended the use of a more advanced turbulence model to improve model predictions and be less dependent on model calibration. In fact, the Standard $k-\varepsilon$ model (SKE) is the most widely applied turbulence model in CFD application to clarifiers. Examples of applications of the Standard $k-\varepsilon$ model can be found in many studies (Armbruster *et al.*, 2001; Gao & Stenstrom, 2018, 2019, 2020a, 2020c; Gong *et al.*, 2011; Lakehal *et al.*, 1999; Liu & García, 2011; Patziger *et al.*, 2012; Ramin *et al.*, 2014; Rodriguez *et al.*, 2021; Stamou *et al.*, 1989; Tamayol *et al.*, 2010; Xanthos *et al.*, 2013; Zhou & McCorquodale, 1992a, 1992c). Gao and Stenstrom (2019) compared simulation results with and without coupling the buoyancy term (G_b) in the Standard $k-\varepsilon$ model. They concluded that the buoyancy term should be included in the model to accurately predict effluent quality.

As a general guideline, the use of the Standard $k-\varepsilon$ model in PST and SST CFD modeling should be appropriate providing the buoyancy term is coupled and the model is calibrated to accurately reproduce the clarifier internal solids profiles. Given some of the limitations of the Standard $k-\varepsilon$ model, the user should consider the application of more advanced models such as the Realizable $k-\varepsilon$, RNG $k-\varepsilon$, Shear Stress Transport (SST) $k-\omega$, and Reynolds Stress Model (RSM).

As a rule of thumb, the user may wish to check the sensitivity of the flow by comparing the results between turbulence models to rule out unphysical simulations. Gao and Stenstrom (2018) compared the influence of the Standard $k-\varepsilon$ model versus the Realizable $k-\varepsilon$ and the RNG $k-\varepsilon$ in a circular SST using a 2D model. They found important differences in the predictions of hydrodynamics and clarifier capacity with the application of the three models, with the Realizable $k-\varepsilon$ and the RNG $k-\varepsilon$ predicting higher capacity than the Standard $k-\varepsilon$ model. However, they did not conclude which of the three turbulence models resulted in more accurate simulation, reinforcing the need for model calibration for the prediction of accurate results.

6.5.5 Calibration and validation of CFD results

Modeling calibration is a critical step for the development of an accurate clarifier CFD model. A clarifier CFD model requires calibration to tune the parameters that affect the effluent suspended solids (ESS), the sludge blanket height (SBH) and the solids profile; this process includes calibration of the settling and flocculation sub-models. The basic data needed for calibration and validation of a clarifier CFD model include:

- effluent flow rate;
- mixed liquor suspended solids (SSTs) or influent total suspended solids (PSTs);
- effluent suspended solids (ESS) concentration;
- RAS flow and concentration (SSTs) or primary sludge flow and concentration (PSTs);
- sludge blanket height and solids profile;
- sludge volume index (SVI, for SSTs);
- solids-settling properties, which depend on the settling sub-model model to be used and may include discrete settling parameters, zone settling parameters and compression settling parameters;
- sludge flocculation parameters, which depend on the flocculation sub-model to be used;
- influent and effluent temperature.

Sampling to acquire this data is best performed at typical operating conditions and during an actual or simulated wet weather event (stressed conditions). A simulated wet weather event can be achieved by taking units offline until the desired level of 'stress' on the system is reached. Operating performance should then be continuously monitored until performance reaches a determined 'failure' point, such as high ESS or SBH, at which point the stress testing should be ended.

Good examples for model calibration of SST CFD models are presented in several studies (Chu *et al.*, 2015; Esping *et al.*, 2015; Gong *et al.*, 2011; Ramalingam *et al.*, 2012; Sawyer *et al.*, 2011; Winkler *et al.*, 2015; Xanthos *et al.*, 2013). Griborio and McCorquodale (2016b) summarized the field data collection and typical procedures for the calibration of a 2D CFD model. A validation step, where the calibration parameters are not adjusted and the model output is compared to an independent set of field data, is typically recommended (Griborio & McCorquodale, 2016b).

Griborio and McCorquodale (2016b) described that the calibration for ESS in their 2Dc model typically involves:

- adjusting the discrete settling fractions and the settling velocity of poorly flocculated solids (see Table 6.4);
- adjusting the vertical diffusivity factor.

They indicated that calibration for the sludge blanket height usually requires:

- adjusting the settling parameter K_c in the compression equation (see Table 6.4);
- adjusting the vertical diffusivity factor.

In addition, increasing the radial (horizontal) diffusivity factor may be required to resolve instabilities in the blanket. Adjustment of the Vesilind r_H (Table 6.4, Equation (6.4)) and/or thresholds between discrete and hindered settling may be needed to adjust the size of the dispersed solids zone.

Other field and modeling techniques can be used to further verify model results including particle size distribution determination (Liu & García, 2011; Rodriguez *et al.*, 2021); dye or tracer studies (Chu *et al.*, 2015); and velocity measurements to validate the hydrodynamic output from the model (see Chapter 4 of this report for an example of the use of a non-invasive acoustic Doppler current profiler (ADCP) to measure velocities in a reactor). Valle-Medina (2019) calibrated and validated her settling sub-model by modeling a batch column in CFD and then transferring the settling coefficients to the clarifier model. Further discussion on validation of CFD models is provided in Chapter 9.

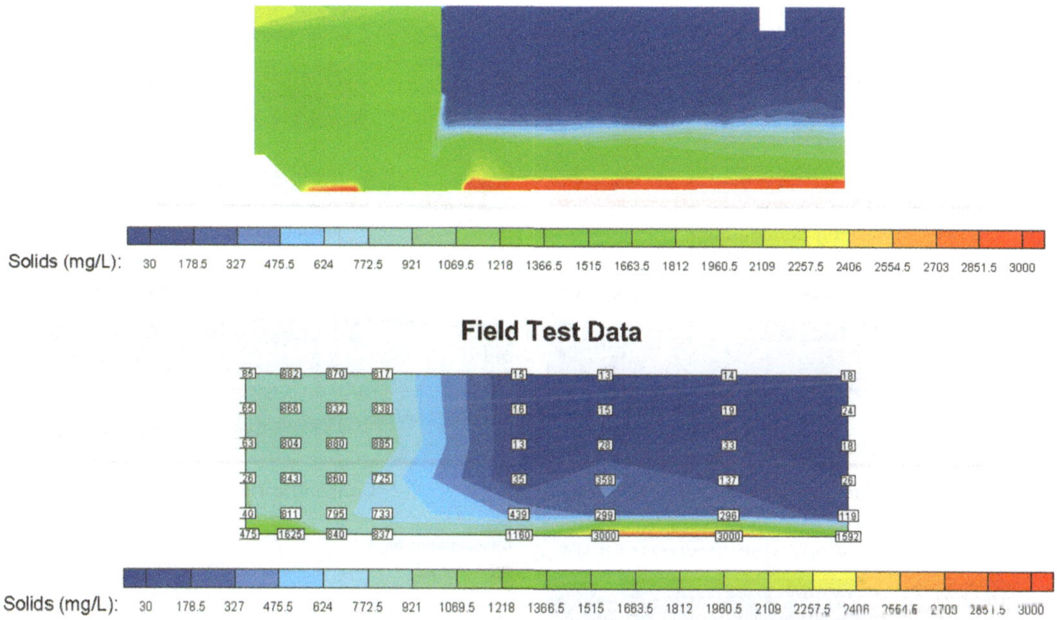

Figure 6.12 Field data CFD model comparison at 8 mgd, 2200 mg.L^{-1} MLSS and SVI of 80 mL.g^{-1} (from Wicklein *et al.*, 2018).

6.6 CASE STUDY

The clarifier shown in Figure 6.7 was part of a study (Wicklein *et al.*, 2018) for the City of Longmont, Colorado to evaluate inlet improvements, including those shown in Figure 6.10, to improve clarifier performance. A 3D model with approximately 1 million hexahedral elements was created of one half of the tank through the central axis with a symmetry boundary (the tangential inlets required a model of the complete tank with ~2 million elements).

Settling velocities were initially set for an estimated sludge volume index (SVI) value using the revised Daigger equation (Daigger, 1995). The settling parameters were then calibrated with data from two field-testing events. Field measurements were taken with a portable solids probe to collect concentration profile data. Table 6.6 summarizes operating conditions during the field testing.

Figure 6.12 presents solids profile contour plots from the field tests compared to the CFD model. This CFD model was based on the Daigger (1995) settling velocity correlation configured for the field

Table 6.6 Sedimentation case study – SST field testing conditions.

	Normal Operation (Test 1)	Simulated Stress Test (Test 2)
Date	May 23, 2012	June 13, 2012
Influent Flow, mgd	8.0	7.7
Number of Secondary Clarifiers Online	2	1
Operating MLSS, mg.L^{-1}	2500	2200
Overflow Rate, gpd.sf^{-1}	350	680
Solids Loading, lb.d^{-1}.sf^{-1}	7.4	12.5
SVI, mL.g^{-1}	76	77

mgd: million gallons per day.

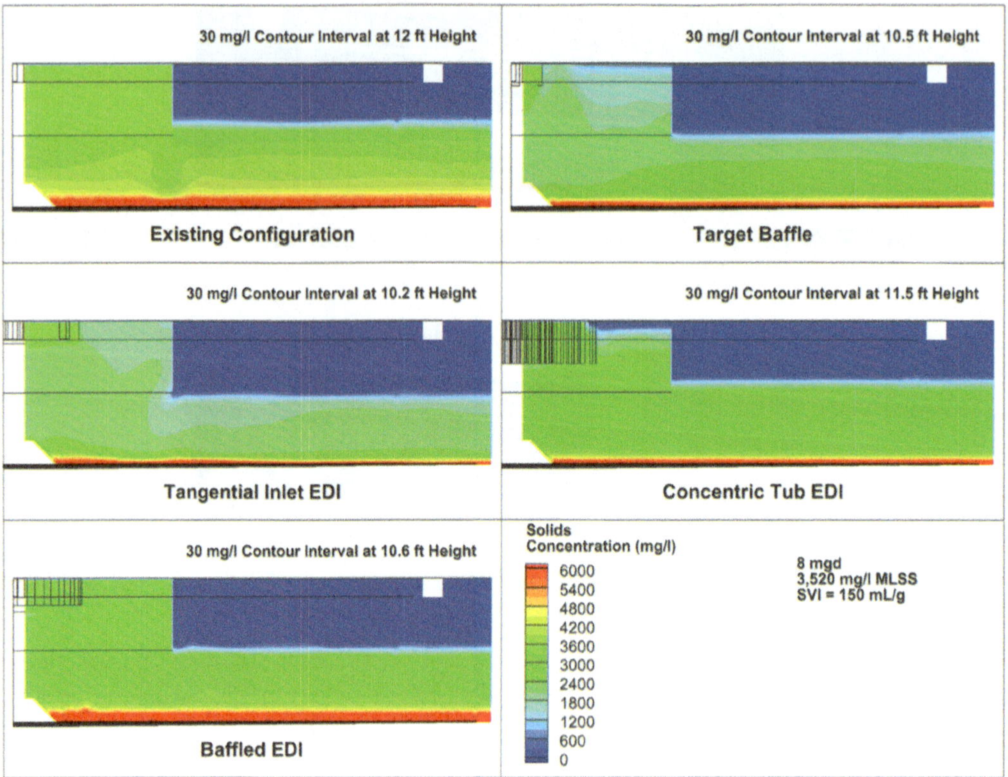

Figure 6.13 Solids profiles at 8 mgd (from Wicklein *et al.*, 2018).

measured SVI value of 80 mL.g^{-1}. The predicted sludge blanket is more expanded than that measured in the field. As the projected SVI was increasing and the model results were more conservative than measured, the setting coefficients were slightly adjusted to match the field test condition.

Once the settling model was calibrated, the initial configuration and four inlet changes were simulated at an MLSS of 3520 mg.L^{-1}, an SVI of 150 mL.g^{-1} and projected flows of 8 and 10 mgd to evaluate performance differences. A section view of the solids concentration profile within the clarifiers is presented in Figure 6.13 for an influent flow rate of 8 mgd and Figure 6.14 for an influent flow rate of 10 mgd. The target baffle, tangential inlet, and baffled EDI show slightly better performance at 8 mgd, with the 30 mg.L^{-1} contour approximately 7–8 feet below the weir, as compared to only 6 feet in the case of the existing configuration. The concentric tub EDI had only a slightly improved solids profile at 8 mgd as compared to the existing configuration, with the 30 mg.L^{-1} contour 6.5 feet below the weir. At an influent flow rate of 10 mgd, the existing configuration is in failure mode, but the four modified EDIs are able to meet the capacity criteria. The concentric tub EDI performs the best at 10 mgd, with a distance of 7 feet between the effluent weir and the top of the 30 mg.L^{-1} solids contour. The target baffle and baffled EDI also performed well, with a distance of 6 feet between the effluent weir and the top of the 30 mg.L^{-1} solids contour. The tangential inlet EDI is right at capacity with the 30 mg.L^{-1} solids contour only 5 feet below the effluent weir. This is supported by the velocity profile, which showed significant zones of higher velocity currents across the clarifier. The results show the existing clarifier capacity can be increased to approximately 10 mgd with any of the proposed modified inlet configurations at an MLSS of 3520 mg.L^{-1} and an SVI of 150 mL.g^{-1}.

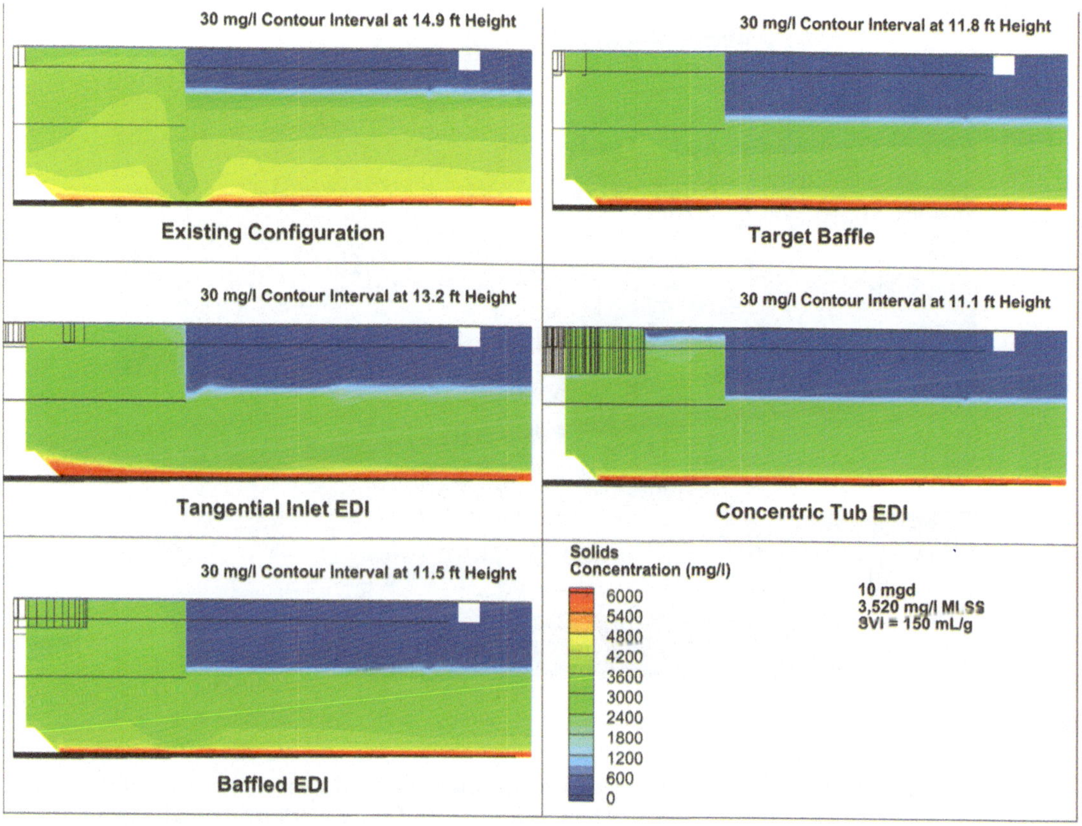

Figure 6.14 Solids profiles at 10 mgd (from Wicklein *et al.*, 2018).

Based on retrofit costs vs model results, the target plates were selected and installed in 2015/2016 and are shown in operation in Figure 6.15. A sample of the performance data for the last several years is summarized in Table 6.7. For similar flows and RAS rates, the effluent suspended solids average was approximately 10 $mg.L^{-1}$ for the original inlets at an average loading of 26.5 $lb.d^{-1}.sf^{-1}$, and dropped to 3.24 $mg.L^{-1}$ once the EDIs were added at a higher loading of 30.6 $lb.d^{-1}.sf^{-1}$. This is a substantial improvement in performance.

The CFD model proved to be a valuable tool in evaluating design changes to improve sedimentation tank performance. Three-dimensional modeling was required to adequately compare different types of EDIs. The model was used to develop a design for a new tank and a design that could be easily retrofitted to existing tanks. The installation of target baffles in the two existing clarifiers provided a 20% increase in capacity at a low cost per tank. Recent data from the retrofit tanks verified improved sedimentation at higher loadings.

6.7 FUTURE RESEARCH NEEDS

There is a need for development of 3D sedimentation models that may include multiple particle distributions for input to the CFD models which would allow for a more accurate representation of effluent quality. Almost all work to date has assumed neutral density conditions for primary SSTs. This is certainly not appropriate for sludge blankets, where the presence of solids contributes to fluid

Figure 6.15 Target plates installed in Longmont, Colorado secondary clarifier (from Wicklein *et al.*, 2018).

density. So density coupling should be included in future work, including these regions either by a drift flux model or by an active scalar approach. There is a need for future work using high-resolution 3D models calibrated to field solids profile and settling tests.

Only the work of Griborio *et al.* (2014), considered here, includes effects of flocculation and this is likely a key issue. Emerging alternatives to primary sedimentation such as rotating belt filters (RBFs) may also benefit from application of CFD, as they involve multiple mechanisms, including sedimentation, mechanical-hydraulic coupling, and multiphase interactions (Paulsrud *et al.*, 2014).

Activated sludge secondary sedimentation tank simulation is the most well-developed area of application for CFD in wastewater treatment. Initial models were only 2D, but almost all included solids transport and settling with coupling of the solids concentration field to tank fluid flow. A significant improvement has been made with incorporation of GSM including discrete, hindered,

Table 6.7 Sedimentation case study – effluent suspended solids pre and post upgrade.

Year	Average Flow (mgd)	RAS (mgd)	Average Effluent TSS ($mg.L^{-1}$)	Max. Effluent TSS ($mg.L^{-1}$)	Min. Effluent TSS ($mg.L^{-1}$)	Standard Deviation Effluent TSS ($mg.L^{-1}$)	SLR ($lb.d^{-1}.sf^{-1}$)	RAS Solids ($mg.L^{-1}$)
2014	7.58	6.03	6.48	10.20	2.80	2.44	22.27	4437.10
2015	7.24	5.75	10.93	17.60	5.00	3.70	24.83	5186.67
2016	7.11	7.14	13.04	54.00	7.20	10.57	32.52	6193.55
2017	7.42	8.27	3.24	4.40	2.20	0.83	30.56	5285.48

and compressive settling regimes and flocculation. The work of Ducoste and Clark (1998) suggests that the relatively widely used Parker *et al.* (1970, 1971) model (for activated sludge flocculation) may not be adequate to capture the influence of flocculation on the behavior of sedimentation tanks, and flocculation and particle interaction has not been included in detail (but is empirically included through scalar compression settling models). Future work should incorporate higher resolution 3D geometric models including GSM and models for flocculation, preferably verified by field tests for solids profile and sludge settling velocity.

Hindered/compression settling is typically represented by empirical equations such as the Vesilind or Takacs exponential equations; future work should attempt to further validate physically-based compression models such as those of Kinnear (2002) and Bürger *et al.* (2011) in 3D modeling. Valle-Medina (2019) incorporated the framework by Bürger *et al.* (2011) in an open-source code based on OpenFOAM and validated this using a 2D axi-symmetric approach.

Additionally, models should incorporate the entire settling domain from the inlet pipe to the effluent and underflow outlets. Knowledge gained from CFD should also be used to improve 1D models for plant-wide simulation as suggested by Laurent *et al.* (2014) and further developed in Guyonvarch *et al.* (2015).

The impact of the clarifier mechanism in the formation of secondary currents or swirl effects has not been fully addressed in current CFD models. This should be included in future 3D models along with the representation of spiral scrapers or suction header mechanisms.

To date, the impact of biological activities and physical/chemical processes has been neglected in current PST and SST 2D/3D CFD models. This area represents a good opportunity for future CFD research.

REFERENCES

Abdel-Gawad S. M. and McCorquodale J. A. (1984a). Strip integral method applied to settling tanks. *Journal of Hydraulic Engineering*, **110**(1), 1–17, https://doi.org/10.1061/(ASCE)0733-9429(1984)110:1(1)

Abdel-Gawad S. M. and McCorquodale J. A. (1984b). Hydrodynamic of circular primary clarifiers. *Canadian Journal of Civil Engineering*, **11**(2), 299–307, https://doi.org/10.1139/l84-043

Abdel-Gawad S. M. and McCorquodale J. A. (1985a). Numerical simulation of rectangular settling tanks. *Journal of Hydraulic Research*, **23**(2), 85–100, https://doi.org/10.1080/00221688509499358

Abdel-Gawad S. M. and McCorquodale J. A. (1985b). Simulation of particle concentration distribution in primary clarifiers. *Canadian Journal of Civil Engineering*, **12**(3), 454–463, https://doi.org/10.1139/l85-053

Ahuja N. and Griborio A. G. (2017). Liquid stream fundamentals: sedimentation. *Water Environment Federation*, 1–7. WSEC-2017-FS-022–Municipal Resource Recovery Design Committee.

Armbruster M., Krebs P. and Rodi W. (2001). Numerical modelling of dynamic sludge blanket behaviour in secondary clarifiers. *Water Science and Technology*, **43**(11), 173–180, https://doi.org/10.2166/wst.2001.0680

Bokil S. D. (1972). Effect of Mechanical Blending on the Aerobic Digestion of the Waste Activated Sludge. PhD Thesis, University of Windsor, Windsor, Canada.

Bokil S. D. and Bewtra J. K. (1975). Rheological characteristics of waste activated sludge. *Indian Journal of Environmental Health*, **17**(1), 1–8.

Brennan D. (2001). The Numerical Simulation of Two-Phase Flows in Settling Tanks. PhD Thesis, University of London.

Brouckaert C. J. and Buckley C. A. (1999). The use of computational fluid dynamics for improving the design and operation of water and wastewater treatment plants. *Water Science and Technology*, **40**(4–5), 81–89, https://doi.org/10.2166/wst.1999.0578

Bürger R., Diehl S. and Nopens I. (2011). A consistent modelling methodology for secondary settling tanks in wastewater treatment. *Water Research*, **45**(6), 2247–2260, https://doi.org/10.1016/j.watres.2011.01.020

Bürger R., Diehl S., Farås S., Nopens I. and Torfs E. (2013). A consistent modelling methodology for secondary settling tanks: A reliable numerical method. *Water Science and Technology*, **68**(1), 192–208, https://doi.org/10.2166/wst.2013.239

Burt D. (2010). Improved Design of Settling Tanks Using an Extended Drift Flux Model. PhD Thesis, University of Bristol, Bristol, UK.

Camp T. (1973). Civil Engineering Classics; Outstanding Papers of Thomas R. Camp. American Society of Civil Engineers, New York, 1973, p. 410.

Camp T. R. (1936). A study of The rational design Of settling tanks. *Sewage Works Jountal*, **8**(5), 742–758.

Camp T. R. (1945). Sedimentation and the design of settling tanks. *Transactions, ASCE*, **111**(1), 895–936, https://doi.org/10.1061/TACEAT.00059

Chu I., Griborio A., Pitt P., Ahmad M. L., Chiu G., Desai J. and Ling E. (2015). Combination of field testing and CFD modeling for optimum clarifier design. 88th Annual Water Environment Federation Technical Exhibition and Conference, WEFTEC 2015.

Dahl C., Larsen T. and Petersen O. (1994). Numerical modelling and measurement in a test secondary settling tank. *Water Science and Technology*, **30**(2), 219–228, https://doi.org/10.2166/wst.1994.0046

Daigger G. T. (1995). Development of refined clarifier operating diagrams using an updated settling characteristics database. *Water Environment Research*, **67**(1), 95–100, https://doi.org/10.2175/106143095X131231

Das S., Bai H., Wu C., Kao J.-H., Barney B., Kidd M. and Kuettel M. (2016). Improving the performance of industrial clarifiers using three-dimensional computational fluid dynamics. *Engineering Applications of Computational Fluid Mechanics*, **10**(1), 130–144, https://doi.org/10.1080/19942060.2015.1121518

De Clercq B. (2003). Computational Fluid Dynamics Of Settling Tanks: Development Of Experiments And Rheological, Settling, And Scraper Submodels. PhD Thesis, Ghent University, Ghent, Belgium.

De Clercq J., Nopens I., Defrancq J. and Vanrolleghem P. A. (2008). Extending and calibrating a mechanistic hindered and compression settling model for activated sludge using in-depth batch experiments. *Water Research*, **42**(3), 781–791, https://doi.org/10.1016/j.watres.2007.08.040

Diehl S. (2015). Numerical identification of constitutive functions in scalar nonlinear convection–diffusion equations with application to batch sedimentation. *Applied Numerical Mathematics*, **95**, 154–172, https://doi.org/10.1016/j.apnum.2014.04.002

Dobbins W. (1944). Effect of turbulence on sedimentation. *Transactions of the American Society of Civil Engineers*, **109**(1), 629–656, https://doi.org/10.1061/TACEAT.0005696

Ducoste J. J. and Clark M. M. (1998). The influence of tank size and impeller geometry on turbulent flocculation: I. Experimental. *Environmental Engineering Science*, **15**(3), 215–224, https://doi.org/10.1089/ees.1998.15.215

Ekama G. A., Barnard J. L., Günthert F. W., Krebs P., McCorquodale J. A., Parker D. S. and Wahlberg E. J. (1997). Secondary Settling Tanks – Theory, Modelling, Design and Operation, Scientific and Technical Report. IWA Publishing, London, UK.

Esping D., Shaposka H., Wahlberg E., Ifft J. and McCorquodale J. A. (2012). Optimizing primary sedimentation and wet weather flow treatment using C f D modeling. WEFTEC 2012 – 85th Annual Technical Exhibition and Conference.

Esping D., Ifft J. and Shaposka H. (2015). Maximizing final settling tank capacity at the northeast ohio regional sewer district wastewater treatment facilities using CFD modeling. 88th Annual Water Environment Federation Technical Exhibition and Conference, WEFTEC 2015, Water Environment Federation, pp. 1605–1624.

Gao H. and Stenstrom M. K. (2018). Evaluation of three turbulence models in predicting the steady state hydrodynamics of a secondary sedimentation tank. *Water Research*, **143**, 445–456, https://doi.org/10.1016/j.watres.2018.06.067

Gao H. and Stenstrom M. K. (2019). Turbulence and interphase mass diffusion assumptions on the performance of secondary settling tanks. *Water Environment Research*, **91**(2), 101–110, https://doi.org/10.1002/wer.1003

Gao H. and Stenstrom M. K. (2020a). The influence of wind in secondary settling tanks for wastewater treatment – A computational fluid dynamics study. Part II: rectangular secondary settling tanks. *Water Environment Research*, **92**(4), 551–561, https://doi.org/10.1002/wer.1244

Gao H. and Stenstrom M. K. (2020b). Development and applications in computational fluid dynamics modeling for secondary settling tanks over the last three decades: A review. *Water Environment Research*, **92**(6), 796–820, https://doi.org/10.1002/wer.1279

Gao H. and Stenstrom M. K. (2020c). The influence of wind in secondary settling tanks for wastewater treatment – A computational fluid dynamics study. Part I: circular secondary settling tanks. *Water Environment Research*, **92**(4), 541–550, https://doi.org/10.1002/wer.1241

Gong M., Xanthos S., Ramalingam K., Fillos J., Beckmann K., Deur A. and McCorquodale J. A. (2011). Development of a flocculation sub-model for a 3-D CFD model based on rectangular settling tanks. *Water Science and Technology*, **63**(2), 213–219, https://doi.org/10.2166/wst.2011.035

Goula A. M., Kostoglou M., Karapantsios T. D. and Zouboulis A. I. (2008). A CFD methodology for the design of sedimentation tanks in potable water treatment. *Chemical Engineering Journal*, **140**(1–3), 110–121, https://doi.org/10.1016/j.cej.2007.09.022

Great Lakes-Upper Mississippi River Board of State Sanitary Engineers (1973). Recommended Standards for Sewage Works. Health Education Service, Albany, N.Y.

Griborio A. G. (2004). Secondary clarifier modeling: A multi-process approach. PhD Thesis, University of New Orleans.

Griborio A. and McCorquodale J. A. (2005). Limiting Solids Flux Analysis: A Study Case Comparing 1-D and 2-D Models. Proceedings Water Environment Federation 78th Annual Conference & Exposition, Washington D.C., pp. 2562–2584.

Griborio A. and McCorquodale J. A. (2016a). Lessons learned from 10 years of clarifier CFD modeling – The 2Dc experience. WEFTEC 2016 – 89th Water Environment Federation Annual Technical Exhibition and Conference, Vol. **5**, pp. 2157–2164.

Griborio A. and McCorquodale J. A. (2016b). Lessons learned from 10 years of clarifier CFD modeling – The 2Dc experience. WEFTEC 2016 – 89th Water Environment Federation Annual Technical Exhibition and Conference.

Griborio A. and McCorquodale J. A. (2006). Optimum design of your center well: Use of a CFD model to understand the balance between flocculation and improved hydrodynamics. *Proceedings of the Water Environment Federation*, **2006**(13), 263–280, https://doi.org/10.2175/193864706783710587

Griborio A., Latimer R., Zhang S., Pitt P., Elmendorf H., Richards T. and Porter R. (2011). Comprehensive approach for the evaluation of rectangular primary clarifiers at the F. Wayne hill water resources center. *Proceedings of the Water Environment Federation*, **2011**(11), 5298–5314, https://doi.org/10.2175/193864711802765796

Griborio A., Alex McCorquodale J. and Rodriguez J. A. (2014). CFD modeling of primary clarifiers: The state-of-the-art. 87th Annual Water Environment Federation Technical Exhibition and Conference, WEFTEC 2014, Water Environment Federation, pp. 1926–1949.

Guyonvarch E., Ramin E., Kulahci M. and Plósz B. G. (2015). iCFD: interpreted computational fluid dynamics – degeneration of CFD to one-dimensional advection-dispersion models using statistical experimental design – The secondary clarifier. *Water Research*, **83**, 396–411, https://doi.org/10.1016/j.watres.2015.06.012

Hazen A. (1904). On sedimentation. *Trans. ASCE*, **53**, 45–88.

Imam E. and McCorquodale J. A. (1983). Simulation of flow in rectangular clarifiers. *Journal of Environmental Engineering*, **109**(3), 713–730, https://doi.org/10.1061/(ASCE)0733-9372(1983)109:3(713)

Imam E., McCorquodale J. A. and Bewtra J. K. (1983). Numerical modeling of sedimentation tanks. *Journal of Hydraulic Engineering*, **109**(12), 1740–1754, https://doi.org/10.1061/(ASCE)0733-9429(1983)109:12(1740)

Isenmann G. (2016). Approche Euler-Lagrange pour la modélisation du transport solide dans les ouvrages de décantation (Euler-Lagrange approach for modeling solids transport in sedimentation tanks). PhD Thesis, Université de Strasbourg, Strasbourg, France.

Isenmann G., Kelly M., Vazquez N., Mosé R. and Fagot C. (2021). Experimental investigation and CFD modelling of settling efficiency in a cylindrical tank. *Journal of Applied Fluid Mechanics*, **14**(3), 733–740.

Jobson H. and Sayre W. (1970). Predicting concentration profiles in open channels. *Journal of the Hydraulics Division*, **96**(10), 1983–1996, https://doi.org/10.1061/JYCEAJ.0002723

Karpinska A. M. and Bridgeman J. (2016). CFD-aided modelling of activated sludge systems – A critical review. *Water Research*, **88**, 861–879, https://doi.org/10.1016/j.watres.2015.11.008

Khalili Garakani A. H., Mostoufi N., Sadeghi F., Hosseinzadeh M., Fatourechi H., Sarrafzadeh M. H. and Mehrnia M. R. (2011). Comparison between different models for rheological characterization of activated sludge. *Iranian Journal of Environmental Health Science and Engineering*, **8**(3), 255–264.

Kinnear D. J. (2002). Biological solids sedimentation: A model incorporating fundamental settling and compression properties. PhD Thesis, University of Utah, Salt Lake City, UT, USA.

Kinnear D. J. (2004). Comparing primary clarifier design and modeling techniques favors traditional techniques over recent water environment research foundation studies. Proceedings of the Water Environment Federation. Water Environment Federation, pp. 586–607.

Kleine D. and Reddy B. D. (2005). Finite element analysis of flows in secondary settling tanks. *International Journal for Numerical Methods in Engineering*, **64**(7), 849–876, https://doi.org/10.1002/nme.1373

Krebs P. (1991). The hydraulics of final settling tanks. *Water Science and Technology*, **23**(4–6), 1037–1046, https://doi.org/10.2166/wst.1991.0555

Krebs P., Vischer D. and Gujer W. (1995). Inlet-Structure design for final clarifiers. *Journal of Environmental Engineering*, **121**(8), 558–564, https://doi.org/10.1061/(ASCE)0733-9372(1995)121:8(558)

Lakehal D., Krebs P., Krijgsman J. and Rodi W. (1999). Computing shear flow and sludge blanket in secondary clarifiers. *Journal of Hydraulic Engineering*, **125**(3), 253–262, https://doi.org/10.1061/(ASCE)0733-9429(1999)125:3(253)

Larsen P. (1977). On the Hydraulics of Rectangular Settling Basins Experimental and Theoretical Studies. Department of Water Resources Engineering, Lund Institute of Technology, University of Lund, Lund, Sweden.

Laurent J., Samstag R. W., Ducoste J. M., Griborio A., Nopens I., Batstone D. J., Wicks J. D., Saunders S. and Potier O. (2014). A protocol for the use of computational fluid dynamics as a supportive tool for wastewater treatment plant modelling. *Water Science and Technology*, **70**(10), 1575–1584, https://doi.org/10.2166/wst.2014.425

Liu X. and García M. H. (2011). Computational fluid dynamics modeling for the design of large primary settling tanks. *Journal of Hydraulic Engineering*, **137**(3), 343–355, https://doi.org/10.1061/(ASCE)HY.1943-7900.0000313

Locatelli F. (2015). Activated sludge batch settling: from experimental investigation using an ultrasonic tranducer to 1D modelling, sensitivity analysis and parameter identification. PhD Thesis, University of Strasbourg, Strasbourg, France.

Marques N. (2015). OpenFOAM® in wastewater applications. Watermatex 2015 Workshop: Computational Fluid Dynamics – Hands-on CFD Modeling for Water Resource Recovery Facilities, Surfers Paradise, Queensland, Australia.

Mazzolani G., Pirozzi F. and D'Antonoi G. (1998). A generalized settling approach in the numerical modeling of sedimentation tanks. *Water Science and Technology*, **38**(3), 95–102, https://doi.org/10.2166/wst.1998.0184

Mccorquodale J. A. and Zhou S. (1993). Effects of hydraulic and solids loading on clarifier performance. *Journal of Hydraulic Research*, **31**(4), 461–478, https://doi.org/10.1080/00221689309498870

McCorquodale J. A., Yuen E. M., Vitasovic Z. and Samstag R. (1991). Numerical simulation of unsteady conditions in clarifiers. *Water Quality Research Journal*, **26**(2), 201–222, https://doi.org/10.2166/wqrj.1991.013

McCorquodale A., Griborio A. and Georgiou I. (2005). A PUBLIC DOMAIN SETTLING TANK MODEL. *Proceedings of the Water Environment Federation*, **2005**(13), 2546–2561, https://doi.org/10.2175/193864705783865668

McCorquodale J. A., Griborio A. and Georgiou I. (2006). Application of a CFD model to improve the performance of rectangular clarifiers. *Proceedings of the Water Environment Federation*, **2006**(13), 310–320, https://doi.org/10.2175/193864706783710794

McCorquodale J. A., Griborio A., Li J., Horneck H. and Biswas N. (2007). Modeling a retention treatment basin for CSO. *Journal of Environmental Engineering*, **133**(3), 263–270, https://doi.org/10.1061/(ASCE)0733-9372(2007)133:3(263)

Morse D., Siczka J. and Nielsen K. (2016). Extending the McCorquodale model for 3D CFD of settling tanks and clarifiers. WEFTEC 2016 – 89th Water Environment Federation Annual Technical Exhibition and Conference.

Nopens I. (2005). Modelling the activated sludge flocculation process: a population balance approach. PhD Thesis, Ghent University, Ghent, Belgium.

Parker D. S., Kaufman W. J. and Jenkins D. (1970). Characteristics of biological flocs in turbulent regime, SERL Report No. 70-5. University of California, Berkeley, CA. Sanitary Engineering Research Laboratory, University of California.

Parker D. S., Kaufman W. J. and Jenkins D. (1971). Physical conditioning of activated sludge floc. *Journal (Water Pollution Control Federation*, **43**(9), 1817–1833.

Patankar S. V. (1980). Numerical Heat Transfer and Fluid Flow. Taylor & Francis, Boca Raton, FL, USA.

Patziger M. and Kiss K. (2015). Analysis of suspended solids transport processes in primary settling tanks. *Water Science and Technology*, **72**(1), 1–9, https://doi.org/10.2166/wst.2015.168

Patziger M., Kainz H., Hunze M. and Józsa J. (2008). Analysing sludge balance in activated sludge systems with a novel mass transport model. *Water Science and Technology*, **57**(9), 1413–1419, https://doi.org/10.2166/wst.2008.221

Patziger M., Kainz H., Hunze M. and Józsa J. (2012). Influence of secondary settling tank performance on suspended solids mass balance in activated sludge systems. *Water Research*, **46**(7), 2415–2424, https://doi.org/10.1016/j.watres.2012.02.007

Patziger M., Günthert F. W., Jardin N., Kainz H. and Londong J. (2016). On the design and operation of primary settling tanks in state of the art wastewater treatment and water resources recovery. *Water Science and Technology*, **74**(9), 2060–2067, https://doi.org/10.2166/wst.2016.349

Paulsrud B., Rusten B. and Aas B. (2014). Increasing the sludge energy potential of wastewater treatment plants by introducing fine mesh sieves for primary treatment. *Water Science and Technology*, **69**(3), 560–565, https://doi.org/10.2166/wst.2013.737

Plósz B. G., Nopens I., Rieger L., Griborio A., De Clercq J., Vanrolleghem P. A., Daigger G. T., Takács I., Wicks J. and Ekama G. A. (2012). A critical review of clarifier modelling: state-of-the-art and engineering practices. Wastewater Treatment Modelling, 3rd IWA/WEF Seminar, Papers, International Water Associationa (IWA); Water Environment Federation (WEF).

Ramalingam K., Fillos J., Xanthos S., Gong M., Deur A. and Beckmann K. (2009). Development and validation of a 3-dimensional computational fluid dynamics (CFD) model for rectangular settling tanks in New York city water pollution control plants. *Water Practice and Technology*, **4**(1), wpt2009015, https://doi.org/10.2166/wpt.2009.015

Ramalingam K., Xanthos S., Gong M., Fillos J., Beckmann K., Deur A. and McCorquodale J. A. (2012). Critical modeling parameters identified for 3D CFD modeling of rectangular final settling tanks for New York city wastewater treatment plants. *Water Science and Technology*, **65**(6), 1087–1094, https://doi.org/10.2166/wst.2012.944

Ramin E., Wágner D. S., Yde L., Binning P. J., Rasmussen M. R., Mikkelsen P. S. and Plósz B. G. (2014). A new settling velocity model to describe secondary sedimentation. *Water Research*, **66**, 447–458, https://doi.org/10.1016/j.watres.2014.08.034

Razmi A., Firoozabadi B. and Ahmadi G. (2009). Experimental and numerical approach to enlargement of performance of primary settling tanks. *Journal of Applied Fluid Mechanics*, **2**(1), 1–12.

Roache P. J. (1972). Computational Fluid Dynamics. Hermosa Publishers, Albuquerque.

Rodi W. (2000). Turbulence Models and Their Application in Hydraulics: A State-of-the-art Review, 3rd ed., CRC Press. Ashgate Pub Co. Farnham , UK, pp. 124.

Rodriguez J., Griborio A. G. and McCorquodale J. A. (2021). Use of three-dimensional CFD model for development of a New circular primary clarifier configuration. *Water Science and Technology*, **84**(2), 333–348, https://doi.org/10.2166/wst.2021.110

Samstag R. W., Dittmar D. F., Vitasovic Z. and McCorquodale J. A. (1992). Underflow geometry in secondary sedimentation. *Water Environment Research*, **64**(3), 204–212, https://doi.org/10.2175/WER.64.3.3

Samstag R. W., Wicklein E. A., Reardon R. D., Leetch R. J., Parks R. M. and Groff C. D. (2012). Field and CFD analysis of Jet aeration and mixing. *Proceedings of the Water Environment Federation*, **2012**(12), 4113–4139, https://doi.org/10.2175/193864712811708301

Samstag R. W., Ducoste J. J., Griborio A., Nopens I., Batstone D. J., Wicks J. D., Saunders S., Wicklein E. A., Kenny G. and Laurent J. (2016). CFD For wastewater treatment: An overview. *Water Science and Technology*, **74**(3), 549–563, https://doi.org/10.2166/wst.2016.249

Sawyer L. K., Merlo R., Faisst B., Thomas B., Maglione S., Wing D. and Gharagozian A. (2011). How Much Can My Plant Treat? Using Models and Stress Testing to Determine Plant Capacity. Proceedings of the Water Environment Federation.

Seddon J. A. (1889). Clearing water by settlement – observations and theory. *Journal of the Association of Engineering Societies*, **8**(10), 477–494.

Siczka J. S., Smith T. F., Hanna E. M., Daigger G. T., Cooper J. P. and Frank D. A. (2015). Using existing facility to maximize the wet weather treatment capacity of the Akron, Ohio, water reclamation facility: Sludge settling theory and state point analysis do not characterize ultimate capacity of shallow clarifiers. 88th Annual Water Environment Federation Technical Exhibition and Conference, WEFTEC 2015.

Stamou A. and Gkesouli A. (2015). Modeling settling tanks for water treatment using computational fluid dynamics. *Journal of Hydroinformatics*, **17**(5), 745–762, https://doi.org/10.2166/hydro.2015.069

Stamou A. I., Adams E. W. and Rodi W. (1989). Numerical modeling of flow and settling in primary rectangular clarifiers. *Journal of Hydraulic Research*, **27**(5), 665–682, https://doi.org/10.1080/00221688909499117

Takács I., Patry G. G. and Nolasco D. (1991). A dynamic model of the clarification-thickening process. *Water Research*, **25**(10), 1263–1271, https://doi.org/10.1016/0043-1354(91)90066-Y

Takamatsu T., Naito M., Shiba S. and Ueda Y. (1974). Effects of deposit resuspension on settling basin. *Journal of the Environmental Engineering Division*, **100**(4), 883–903, https://doi.org/10.1061/JEEGAV.0000240

Tamayol A., Firoozabadi B. and Ahmadi G. (2008). Effects of inlet position and baffle configuration on hydraulic performance of primary settling tanks. *Journal of Hydraulic Engineering*, **134**(7), 1004–1009, https://doi.org/10.1061/(ASCE)0733-9429(2008)134:7(1004)

Tamayol A., Firoozabadi B. and Ashjari M. A. (2010). Hydrodynamics of secondary settling tanks and increasing their performance using baffles. *Journal of Environmental Engineering*, **136**(1), 32–39, https://doi.org/10.1061/(ASCE)EE.1943-7870.0000126

Tekippe R. J. (1987). Critical Literature Review and Research Needed on Activated Sludge Secondary Clarifiers. Water Engineering Research Laboratory, Office of Research and Development, U.S. EPA, Cincinnati, Ohio.

Torfs E., Locatelli F., Balemans S., Diehl S., Bürger R., François P., Laurent J. and Nopens I. (2015). Impact of the flocculation state on hindered and compression settling: experimental evidence and overview of available modelling frameworks. Systems Analysis and Integrated Assessment, 9th IWA Symposium, Papers, International Water Association (IWA).

Torfs E., Balemans S., Locatelli F., Diehl S., Bürger R., Laurent J., François P. and Nopens I. (2017a). On constitutive functions for hindered settling velocity in 1-D settler models: selection of appropriate model structure. *Water Research*, **110**, 38–47, https://doi.org/10.1016/j.watres.2016.11.067

Torfs E., Martí M. C., Locatelli F., Balemans S., Bürger R., Diehl S., Laurent J., Vanrolleghem P. A., François P. and Nopens I. (2017b). Concentration-driven models revisited: towards a unified framework to model settling tanks in water resource recovery facilities. *Water Science and Technology*, **75**(3), 539–551, https://doi.org/10.2166/wst.2016.485

Valle-Medina M. E. (2019). Modélisation des décanteurs secondaires : étude en mécanique des fluides numériques de la dynamique de la sédimentation des boues activées (Secondary settling tanks modeling: computational fluid dynamics study of activated sludge settling). PhD Thesis, Université de Strasbourg, Strasbourg, France.

Valle Medina M. E. and Laurent J. (2020). Incorporation of a compression term in a CFD model based on the mixture approach to simulate activated sludge sedimentation. *Applied Mathematical Modelling*, **77**, 848–860, https://doi.org/10.1016/j.apm.2019.08.008

Vesilind P. A. (1968). Design of prototype thickeners from batch settling tests. *Water Sewage Works*, **115**(7), 302–307.

Wahlberg E. J., Sheridan C. and Koho S. (2003). Determine the Effect of Individual Wastewater Characteristices and Variances on Primary Clarifier Performance. Proceedings of the Water Environment Federation, pp. 275–294.

Wang L. P. and Maxey M. R. (1993). Settling velocity and concentration distribution of heavy particles in homogeneous isotropic turbulence. *Journal of Fluid Mechanics*, **256**(3), 27–68, https://doi.org/10.1017/S0022112093002708

Water Environment Federation (2005). Clarifier Design: WEF Manual of Practice No. FD-8.

Weiss M., Plósz B. G., Essemiani K. and Meinhold J. (2007). Suction-lift sludge removal and non-Newtonian flow behaviour in circular secondary clarifiers: numerical modelling and measurements. *Chemical Engineering Journal*, **132**(1–3), 241–255, https://doi.org/10.1016/j.cej.2007.01.004

Wicklein E. A. and Samstag R. W. (2009). Comparing commercial and transport CFD models for secondary sedimentation. *Proceedings of the Water Environment Federation*, **2009**(10), 6066–6081, https://doi.org/10.2175/193864709793952765

Wicklein E., Batstone D. J., Ducoste J., Laurent J., Griborio A., Wicks J., Saunders S., Samstag R., Potier O. and Nopens I. (2016). Good modelling practice in applying computational fluid dynamics for WWTP modelling. *Water Science and Technology*, **73**(5), 969–982, https://doi.org/10.2166/wst.2015.565

Wicklein E., Rauch-Williams T., Luna B., Samstag R. and Michalski J. (2018). 3D CFD modeling helped develop geometric improvements that increased an existing plant's secondary clarifier capacity. *Proceedings of the Water Environment Federation*, **2018**(8), 4729–4738, https://doi.org/10.2175/193864718825139618

Winkler M., Kelly R., Schauer P. and Menniti A. (2015). Using CFD and cost-benefit analysis for selection of improvements to secondary clarifiers at the rock creek AWWTF in Oregon. 88th Annual Water Environment Federation Technical Exhibition and Conference, WEFTEC 2015.

Xanthos S., Gong M., Ramalingam K., Fillos J., Deur A., Beckmann K. and McCorquodale J. A. (2011). Performance assessment of secondary settling tanks using CFD modeling. *Water Resources Management*, **25**(4), 1169–1182, https://doi.org/10.1007/s11269-010-9620-1

Xanthos S., Ramalingam K., Lipke S., McKenna B. and Fillos J. (2013). Implementation of CFD modeling in the performance assessment and optimization of secondary clarifiers: the PVSC case study. *Water Science and Technology*, **68**(9), 1901–1913, https://doi.org/10.2166/wst.2013.280

Zhou S. and McCorquodale J. A. (1992a). Modeling of rectangular settling tanks. *Journal of Hydraulic Engineering*, **118**(10), 1391–1405, https://doi.org/10.1061/(ASCE)0733-9429(1992)118:10(1391)

Zhou S. and McCorquodale J. A. (1992b). Influence of skirt radius on performance of circular clarifier with density stratification. *International Journal for Numerical Methods in Fluids*, **14**(8), 919–934, https://doi.org/10.1002/fld.1650140804

Zhou S. and McCorquodale J. A. (1992c). Mathematical modelling of a circular clarifier. *Canadian Journal of Civil Engineering*, **19**(3), 365–374, https://doi.org/10.1139/l92-044

doi: 10.2166/9781780409030_0153

Chapter 7
Disinfection

Christopher DeGroot[1], Edward Wicklein[2] and Stephen Saunders[3]

[1]Department of Mechanical and Materials Engineering, Western University, London, ON, Canada; Email: cdegroo5@uwo.ca
[2]Carollo Engineers, Seattle, WA, USA; Email: ewicklein@carollo.com
[3]IBIS Group Inc., Chamcook, NB, Canada; Email: steve@ibisgroupcfd.com

7.1 INTRODUCTION

Disinfection is an important component of an effective wastewater treatment process (Figure 7.1), since it is responsible for protecting human health and the environment through inactivation or destruction of pathogens such as bacteria, viruses, and protozoa. Disinfection refers specifically to processes that are used to achieve a given level of pathogen reduction. This is distinct from sterilization where the objective is to destroy all organisms present. Broadly, disinfection processes can by classified as being either chemical or physical in nature. Chemical disinfection processes use an oxidant (e.g., chorine, peracetic acid, or ozone) to cause a loss of electrons from a pathogen's cell thereby leading to subsequent damage to its structure. Ultraviolet (UV) disinfection uses germicidal wavelengths of UV light that are readily absorbed by cellular DNA and RNA, thus causing damage that prevents the pathogens from further replication.

A common element between all disinfection technologies is that they are based on the presence of a disinfectant and a contact time between the disinfectant and the microorganisms that are inactivated by the disinfectant. In the case of chemical disinfection, the disinfectant is the chemical compound, while in UV disinfection, the disinfectant is the UV radiation. The rate at which pathogens are inactivated depends on the strength of the disinfectant. For chemical disinfectants, this is its concentration. For UV disinfection, the strength is the fluence rate (the total radiant energy from all directions passing through an infinitesimal sphere, per unit area of the sphere cross-section). Each type of pathogen displays a unique relationship between disinfectant strength, contact time, and inactivation. Therefore, it is important to characterize thoroughly the *dose-response curve* for each organism. In most reactors, the contact time is not uniform due to non-ideal hydraulic behavior. For example, flow short circuiting may cause some particles to experience much shorter contact times, while dead zones in the flow may cause some particles to experience very long contact times. To complicate matters further, the disinfectant may not be uniformly distributed in the reactor, leading to inconsistent exposure. In the case of chemical disinfectants, this could be due to inadequate mixing, while in the case of UV disinfection it can be brought about by non-uniform irradiance due to absorption and scattering by various constituents in the wastewater.

Editor(s): Julien Laurent, Randal Samstag, Jim Wicks and Ingmar Nopens

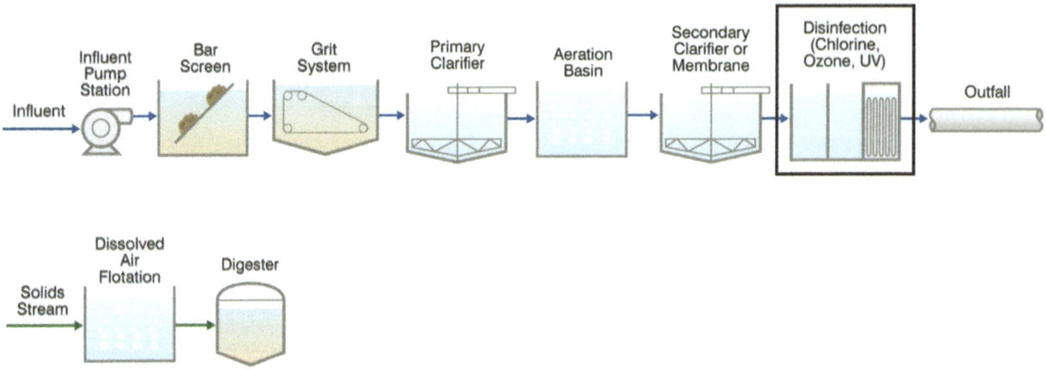

Figure 7.1 Placement of the disinfection step in a wastewater resources recovery facility.

Considering the profound effects of reactor hydraulic behavior on the resulting disinfection performance, computational fluid dynamics (CFD) modeling has emerged as an important tool contributing to the effective design of disinfection unit processes for wastewater effluents. It has proven to be useful both as a complement to physical evaluation and as a stand-alone analysis tool. Contact tanks and UV reactors used for effluent disinfection at wastewater treatment plants are particularly suitable subjects where CFD has become the first choice for preliminary evaluation and diagnostics. In each case, CFD can solve for the flow field in the reactor, the resulting distribution of the disinfectant in the reactor, and the disinfection kinetics for parcels of fluid containing pathogens as they flow through the domain. Each of these analyses are readily accomplished by existing CFD solution algorithms. Also significant is the ability to introduce virtual tracers into a CFD solution and collect statistical data on their progression through the reactor or contact tank (see also Chapter 10). Ultimately, the goal of the CFD analysis is to aid in the arrival at an efficient and cost-effective means of dispersing a disinfectant into an effluent stream such that the facility meets appropriate public health standards and has manageable life cycle costs.

7.2 PROCESS BACKGROUND: DISINFECTION KINETICS

The origins of applying quantitative analysis to chemical disinfection predate practical CFD modeling by several decades. The pioneers Chick and Watson were colleagues at the Lister Institute of Preventive Medicine in London, England, when their respective papers were published in the Journal of Hygiene in 1908 (Chick, 1908; Watson, 1908). Chick and Watson were among the first to publish empirical relationships for the rates of microbial destruction which remain the basis for the kinetic equations used within CFD studies to this date.

All disinfection processes rely on the concept of a *contact time* to quantify the typical amount of time for which the disinfectant has come into contact with the pathogens that are targeted for inactivation. It was first discovered by Chick (1908) that increased contact time resulted in greater destruction of pathogens, which is expressed in differential form using Chick's law:

$$\frac{dN}{dt} = -KN \tag{7.1}$$

where N is the number of microorganisms, K is the inactivation rate constant, and t is time. Integrating Equation (7.1) from the initial condition, time $t=0$, N_0, results in the expression:

$$\ln\left(\frac{N}{N_0}\right) = -Kt \tag{7.2}$$

The influence of disinfectant concentration (or equivalently, irradiance in the case of UV disinfection) on the inactivation rate constant was reported by Watson (1908) to follow a relationship given by:

$$K = \Lambda C^n \tag{7.3}$$

where Λ is a coefficient representing the specific lethality of a disinfectant with respect to a specific pathogen, and n is an empirical constant.

The combined influences of contact time and disinfectant concentration are described by the Chick-Watson equation, given as:

$$\log\left(\frac{N}{N_0}\right) = -\Lambda_{10} C^n t \tag{7.4}$$

It is noted that Equation (7.4) has been rewritten in terms of a base-10 logarithm, rather than the natural logarithm, as it is more common to report inactivation data in this way. It is noted that the coefficient of lethality has been adjusted and given the symbol Λ_{10} to denote the change of base in the logarithm. The left side of Equation (7.4) is commonly referred to as the *log inactivation*, which is found to be proportional to both the contact time t and the concentration C, for a given value of Λ_{10}, representing the specific susceptibility of a given microorganism to a particular disinfectant. The ideal relationship between log inactivation as a function of contact time and disinfectant concentration is illustrated in Figure 7.2.

The empirical relationships of Chick and Watson are of significant value for quantifying the relationship between inactivation of pathogens and the independent variables of disinfectant

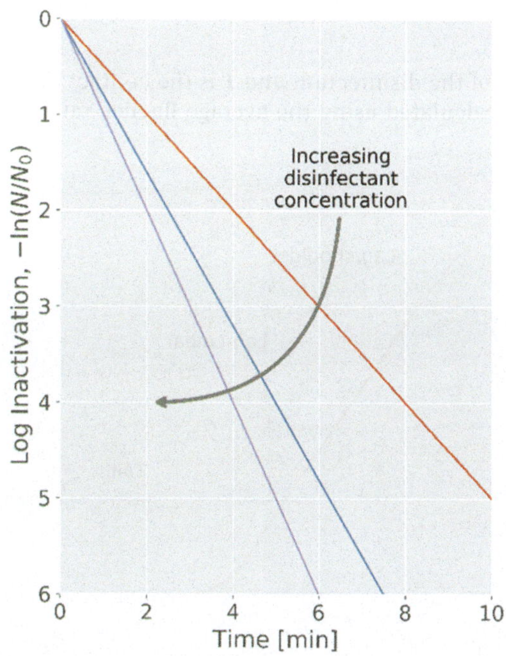

Figure 7.2 Illustration of the ideal Chick-Watson relationship between contact time and log inactivation for different disinfectant concentrations.

concentration and contact time. However, they neglect the influence of other constituents in the suspending liquid, which may have significant effect in the case of wastewater disinfection. Specifically, interactions are known to occur between disinfectants and common constituents of wastewater. For example, suspended matter can reduce the effectiveness of disinfectants both by absorption of the disinfectant and by shielding the target microorganisms from exposure. Further, various organic and inorganic materials present in wastewater can react with the disinfectant, reducing its availability for reaction with the target pathogen(s). These factors lead to a departure from the classical Chick-Watson curves. A *lag* or *shoulder* effect results when the disinfectant initially reacts with other constituents in the wastewater, such that there is a negligible contribution to inactivation of the target microorganisms. In contrast, a *tailing* effect can result when some of the microorganisms become shielded by larger particles in the wastewater, reducing their exposure to the disinfectant. Between the lag and tailing portions of the inactivation curves, a log-linear region can typically be observed, where the Chick-Watson relationship is obeyed. In the log-linear region, it can be assumed that there is negligible reaction between the disinfectant and other constituents of wastewater and that non-shielded microorganisms are being effectively inactivated. An illustration of these departures from Chick's law (assuming a constant disinfectant concentration) is shown in Figure 7.3.

Returning to the Chick-Watson equation (Equation 7.4), it is observed that the log inactivation is proportional to both the concentration and the contact time. The relative importance of concentration and contact time is quantified by the exponent n. If $n=1$, it is conjectured that concentration and time have equal importance when determining the disinfection performance. Likewise, if $n<1$ then it can be said the contact time has a more significant effect than concentration. If $n>1$, then concentration is more significant than contact time. On the basis that $n \approx 1$, a concept for predicting disinfection performance based on the product of concentration and time, that is a *CT concept*, has been proposed (US EPA, 1986). In this case, it is common to express the product of concentration and contact time as the *dose* that is delivered. The dose, D, is expressed as:

$$D = CT \tag{7.5}$$

where C is the concentration of the disinfectant and T is the contact time. It should be noted that for UV disinfection, the dose is calculated using the average fluence rate for the reactor in place of the concentration.

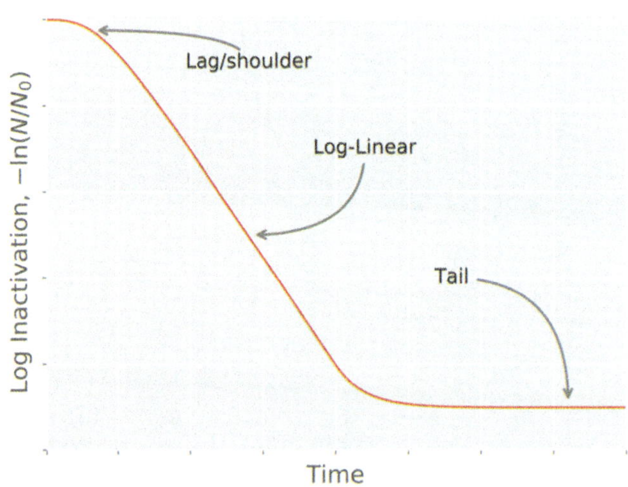

Figure 7.3 Illustration of the typical departures from Chick's law, including the lag/shoulder and tailing effects.

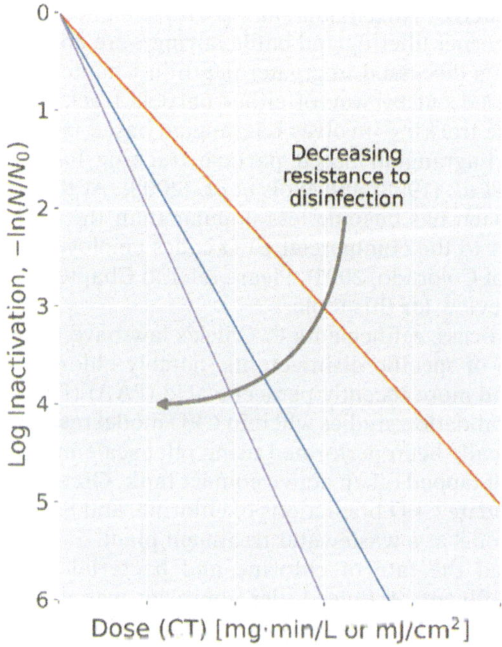

Figure 7.4 Illustration of a typical dose-response curve for three hypothetical microorganisms with different resistances to the disinfectant.

The germicidal dose, defined by the CT concept, is important for evaluating the effectiveness of treatment across a range of operating conditions (e.g., different dosing concentrations, different reactor geometries, etc.). Typically, CT values for a particular disinfectant and microorganism pairing are determined under carefully controlled laboratory conditions and are reported on the basis of the dose required per log reduction. The relationship between the dose and the log-reduction is described by the dose-response curve, as shown in Figure 7.4. Then, the dose distribution that is calculated by CFD can be compared with the dose-response curve to predict the log-reduction for the reactor under consideration.

7.3 LITERATURE REVIEW

7.3.1 Chemical disinfection

Research into the application of CFD in chemical disinfection has fallen into three distinct fields: flow simulation, tracer transport and reaction process.

Reliable flow simulation is the lynchpin of disinfection simulation as it forms the continuum in which the tracer and reaction process models operate. To that end, extensive validation studies have been carried out to demonstrate the predictive capabilities of CFD simulation in this area. Khan *et al.* (2006) conducted extensive experiments with a 1/8-scale baffled contact tank where they verified CFD results against velocity and turbulence intensity profiles that had been recorded by a laser Doppler velocimeter (LDV). Wilson and Venayagamoorthy (2010) used three different contact systems to verify CFD results against lab data. These three contact systems comprised a pilot pipe-loop system tapped into the Fort Collins, Colorado Municipal Water Treatment Facility, a series of consecutive pressurized tanks system at Colorado State University, and a pilot-scale baffled contact tank based on one used for turbulence measurement studies carried out by Shiono and Teixeira (2000). Modifications to tank geometry were evaluated by Angeloudis *et al.* (2016) in a pilot-scale baffled contact tank where the predicted effects of end corner filleting and baffle fairing were compared against lab data.

The simulation of tracers is discussed in greater depth in Chapter 10 of this report. In summary, simulation of tracers is carried out by way of either particle tracking or the solution of transport equations for scalars. Particle tracking involves Lagrangian based methods whereas scalar transport is Eulerian method based. Lagrangian based particle tracking has been applied successfully by Stropky *et al.* (2007), Thýn *et al.* (1998) and Wols *et al.* (2008). At the time of this writing, however, the Lagrangian based approach has become less popular than the Eulerian based scalar transport approach. This is due in part to the commercial CFD codes employed in water industry simulations being Eulerian based (State of Colorado, 2021). Please refer to Chapter 10 wherein the reader will find a more extensive literature review for this topic.

In the field of reaction process, refinements to Chick's law have been made to accommodate the performance characteristics of specific disinfectants, notably chlorine (Haas & Joffe, 1994; Hom, 1972; Selleck *et al.*, 1978) and more recently peracetic acid (PAA) (Haas & Finch, 2001; Rossi *et al.*, 2007; Santoro *et al.*, 2007). Validation studies wherein CFD model results have been compared against experimental data have typically been performed using pilot-scale models either in labs or on-site at treatment plants with effluent tapped off an active contact tank. Greene (2002) describes a serpentine lab-scale contact tank to validate CFD predictions of chlorine, and Santoro *et al.* (2007) studied PAA performance in an on-site model at a wastewater treatment plant.

Greene (2002) researched the fate of chlorine and bacterial pollutants in chlorine contact tanks treating wastewater effluent. A Euler-Euler approach was used for simulation of microbial inactivation with the spatial distribution of microorganisms in the reactor viewed as a continuous field, similar to a dissolved species. Disinfection models were implemented in the general purpose commercial CFD software. Reactor tanks (3.5 m length, 0.75 m width, 0.75 m height) were modeled in 3D using a k-epsilon turbulence model. Buoyant forces were assumed to be negligible. Grid independence checks were undertaken with a mesh of 132 000 cells selected for the influent piping and 790 000 cells for the reactor. Santoro *et al.* (2015) used CFD as a deterministic tool to identify probabilistic curves for pathogen deactivation by peracetic acid, identifying the value of CFD for prototyping of systems where there are complex mechanisms present, and enabling simple tools for end-user application.

7.3.2 Ultraviolet disinfection

UV radiation refers to a specific portion of the electromagnetic spectrum in the wavelength range of 100–400 nm (U.S. EPA, 2006). The UV spectrum can be further classified as vacuum UV (100–200 nm), UVC (200–280 nm), UVB (280–315 nm), and UVA (315–400 nm). UV radiation has a germicidal effect on microorganisms due to photochemical reactions that are induced by the absorption of photons by proteins, RNA, and DNA in cells. The concept of a *UV dose* (also known as *fluence*) is important in quantifying the disinfection performance of a UV reactor. Classically, the UV dose is calculated as the product of an average fluence for the reactor, multiplied by an average contact time. However, this concept is overly simplified, since some particles will receive a dose that is much less than the average, while others will receive a dose that is much more than the average. In characterizing the performance of a UV reactor it is important to consider the *dose distribution*, which can be done conveniently using CFD simulations. Typically, the flow and radiation fields can be solved first, then particles are tracked through the domain to determine the dose they receive; by tracking a large number of particles, the UV dose distribution can be recovered (Lagrangian approach, see also Chapter 2).

Before discussing the details of how to use CFD simulations to assess UV reactor performance, it is important to define the correct radiometric nomenclature, since much of the terminology is commonly misused in the literature. There are four fundamental quantities that enable the description of radiation (McCluney, 2014), which are summarized in Table 7.1. Reviewing the literature, especially

Table 7.1 Description of the four fundamental radiometric quantities.

Quantity	Symbol	Units	Description
Radiant flux	Φ	W	Flow of radiant energy through a surface or region of space per unit time
Irradiance	E	$W{\cdot}m^{-2}$	Radiant flux per unit area indecent on, passing through, or emitted from a surface at a specific point
Radiant intensity	I	$W{\cdot}sr^{-1}$	Radiant flux per unit solid angle indecent on, passing through, or emitted from a surface in a specific direction
Radiance	L	$W{\cdot}sr^{-1}{\cdot}m^{-2}$	Radiant flux per unit solid angle and flux per unit area indecent on, passing through, or emitted from a surface in a specific direction and at a specific point

in engineering fields, it becomes apparent that radiance is commonly referred to as 'intensity', which may cause some confusion when reviewing literature from different fields. Each of the quantities listed in Table 7.1 can also be expressed as spectral quantities, denoted with the subscript λ (i.e., Φ_λ, E_λ, I_λ, and L_λ), meaning that they are also expressed on a per wavelength basis.

To fully understand these quantities, it is important to introduce the concept of the *solid angle*, which is measured in *steradians* (sr). The definition of the solid angle is defined for a given closed curve as the area contained within that curve, projected into a unit sphere. It is most convenient to express radiometric quantities in spherical coordinates, as shown in Figure 7.5. The differential solid angle is denoted $d\omega$, which can be expressed in spherical coordinates in terms of the polar angle, θ, and azimuthal angle, φ:

$$d\omega = \sin\theta d\theta d\varphi \tag{7.6}$$

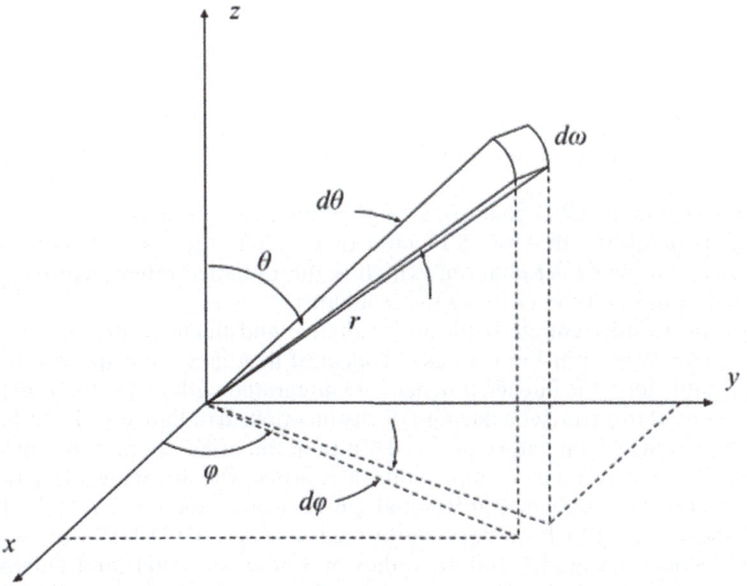

Figure 7.5 Definition of the solid angle in spherical coordinates.

With these definitions in place, it becomes possible to mathematically define the radiometric quantities defined in Table 7.1 in terms of the radiant flux:

$$E = \frac{d\Phi}{dA} \quad \text{or} \quad E_{\nu} = \frac{d^2\Phi}{dA\,d\nu} \tag{7.7}$$

$$I = \frac{d\Phi}{d\omega} \quad \text{or} \quad I_{\nu} = \frac{d^2\Phi}{d\omega\,d\nu} \tag{7.8}$$

$$L = \frac{d^2\Phi}{dA\,d\omega} \quad \text{or} \quad L_{\nu} = \frac{d^3\Phi}{dA\,d\omega\,d\nu} \tag{7.9}$$

Many different types of radiation models exist, including moment-based methods such as P-1 (Cheng, 1964), Monte Carlo (Toor & Viskanta, 1968), and finite volume discrete ordinates models (Murthy & Mathur, 1998; Raithby & Chui, 1990). For the wastewater disinfection applications, the finite volume discrete ordinates method (FVDOM) is most commonly used since it provides greater accuracy as compared to the P-1 model. This is because the FVDOM model is capable of modeling several physical phenomena that are important for UV reactors, including specular reflection and refraction at semi-transparent interior surfaces (e.g., lamp sleeves), specular reflection at opaque walls, and non-gray radiation (i.e., radiation with multiple wavelength bands). It is also noted that the FVDOM model provides greater computational efficiency compared to Monte Carlo methods, although Monte Carlo methods are not typically included within existing CFD packages.

CFD codes implementing the FVDOM solve the radiative transfer equation, which results in a solution for the radiance, L, or spectral radiance, L_{ν}, depending on whether multiple wavelength bands are being modeled. The radiance is a function of both position within the domain and direction; therefore in addition to discretization of the physical domain into grid cells, the angular space must be discretized into a finite number of solid angles which fill the angular space. The relationship between radiance and irradiance is:

$$E = \int_{4\pi} L(\theta,\phi)\cos\theta\, d\omega \tag{7.10}$$

which represents the integration of radiance over all solid angles to find the irradiance at a specific point in the domain.

Again, it is noted that many CFD codes refer to the radiance as 'intensity', so care should be taken when interpreting all quantities. In ANSYS Fluent (Ansys, 2011), the associated quantity available for post-processing is called 'incident radiation,' which is the radiance integrated over all solid angles, which is recognized from Equation (7.10) as the irradiance.

It is also important to differentiate between irradiance and fluence rate, both of which have the same units of $W{\cdot}m^{-2}$ (or $W{\cdot}m^{-2}{\cdot}nm^{-1}$ in the case of spectral irradiance and spectral fluence rate). This distinction is important since it is fluence (fluence rate integrated with respect to time) that defines the germicidal dose delivered to pathogens during UV disinfection. Irradiance is the radiant flux incident on a surface, per unit area. By the nature of this definition, the orientation of the surface with respect to the direction of the incident ray(s) is important. In contrast, the fluence is the total radiant energy passing from all directions through an infinitesimal sphere of cross-sectional area dA, per unit area dA.

A number of research studies have been performed to assess CFD UV process performance as a function of the fluence rate model. Initial studies by Liu *et al.* (2004) and Ducoste *et al.* (2005) compared the following models: line source integration (LSI), radial line source integration (RADLSI), multiple points source summation (MPSS), multiple segment source summation (MSSS), view factor, and discrete ordinates (DO). The study of Liu *et al.* (2004) compared different fluence rate models with experiments, independent of the flow through the reactor. One of the conclusions of this work was that the DO model significantly over-predicted the fluence rate near the lamp surface and under-predicted the fluence rate in regions far from the lamp surface, since the DO mode did not incorporate the effects of refraction due to the lamp sleeve. However, more recent implementations of the DO model, including FVDOM implementations in many CFD codes, include the ability to model refraction. The study of Ducoste *et al.* (2005) considered the effect of both the fluence rate model (RADSLI and MSSS) as well as the model that is used to calculate the delivered UV dose (Eulerian or Lagrangian). The Eulerian approach implements a scalar transport equation to compute the concentration of microorganisms with a source term describing the inactivation kinetics as a function of the local fluence rate. The Lagrangian approach implements a particle-tracking algorithm to derive particle pathlines over which the fluence rate is integrated to calculate the UV dose. Random perturbations to the particle velocity, can be added to account for turbulent velocity fluctuations. It was concluded by Ducoste *et al.* (2005) that the results of the Eulerian and Lagrangian models were comparable, but noted that the Eulerian method was sensitive to the microbial inactivation rate constant as well as selection of the outlet plane, when downstream flow structures were present.

In drinking water treatment applications, a significant effort was made to demonstrate the accuracy of these models for several reactor designs and different microbial disinfection kinetics (Ducoste *et al.*, 2005; Wols *et al.*, 2010). In addition, the UV dose distribution, which was exclusively an output from the CFD model by simulating the transport of particles (i.e., Lagrangian approach) within the reactor and monitoring their exposure to local UV light intensity along the particle path, was only validated indirectly through the model's ability to predict the experimental effluent log inactivation. Researchers were able to develop an experimental Lagrangian approach that involved tracking fluorescent polystyrene microspheres through a UV reactor and monitoring the change in fluorescence due to their cumulative exposure to UV light (Blatchley *et al.*, 2006; Bohrerova *et al.*, 2005; Zhao *et al.*, 2008). While this approach has its limitation, it has provided additional information about the accuracy of CFD UV disinfection models and, in combination with experimental log inactivation, improves the validation process of UV reactors.

A particular concern in wastewater treatment is the disinfection of water containing particle aggregates. Researchers have shown that bioflocs that shield or harbor microorganisms can reduce the UV disinfection performance (Caron *et al.*, 2007; Kollu & Örmeci, 2012; Mamane & Linden, 2006). Experimental results suggest that floc size, number of microbes incorporated into the floc, and the spatial location of these organisms within the floc can influence the UV dose to achieve a high log inactivation (Mamane & Linden, 2006). Mamane *et al.* (2006) explored the effect of light scattering on measurement of UV absorbance and penetration of germicidal UVC irradiance in a UV reactor. As part of that study, CFD was used to evaluate the radiative transfer equation in predicting the UV fluence rate within a reactor. Mamane *et al.* (2006) results showed that the model over-predicted the fluence rate in wastewater effluent augmented with particles. The researchers hypothesized the higher model results were due to the type and size of particles in the augmented wastewater solution, which was augmented with irregular, 2–5 micron particles that scattered light anisotropically. The anisotropic nature of the scattered light was not modeled in their study.

A good starting point for reading about the application of CFD modeling to disinfection is the review compiled by Zhang *et al.* (2014). In their paper, *Developments in Computational Fluid Dynamics-Based Modeling for Disinfection Technologies Over the Last Two Decades: A Review*, the authors discuss elements of CFD disinfection simulation including hydraulics, practical turbulence modeling, mesh resolution and disinfection kinetics. Their reference list is extensive with over 180 entries.

7.3.3 Hydraulic efficiency

The baffling factor (BF) is a measure of a disinfection system's hydraulic efficiency and effectiveness. It defines the degree to which short circuiting occurs within a contact as the ratio of time it takes for

10% of the entering water (by volume) to move through a contact tank (T10) to the hydraulic retention time (HRT). In other words, the BF is calculated by Equation (7.11):

$$BF = T10/HRT \tag{7.11}$$

BF values range from 0 to 1.0, which are represented as follows:

- 0: complete short circuiting, that is some water entering the contact tank moves directly to the outlet while water in other areas is not exchanged;
- 0.1: complete mixing such as continuous stirred tank reactor (CSTR);
- 1.0: perfect plug flow and best disinfection efficiency;
- Any value in-between: varying levels of mixing and short circuiting.

The higher the value of the BF, the more effective a disinfection tank becomes through a longer and more consistent contact time. The CT value is modified by the BF to generate the following formula:

$$CT = C * HRT * BF \tag{7.12}$$

CFD modeling has been used to evaluate the hydraulic performance of contact tanks and is one of the oldest applications of water and wastewater CFD modeling, going back to studies conducted by Hannoun and Boulos (1997). Several years later, Crozes *et al.* (1999) conducted an extensive research effort for the American Water Works Association Research Foundation (AWWARF), now known as the Water Research Foundation (WRF), which compared the results of field tracer tests against those of numerical modeling. Their early CFD modeling found that the numerical tools could replicate the physical tracer tests reasonably well and were slightly conservative. The study found a relationship between BF and length to width ratio, improving estimation of where a tank would operate between a CSTR and plug flow. Their curve is presented in Figure 7.6.

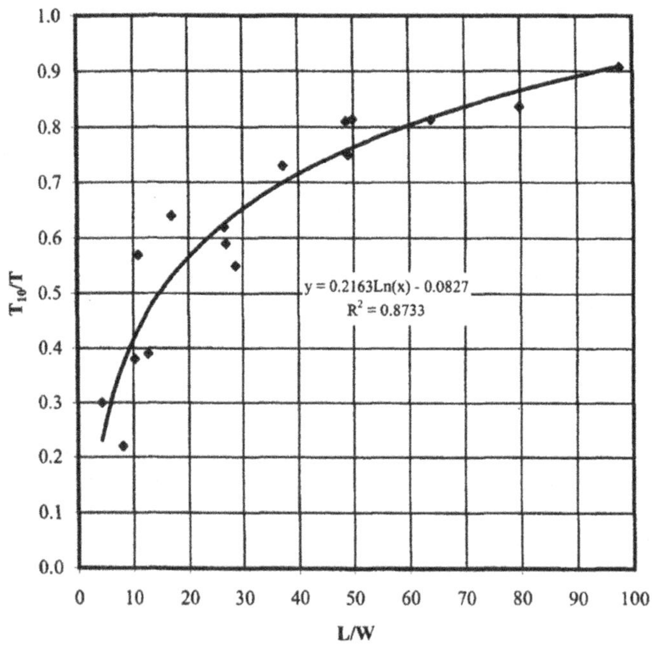

Figure 7.6 Impact of L/W on T10/T ratio (Figure ES.2 from Crozes *et al.*, 1999).

7.4 CFD APPROACH

Accurate modeling of the flow field in disinfection applications is essential, since this determines the contact time with the disinfectant or, in the case of UV disinfection, the time of exposure to UV light. Simplified hydrodynamic models are possible but may not properly characterize features such as dead zones or behaviors that do not resemble idealized flows such as plug flow or completely mixed reactors. CFD is able to fully resolve the details of the flow field such that the history of exposure to the disinfectant or irradiation can be accurately modeled. Flow in contact tanks and UV reactors is expected to be turbulent, necessitating the inclusion of a turbulence model within the CFD simulation of the flow field. While this is typically done using two-equation turbulence models (varieties of the k-ε and k-ω models), it is also possible to consider more advanced turbulence models, such as the Reynolds stress model or large eddy simulation, to better account for anisotropy in the turbulence.

7.4.1 Hydraulic efficiency

The hydraulic efficiency of a contact tank is a set of parameters that indicate the relationship between the actual flow patterns within the tank and an idealized plug flow condition. That is, the closer the flow within a tank is to plug flow, the better its efficiency. However, in reality, most tanks are likely to have instances of short circuiting and recirculation zones. The current practice for evaluating the hydraulic efficiency of existing contact systems in the field is based on tracer studies performed with a salt such as lithium chloride (LiCl). This is done by introducing a LiCl brine solution at the inflow of a tank and measuring its concentration at the tank exit over time by means of an array of conductivity probes.

It is possible to emulate this procedure in a CFD simulation by using a Eulerian approach wherein a passive scalar is introduced into a 'frozen' flow field. The assumptions made in using this technique are that the presence of the scalar has no influence on its surrounding flow field and that the flow field in the tank has arrived at a steady state. In this approach, the flow field in the tank with no tracer present is simulated until it reaches a steady state. At this point, the flow and turbulence equations are turned off and the advection-diffusion equation for the tracer is turned on, which may also include turbulent diffusion. In a CFD simulation, a tracer is introduced at a constant inflow rate to the contact system inlet starting at a time designated as $t = 0$. The tracer concentration is then measured over time at the tank exit, emulating the measurements of conductivity probes that would be used in a physical tracer test. These tracer data are used to generate a residence time distribution (RTD) curve for the reactor that is evaluated against the theoretical detention time (τ) of a contact system. The theoretical detention time is simply the liquid volume of the contact tank, divided by the volumetric inflow rate.

An example of a typical RTD curve for a contact system is shown in Figure 7.7, which is interpreted as follows:

- The vertical axis is the measured tracer concentration at the system outlet normalized against the target steady state concentration;
- The horizontal axis is the elapsed time normalized against theoretical detention time. Values of interest are t_{10} and t_{90}, where t_{10} is the time for the tracer to reach 10% of the target value and t_{90} is the time for the tracer to reach 90% of target value;
- The slope and inflection points of the RTD curve are indicative of the hydraulic performance of the contact system;
- The baffle factor (BF) is defined at t_{10}/τ. Values near 1.0 indicate good plug flow while values lower than 0.3 indicate significant short circuiting is taking place;
- The Morrill index (MI) is defined as t_{90}/t_{10}. Values greater than 4.0 indicate a significant portion of flow is getting held up in recirculation zones.

Through CFD simulations, it is possible to emulate physical tracer studies by introducing an inert scalar to the solution equations that behaves like a real-world tracer. Comparisons with field data over

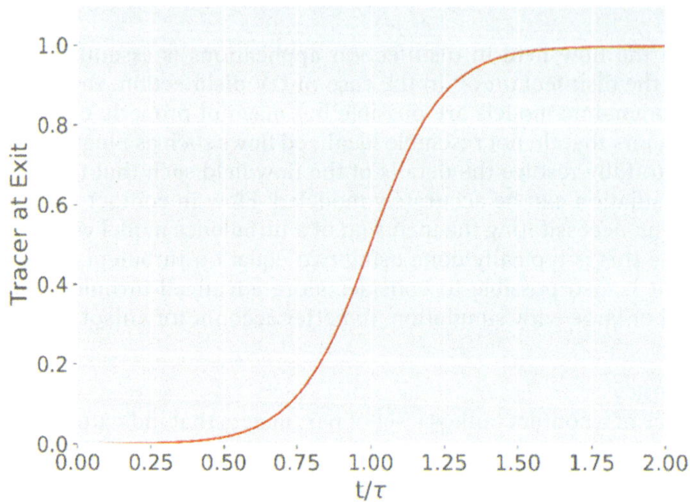

Figure 7.7 Example tracer curve for a system.

the past 20 years show excellent agreement between the physical and numerical methods (Angeloudis *et al.*, 2016; Greene, 2002; Khan *et al.*, 2006).

Chapter 10 of the present report provides an extensive discussion of the RTD concept and CFD modeling.

7.5 CASE STUDIES

Two case studies are presented as different approaches are used for contact tanks vs UV systems. In contact tanks the goal is maximizing hydraulic efficiency so all paths through the tank take sufficient time. In UV systems the goal is for each flow path to accumulate light exposure from a fixed light field.

7.5.1 UV case study

A typical wastewater treatment plant UV system was evaluated using the discrete ordinance disinfection approach presented by Wicklein (2016) and discussed in much greater detail in Ho *et al.* (2011), and at https://www.sandia.gov/cfd-water/uvdisinfection.htm. A representative facility had three UV reactor channels with a Trojan 4000 system rated for treating flows up to 157 mgd. The UV system had 3 trains with 2 banks in each train and 90 lamps per bank, for 540 total system lamps.

The hydraulics were simulated using Ansys Fluent using a free surface model. The system had high head loss, approximately 1.5 feet as shown in the left image in Figure 7.8, which helped provide a relatively even flow split for this system. The geometry of the system led to a non-uniform velocity profile thorough the lamps, with velocity generally higher through the central region and lower near the walls, as shown in the right image in Figure 7.8.

Following hydraulic simulation, the discrete ordinance (DO) model in Fluent was run to create a light field within the fluid domain, radiating out from the lamps, and decreasing as a function of distance and fluid adsorption. The left image of Figure 7.9 shows an example of the light radiation field at a plane through the lamps. Then a statistically significant number of particles were released in the fluid, which carried them through the light field. The particles accumulate light exposure based on how long they spend being exposed to the variable light filed, as shown in the right image of Figure 7.9.

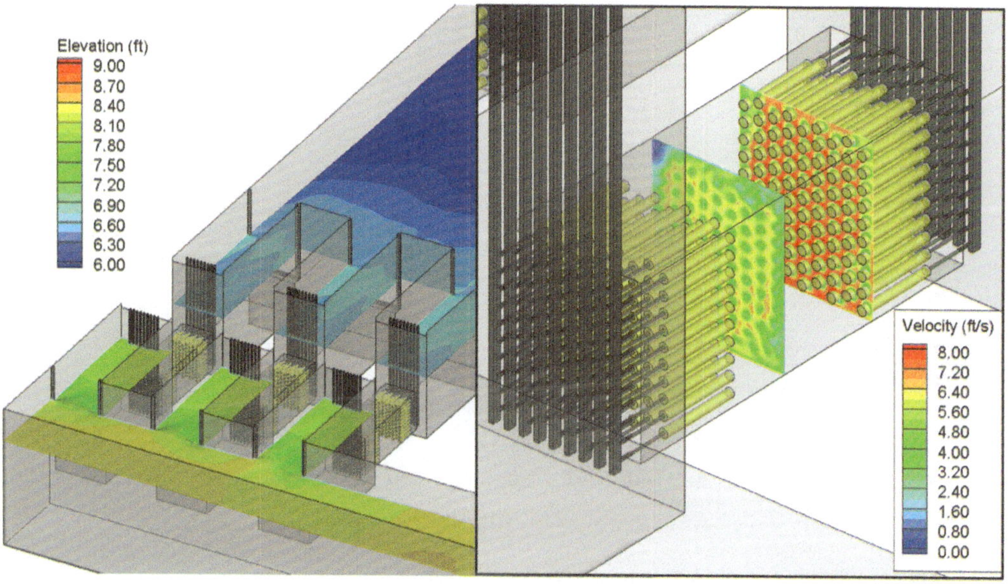

Figure 7.8 The water surface elevation and velocity profile through the UV reactor (Wicklein, 2016).

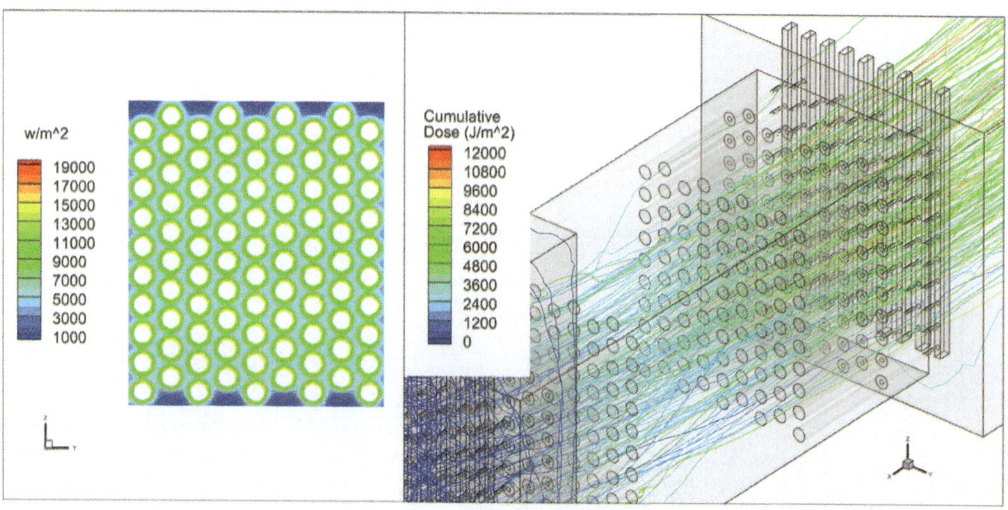

Figure 7.9 Particles collect UV light exposure from the lamps as they travel through the UV reactor (Wicklein, 2016).

The dose distribution plot of the particle tracks is plotted in Figure 7.10. This information is used to estimate log inactivation of target species based on known kinetics. The plot can also show the severity of hydraulic short circuiting through shaded or low-light areas of the reactor.

7.5.2 Contact tank case study

A good case study of a contact tank was presented by Khan *et al.* (2006), using the Star-CD CFD model to replicate a laboratory-scale contact time modeled by Shiono and Teixeira (2000). The

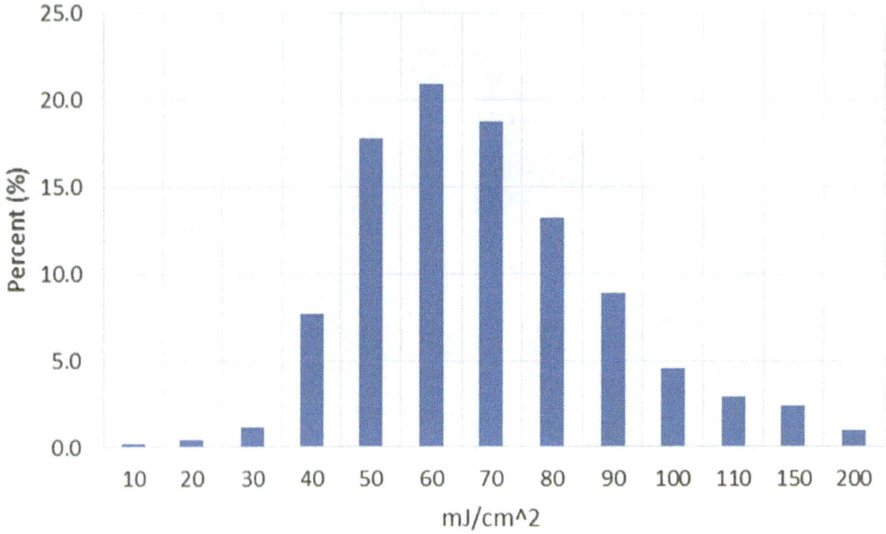

Figure 7.10 UV dose distribution of particle tracks (Wicklein, 2016).

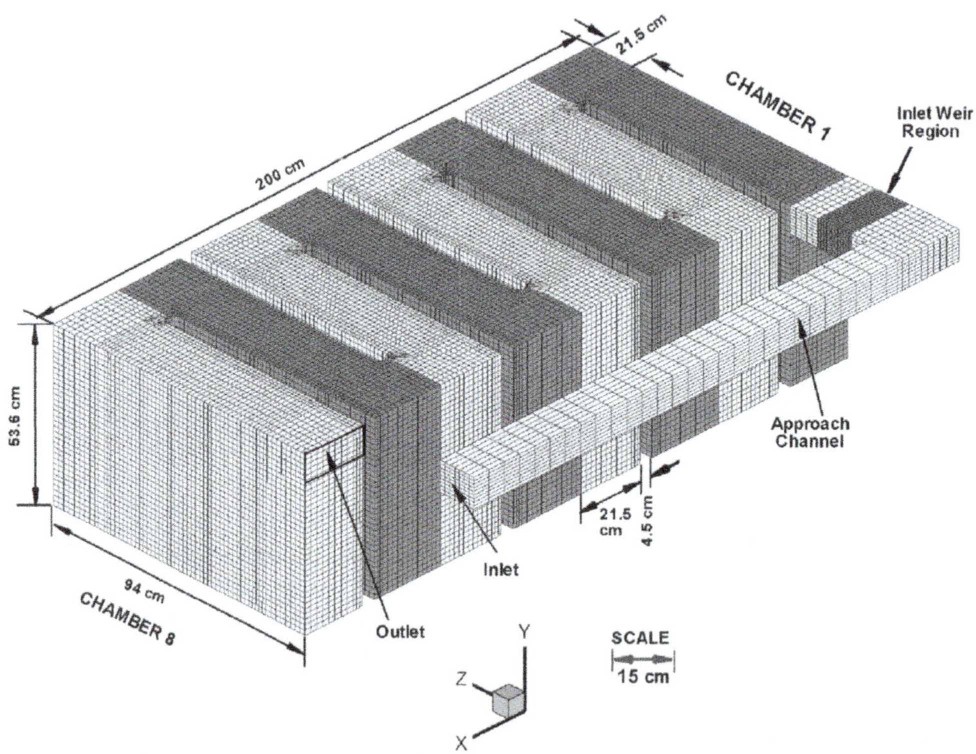

Figure 7.11 Contact tank geometry and computation mesh (Khan *et al.*, 2006).

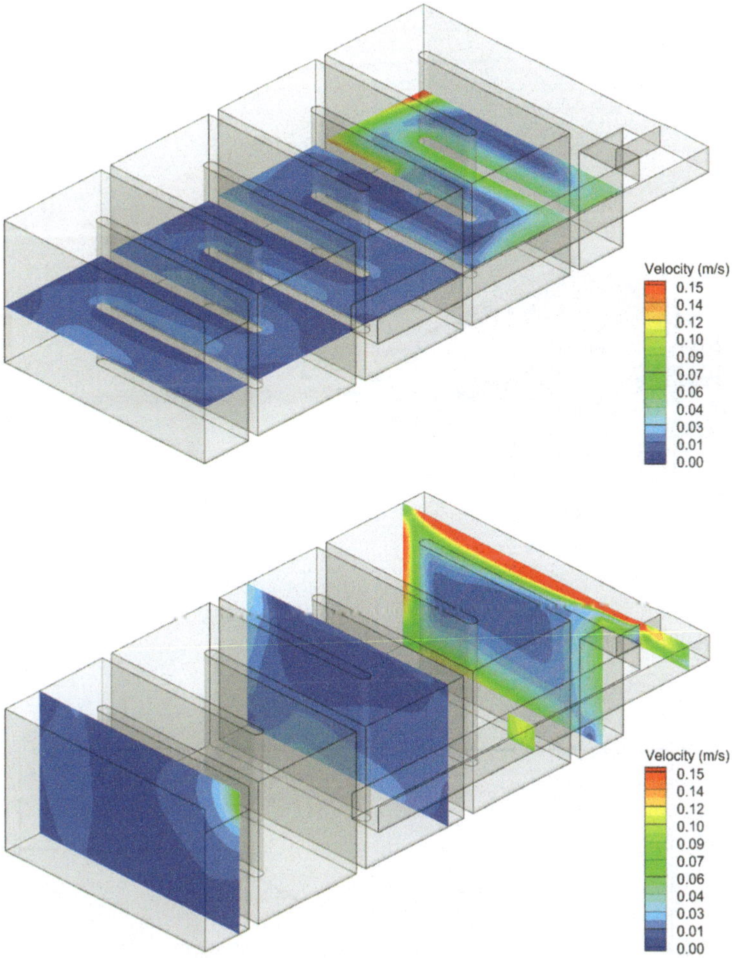

Figure 7.12 Velocity through contact tank.

standard frozen fluid method was used for this example case. Figure 7.11 shows the modeled contact tank geometry and computation mesh. A mesh independence study was conducted, the final mesh is hexahedral dominant with 228 888 computational elements in the contact tank.

The hydraulics were simulated for a 1.17 $l \cdot s^{-1}$ flow to match a laboratory flow condition, using the k-ε turbulence model. Measured velocity profiles were compared with the modeled flow field, showing good agreement. Figure 7.12 shows the velocity at a mid-depth section and through several of the cells.

Following hydraulic simulation, a slug dose tracer test was simulated using a non-reacting neutrally buoyant tracer. The velocity and turbulence fields were held steady from the hydraulic solution, while the model solved a transport equation using a time marching scheme to calculate the advection and dispersion of the tracer through the tank. A total of 1755 mg of tracer was uniformly added over 15 seconds of simulation. A 0.05-second time step was used to simulate 25 minutes of tank operation while monitoring the concentration at the outlet. The tracer concentration is shown at 150 seconds, or 20% of the HRT (Figure 7.13). Figure 7.14 shows the normalized outlet concentration measurement from both the experiment and CFD simulation, as well as the cumulative tracer at the CFD simulated

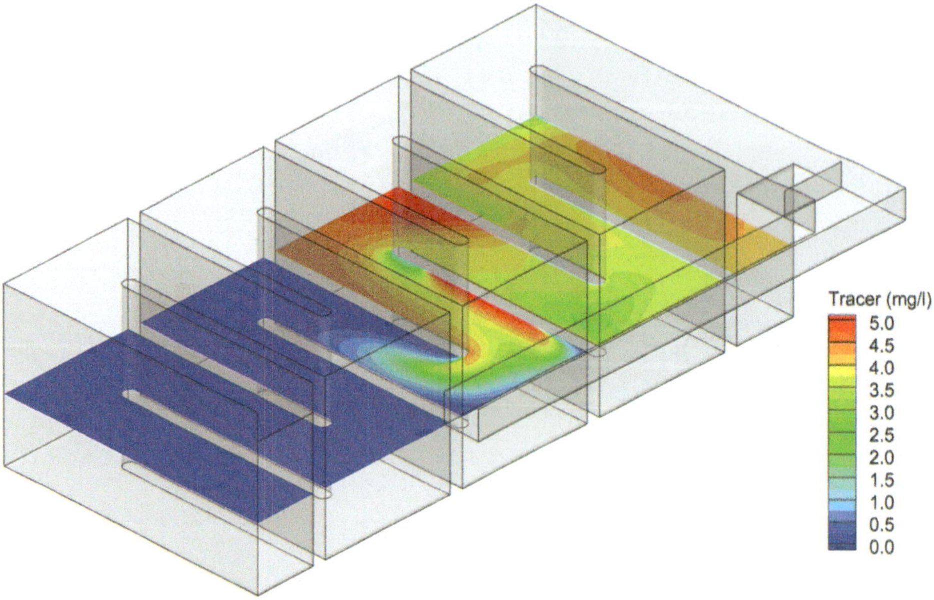

Figure 7.13 Tracer concentration at 150 seconds (~20% of HRT).

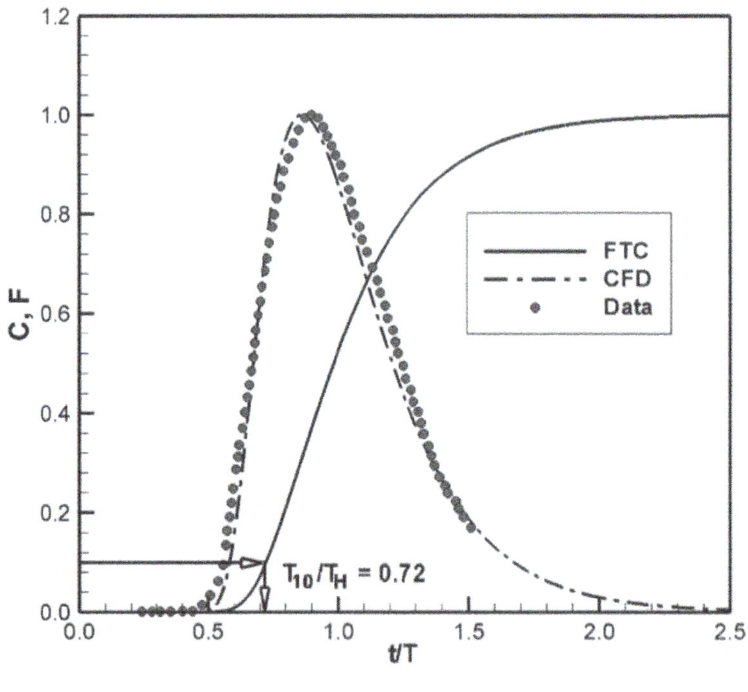

Figure 7.14 Simulated and measured tracer comparison (from Khan *et al.*, 2006). FTC = flow through curve.

tank outlet. The CFD results closely match the experimental results. Based on the cumulative curve, the T10, or BF of the tank is 0.72.

7.6 RESEARCH NEEDS

Application of CFD to disinfection is relatively well developed. It is normally assumed that disinfection flows are not affected by solids density or other buoyant forces. Hence neutral density simulations are typical, and this has seemed to be a reasonable assumption. To date, no CFD model has been developed to explore the performance of UV disinfection in wastewater containing particle aggregates. The ability to model such a system may require the simulation of continued aggregation within a UV reactor, the floc size distribution, and UV light path length, absorbance, and potential scattering due to the presence of these aggregates.

REFERENCES

Angeloudis A., Stoesser T., Gualtieri C. and Falconer R. A. (2016). Contact tank design impact on process performance. *Environmental Modeling & Assessment*, **21**(5), 563–576, https://doi.org/10.1007/s10666-016-9502-x

Ansys I. (2011). ANSYS FLUENT Theory Guide. Southpointe, 275 Technology Drive, Canonsburg, PA 15317, USA.

Blatchley E. R., Shen C., Naunovic Z., Lin L.-S., Lyn D. A., Robinson J. P., Ragheb K., Grégori G., Bergstrom D. E., Fang S., Guan Y., Jennings K. and Gunaratna N. (2006). Dyed microspheres for quantification of UV dose distributions: photochemical reactor characterization by lagrangian actinometry. *Journal of Environmental Engineering*, **132**(11), 1390–1403, https://doi.org/10.1061/(ASCE)0733-9372(2006)132:11(1390)

Bohrerova Z., Bohrer G., Mohanraj S. M., Ducoste J. and Linden K. G. (2005). Experimental measurements of fluence distribution in a UV reactor using fluorescent microspheres. *Environmental Science and Technology*, **39**(22), 8925–8930, https://doi.org/10.1021/es050034c

Caron E., Chevrefils G., Barbeau B., Payment P. and Prévost M. (2007). Impact of microparticles on UV disinfection of indigenous aerobic spores. *Water Research*, **41**(19), 4546–4556, https://doi.org/10.1016/j.watres.2007.06.032

Cheng P. (1964). Two-dimensional radiating gas flow by a moment method. *AIAA Journal*, **2**(9), 1662–1664, https://doi.org/10.2514/3.2645

Chick H. (1908). An investigation of the laws of disinfection. *Epidemiology & Infection*, **8**(1), 92–158.

Crozes G. F., Hagstrom J. P., Clark M. M., Ducoste J. and Burns C. (1999). Improving Clearwell Design for CT Compliance. AWWA Research Foundation, Denver, CO, USA.

Ducoste J. J., Liu D. and Linden K. (2005). Alternative approaches to modeling fluence distribution and microbial inactivation in ultraviolet reactors: lagrangian versus eulerian. *Journal of Environmental Engineering*, **131**(10), 1393–1403, https://doi.org/10.1061/(ASCE)0733-9372(2005)131:10(1393)

Greene D. J. (2002). Numerical simulation of chlorine disinfection processes. PhD thesis, Drexel University, Philadelphia, PA, USA.

Haas C. N. and Finch G. R. (2001). Methodologies for the Determination of Disinfection Effectiveness. AWWA Research Foundation and American Water Works Association, Denver, CO, USA.

Haas C. N. and Joffe J. (1994). Disinfection under dynamic conditions: modification of Hom's model for decay. *Environmental Science & Technology*, **28**, 1367–1369, https://doi.org/10.1021/es00056a028

Hannoun I. A. and Boulos P. F. (1997). Optimizing distribution storage water quality: a hydrodynamic approach. *Applied Mathematical Modelling*, **21**(8), 495–502, https://doi.org/10.1016/S0307-904X(97)00043-7

Ho C. K., Khalsa S. S., Wright H. B., Wicklein E. A. and Khalsa S. S. (2011). Computational Fluid Dynamics Based Models for Assessing UV Reactor Design and Installation, Water Research Foundation. Denver, CO, USA.

Hom L. W. (1972). Kinetics of chlorine disinfection in an ecosystem. *Journal of the Sanitary Engineering Division*, **98**(1), 183–194, https://doi.org/10.1061/JSEDAI.0001370

Khan L. A., Wicklein E. A. and Teixeira E. C. (2006). Validation of a three-dimensional computational fluid dynamics model of a contact tank. *Journal of Hydraulic Engineering*, **132**(7), 741–746, https://doi.org/10.1061/(ASCE)0733-9429(2006)132:7(741)

Kollu K. and Örmeci B. (2012). Effect of particles and bioflocculation on ultraviolet disinfection of Escherichia coli. *Water Research*, **46**(3), 750–760, https://doi.org/10.1016/j.watres.2011.11.046

Liu D., Ducoste J., Jin S. and Linden K. (2004). Evaluation of alternative fluence rate distribution models. *Journal of Water Supply: Research and Technology – AQUA*, **53**(6), 391–408, https://doi.org/10.2166/aqua.2004.0031

Mamane H. and Linden K. G. (2006). Impact of particle aggregated microbes on UV disinfection. I: evaluation of spore–clay aggregates and suspended spores. *Journal of Environmental Engineering*, **132**(6), 596–606, https://doi.org/10.1061/(ASCE)0733-9372(2006)132:6(596)

Mamane H., Ducoste J. J. and Linden K. G. (2006). Effect of particles on ultraviolet light penetration in natural and engineered systems. *Applied Optics*, **45**(8), 1844, https://doi.org/10.1364/AO.45.001844

McCluney W. R. (2014). Introduction to Radiometry and Photometry, 2nd edn. Artech House, Boston, MA, USA.

Murthy J. Y. and Mathur S. R. (1998). Finite volume method for radiative heat transfer using unstructured meshes. *Journal of Thermophysics and Heat Transfer*, **12**(3), 313–321, https://doi.org/10.2514/2.6363

Raithby G. D. and Chui E. H. (1990). A finite-volume method for predicting a radiant heat transfer in enclosures with participating media. *Journal of Heat Transfer*, **112**(2), 415–423, https://doi.org/10.1115/1.2910394

Rossi S., Antonelli M., Mezzanotte V. and Nurizzo C. (2007). Peracetic acid disinfection: a feasible alternative to wastewater chlorination. *Water Environment Research*, **79**(4), 341–350, https://doi.org/10.2175/106143006X101953

Santoro D., Gehr R., Bartrand T. A., Liberti L., Notarnicola M., Dell'Erba A., Falsanisi D. and Haas C. N. (2007). Wastewater disinfection by peracetic acid: assessment of models for tracking residual measurements and inactivation. *Water Environment Research*, **79**(7), 775–787, https://doi.org/10.2175/106143007X156817

Santoro D., Crapulli F., Raisee M., Raspa G. and Haas C. N. (2015). Nondeterministic computational fluid dynamics modeling of escherichia coli inactivation by peracetic acid in municipal wastewater contact tanks. *Environmental Science and Technology*, **49**(12), 7265–7275, https://doi.org/10.1021/es5059742

Selleck R. E., Saunier B. M. and Collins H. F. (1978). Kinetics of bacterial deactivation with chlorine. *Journal of the Environmental Engineering Division*, **104**(6), 1197–1212, https://doi.org/10.1061/JEEGAV.0000842

Shiono K. and Teixeira E. C. (2000). Turbulent characteristics in a baffled contact tank. *Journal of Hydraulic Research*, **38**(6), 403–416, https://doi.org/10.1080/00221680009498294

State of Colorado. (2021). Technical Guidance Manual on Computational Fluid Dynamics Modeling of Chlorine Contact Basins. Colorado Department of Public Health and Environment Water Quality Division, Glendale, Colorado, USA.

Stropky D., Pougatch K., Nowak P., Salcudean M., Pagoria P., Gartshore I. and Yuan J. (2007). RTD (residence time distribution) predictions in large mechanically aerated lagoons. *Water Science and Technology*, **55**(11), 29–36, https://doi.org/10.2166/wst.2007.346

Thýn J., Ha J. J., Střasák P. and Žitný R. (1998). RTD Prediction, modelling and measurement of gas flow in reactor. *Nukleonika*, **43**(1), 95–114.

Toor J. S. and Viskanta R. (1968). A numerical experiment of radiant heat interchange by the monte carlo method. *International Journal of Heat and Mass Transfer*, **11**(5), 883–897, https://doi.org/10.1016/0017-9310(68)90131-2

US EPA. (1986). Design Manual: Municipal Wastewater Disinfection. Cincinnati, Ohio, USA.

U.S. Environmental Protection Agency. (2006). Ultraviolet Disinfection Guidance Manual for the Final Long Term 2 Enhanced Surface Water Treatment Rule. Washington, D.C., USA.

Watson H. E. (1908). A note on the variation of the rate of disinfection with change in the concentration of the disinfectant. *Epidemiology & Infection*, **8**(4), 536–542, https://doi.org/10.1017/S0022172400015928

Wicklein E. A. (2016). Aspects of Modeling UV Systems. WEFTEC 2016 Workshop – CFD for WRRF, WEF, New Orleans.

Wilson J. M. and Venayagamoorthy S. K. (2010). Evaluation of hydraulic efficiency of disinfection systems based on residence time distribution curves. *Environmental Science & Technology*, **44**(24), 9377–9382, https://doi.org/10.1021/es102861g

Wols B. A., Uijttewaal W. S. J., Rietveld L. C., Stelling G. S., van Dijk J. C. and Hofman J. A. M. H. (2008). Residence time distributions in ozone contactors. *Ozone: Science & Engineering*, **30**(1), 49–57, https://doi.org/10.1080/01919510701759538

Wols B. A., Shao L., Uijttewaal W. S. J., Hofman J. A. M. H., Rietveld L. C. and van Dijk J. C. (2010). Evaluation of experimental techniques to validate numerical computations of the hydraulics inside a UV bench-scale reactor. *Chemical Engineering Science*, **65**(15), 4491–4502, https://doi.org/10.1016/j.ces.2010.04.013

Zhang J., Tejada-Martínez A. E. and Zhang Q. (2014). Developments in computational fluid dynamics-based modeling for disinfection technologies over the last two decades: a review. *Environmental Modelling and Software*, **58**, 71–85, https://doi.org/10.1016/j.envsoft.2014.04.003

Zhao X., Alpert S. M. and Ducoste J. J. (2008). Assessing the impact of upstream hydraulics on the dose distribution of ultraviolet reactors using fluorescence microspheres and computational fluid dynamics. *Environmental Engineering Science*, **26**(5), 947–959, https://doi.org/10.1089/ees.2008.0139

doi: 10.2166/9781780409030_0171

Chapter 8
Anaerobic digestion

David Fernandes Del Pozo[1], Damien Batstone[2] and Jim Wicks[3]

[1]BIOMATH, Department of Data Analysis and Mathematical Modelling, Ghent University, Gent, Belgium; Email: david.fernandesdelpozo@ugent.be/fernandesdelpozodavid@gmail.com
[2]Australian Centre for Water and Environmental Biotechnology, University of Queensland, Brisbane, Australia; Email: d.batstone@uq.edu.au
[3]The Fluid Group, Oxford, UK; Email: jim.wicks@thefluidgroup.com

8.1 INTRODUCTION

Anaerobic digestion (AD) is a widespread process for stabilisation of sewage solids and production of renewable energy from low value or waste biomass. It is primarily used as a treatment process in a mixed digester configuration to treat high-solids liquid streams (1–10% w/w) (Figure 8.1). However, other technologies such as high-rate granular and biofilm systems have also been developed for highly soluble industrial streams. Additionally, solid phase and plug-flow digesters have been developed for higher solids feeds (Batstone & Jensen, 2011). Although there is a large variety of available digestion technologies, this chapter focuses on the mixed digester application since it is the most widespread application in anaerobic digestion. High-rate biofilm systems are more correctly considered as biofilm reactors, and emerging technologies such as plug-flow and solid phase are considered in research needs.

Methane (CH_4), the main component in biogas, is an energy intensive by-product that can conserve the energetic value of the organic substrate, and it can be further valorised into heat and electricity. Effective digestion of feed sludge, often incorporating co-feeds such as waste fats and highly degradable carbohydrates (including glycerol), can meet a substantial fraction of, or even completely meet, the needs of the parent activated sludge process (Mata-Alvarez *et al.*, 2011).

AD involves substrate coupled oxidation and reduction reactions, in the absence of molecular oxygen or another major electron acceptor to generate an approximately even amount of the most oxidised, and most reduced forms of carbon: CO_2 and CH_4 (Metcalf & Eddy *et al.*, 2014). The complete process consists of a complex cascade of reaction pathways performed by different microorganisms. Physically, these reactions take place in a complex multiphase matrix consisting of various sized solids, a liquid slurry, and biogas bubbles.

The design of mixed-liquid anaerobic digesters depends heavily on feed concentration and substrate profile and properties. The digester must (1) be of sufficient size to remove degradable organics and generate a stabilised product (a function of the speed of degradation), (2) be able to mix the contents

Editor(s): Julien Laurent, Randal Samstag, Jim Wicks and Ingmar Nopens

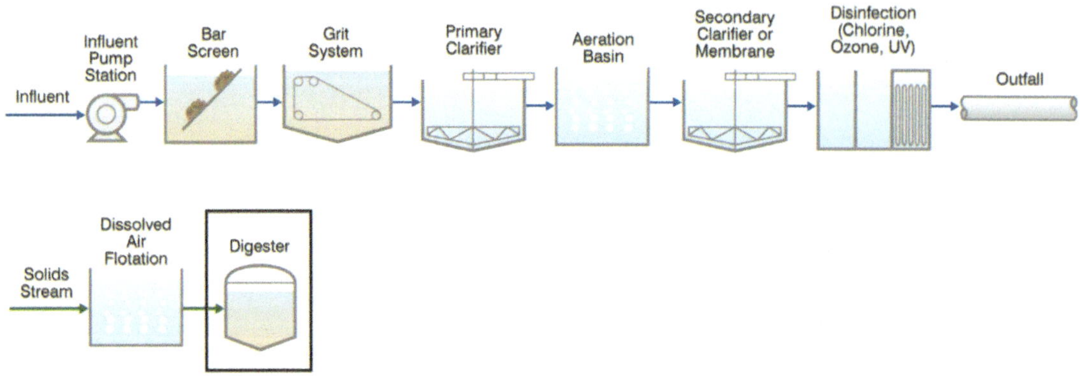

Figure 8.1 Placement of the anaerobic digestion in a wastewater resources recovery facility.

effectively (a function of liquid viscosity, mixing system design, and power), (3) be able to effectively capture gas (a function of hydraulics and mechanical equipment), and (4) retain the thermal setpoint. Trade-offs between the different objectives are common. For instance, increasing feed total solids (TS%) concentration will increase the retention time, and reduce the fraction of degradable material in the effluent. However, this will increase the fluid viscosity and hence reduce mixing effectiveness for a given power input and design.

The critical substrate properties can be summarised as (1) the substrate degradability speed, normally expressed as a first order hydrolysis coefficient (k_{hyd}), and (2) substrate degradability extent, normally expressed as a degradable fraction (f_d), or a methane potential (B_0, mL CH_4·g VS_{fed}) (Batstone & Jensen, 2011). For a given COD:VS (chemical oxygen demand:volatile solids) ratio, the f_d and B_0 are fully correlated (and calculable from each other). COD:VS, another critical fundamental parameter, influences the energy content and hence effective methane yield per volatile organic solids fed. These parameters are estimable in batch or continuous laboratory-scale anaerobic digesters (though batch tests tend to be conservative since they impose effective mixing through the digester) (Jensen *et al.*, 2011). However, achieving homogeneous conditions in a scaled-up reactor is not a trivial task, and this criterion is rarely fulfilled. While non-homogeneous conditions do not necessarily mean diminished performance, this is the most common consequence, and ineffective mixing is one of the most common causes of poor or unintended operational modes (including spilling, foaming, and inconsistent digestion).

Mixing has been classically designed using chemical engineering principles, enabling effective turbulent energy dissipation for a given geometry. The mixing strategy depends on the digester type, broadly egg-shaped or cylindrical. Egg-shaped digesters have a higher vessel capital cost, but are designed for effective mixing, and are generally mechanically mixed, through either draft tube or impeller-based mixing (Wu, 2010a). Cylindrical digesters may be mixed mechanically (generally by multiple impellers), by a liquid draft tube, by hydraulic means (recirculated liquid injected), or by recirculated gas (through either lances or draft tubes). Particularly in cylindrical digesters, mixing systems are overdesigned to account for variations in material properties and hydraulic profile (Metcalf & Eddy *et al.*, 2014).

Computational fluid dynamics (CFD) is an ideal tool to investigate mixing in anaerobic digesters, as it can be used throughout the design and application cycle to produce effective and energy-efficient designs, as well as investigate existing digesters under normal operation and in response to specific incidents. However, there are several challenges that need to be considered compared to other processes. These may include (1) non-Newtonian rheology, (2) the need to consider multiple phases (solid, liquid, and gas), (3) the relatively long pressure field, (4) the large dimensional scales involved in full-scale digesters, and (5) the mixing impact of native vs applied gas phases (i.e., some gas is imposed through mixing systems, while some gas is self-generated). CFD is an ideal tool to explore the interaction when different processes simultaneously occur. Consequently, there are multiple studies that range from considering sludge as a Newtonian single-phase fluid to more advanced studies including biokinetic integration, three-phase modelling, and non-Newtonian rheology. Given that digesters are expensive to build, and the cost of failure is high (often resulting in reseeding or clean-out operations), understanding the hydrodynamics is critical. This is reflected in environmental engineering were CFD applied to digesters is a rapidly emerging field, in both research and applied areas.

8.2 LITERATURE REVIEW

Within the past decade, several CFD studies relevant to digestion have been published. While a significant amount of work has been completed, the literature is generally lightweight, particularly in comparison with its industrial relevance.

Wu (2013) reviewed the application of CFD in bioenergy systems, which included digesters. The focus in anaerobic digestion was mainly on mixing efficiency and commonly used single phase techniques, and presented a mix of Newtonian and non-Newtonian viscosity assumptions. Several works have provided an extended review of AD CFD, focusing more on the recommended mechanisms and implementation strategy, rather than a state-of-the-art assessment of quality and accuracy (Sadino-Riquelme *et al.*, 2018; Samstag *et al.*, 2016; Wang *et al.*, 2018).

The key challenge, often noted, is the need to incorporate biochemical reactions in a CFD model to predict biogas production from first principles.

8.2.1 Single phase

The vast majority of CFD work in literature (and the bulk of the commercial analysis to date) has been performed assuming the digester contents as a single phase, incompressible, steady state, and isothermal fluid. Sludge properties have been modelled assuming either Newtonian or non-Newtonian fluid physics over a wide range of solids concentrations (Wu, 2013).

Although it is well known that Newtonian dynamics (as we show in Section 1.1.4.2) is not a reasonable modelling assumption when the dry solids content exceeds 2%, this continues to be applied in CFD modelling of digesters. Different non-Newtonian models have been considered for higher solids concentrations, but often with little or no justification. This is particularly important, since a power law (PL) model is fundamentally different from a Bingham (BN) plastic, or a Herschel-Bulkley model (HB) (Baudez *et al.*, 2011).

Jayranaiwachira and James (1999) made one of the earliest attempts to systematically study the non-Newtonian effect in anaerobic digesters using CFD (in a purely theoretical study). The main conclusion from this article was that non-Newtonian behaviour could potentially have a significant impact in full-scale digesters and should be included. The work of Fleming (2002) described the main processes that occur in anaerobic lagoons, but are applicable to any AD modelling with CFD. The author identified processes such as sedimentation, heat transfer, buoyancy and biokinetics, and provided simple closure modelling for each.

Eftekharzadeh *et al.* (2006) assessed a pilot CFD system in a single-phase non-Newtonian (PL) model to determine the influence of the design on a liquid JML (jet mixing loop) at several sludge concentrations. The authors used xantham gum (XG) as a clear model fluid (with a dye to assess mixing performance) to mimic the rheology of sludge for validation purposes. The authors surmised that to maintain effective mixing when the inlet feed flow rate has increased, it is beneficial to use an intermittent operation of mixing nozzles, and to use top and bottom suction headers to avoid any eventual stratification (in temperature and concentration) to promote vertical mixing.

Terashima *et al.* (2009) simulated an existing full-scale digester (modelled in laminar flow, PL model) and validated the CFD model with a tracer test. The results of the tracer test indicated short circuiting with a mean internal recirculation time of 1.25 h and were replicated by the CFD model. Interestingly, mixing was assessed with a uniformity index metric, which showed that increasing TS concentration led to poorer mixing indexes.

Langner (2009) provided important manure physical characterisation data, as well as velocity profile validation studies of their 3D model with particle image velocimetry (PIV) data. Martínez Mendoza *et al.* (2011) used a CFD model to represent the hydraulic behaviour of an existing full-scale plant. A novel criterion was used to determine dead zones based on local terminal sedimentation velocities of sludge particles. A Newtonian fluid assumption was applied ($h = 0.065$ Pa·s). Yu *et al.* (2011) used a CFD model (non-Newtonian power law) to demonstrate the impact of impeller geometry on the flow field when different high solids concentrations were considered.

Bridgeman (2012) performed CFD simulations of a lab-scale mechanical mixer to obtain information about the effect of mixing with respect to experimental biogas yield. The author concluded that lab-scale performance was not critically dependent on mixing intensity (this statement not being applicable to large-scale digesters). The author proposed the use of shear-rate distribution values to analyse the mixing state in the digester. Later, this concept was linked with biogas digester performance in Sindall and Bridgeman (2013). Kamarád *et al.* (2013) took a very similar approach (PL model) in assessing full-scale manure digesters. An important conclusion from this work is that the overall flow pattern was governed by the slow-motion agitators rather than the fast-motion mixing impellers. Craig *et al.* (2013) assessed the effect of different viscosity models (HB and BM) in a liquid draft-tube reactor. The authors stressed the need for a non-Newtonian viscosity model to represent the sludge behaviour.

Hurtado *et al.* (2015) developed a CFD model of a full-scale system to assess the influence of digester operation with respect to active volume and mixing time. Both Newtonian and non-Newtonian (PL and HB) models were compared. It was concluded that the Newtonian assumption was sufficiently valid at the solids concentration evaluated (2% TS), and that intermittent mixing could effectively mix the digester. Hernández-Aguilar *et al.* (2016) developed a lab-scale egg-shaped system and modelled it with CFD, assuming sludge as a Newtonian fluid with laminar flow. Different digestion stages were modelled by linking dynamic viscosity to sludge concentration and time.

Sajjadi *et al.* (2016) used a CFD model of a lab-scale rheologically mimicking digester (XG) with submerged recirculating jets. It was found that the rheological properties (applying a PL model) dictate the presence of cavern structures, and different mixing modes were studied to optimise active volume with respect to energy input.

During the last five years, AD CFD modelling literature has notably shifted towards demonstrating the influence of physical properties, geometries, and mixing efficiency with respect to biogas performance.

Lebranchu *et al.* (2017) used a CFD model (PL) to simulate sludge viscosity (8.8% TS) while considering both helical and Rushton impellers in a lab-scale digester to assess methane production efficiency. A reduction of the digester dead volume, calculated by the model, was obtained by an effective dispersion of the inlet feed by the helical impeller. This configuration led to an experimental increase in the maximum methane production rates of 50%.

Meister *et al.* (2018) simulated a full-scale egg-shaped digester mixed with an internal draft tube. For the CFD model, a PL fluid model was adopted involving different concentrations up to 12% TS. Emphasis was placed on the importance of the inlet and outlet in the hydrodynamic patterns. The CFD model was used to study different operational modes and the effect of higher TS concentrations in the development of the flow pattern.

Mohammadrezaei *et al.* (2018) used the rheological model of Yu *et al.* (2011) to study the influence of different rotational speeds in a multiple mechanically stirred digester. The authors demonstrated the existence of an optimal stirring speed to maximise total biogas production. Wiedemann *et al.* (2018) presented a study comprising the validation and optimisation of a CFD model of an 800 litre tank equipped with submerged propellers (using cellulose as a sludge surrogate fluid). Optimal placement and angle orientation of submerged propellers was based on maximising different criteria related to velocity uniformity indexes in the tank. Other relevant work can be found in Conti *et al.* (2019), Leonzio (2018), Mao *et al.* (2019) and Zhu and Jing (2019).

8.2.2 Eulerian multiphase models

Karim *et al.* (2004) provided one of the first hydrodynamic data sets to study the hydrodynamics of a gas-lift reactor (GLR). This configuration was later used in Karim *et al.* (2005) to evaluate the performance of different digester configurations. Within the latter study, small differences were found in terms of biogas production between mixed and unmixed configurations (possibly due to native production). The data set from Karim *et al.* (2004) included measurements at different sparger flow rates achieved with CT (computed tomography) and CARPT (computer-automated radioactive particle tracking). This data set was later used by multiple authors in the AD CFD modelling community. Vesvikar and Al-dahhan (2005) built a two-phase implementation CFD model of the same GLR and provided validation of the model in terms of liquid velocity profiles. Karim *et al.* (2007) modelled this GLR too, but the outlet draft-tube slurry velocity boundary condition was calculated from a correlation generated by Kojima *et al.* (1999) to include the effect of the uprising biogas injection on the liquid recirculated through the draft-tube. Later, both CFD models were used to explore different digester configurations.

A well-known author in the AD CFD community (Wu, 2010a, 2012a, 2012b, 2014; Wu & Chen, 2008; Wu *et al.*, 2009) conducted an extensive analysis of gas-fed systems with CFD modelling, and further published a collection of relevant data. The author analysed the influence of digester geometry, mixing mode and turbulence model selection in the context of wet anaerobic, single and multiphase digestion for laboratory, pilot, and full-scale digesters. The approach adopted was generally a multiphase Eulerian using a PL viscosity model. An analysis of gas fed systems with CFD identified that the US EPA (United States Environmental Protection Agency) velocity-gradient approach to mixing efficiency (US EPA, 1987) was not particularly relevant to non-Newtonian fluids (Wu, 2014). It was also demonstrated in these studies (by CFD) that confined gas mixing was more efficient than unconfined mixing.

An Eulerian multiphase approach was also applied by Maier *et al.* (2010) to model the hydrodynamics of a full-scale stratified digester. Mechanical mixing was modelled by CFD and included the sludge as three phases, each with its own rheological behaviour.

Based on the cited works, Eulerian approaches have typically assumed a uniform distribution of gas bubble size, mainly due to modelling simplicity. CFD coupled with Population Balance Models (PBMs), where the moments of a gas size distribution are included as state variables (Nopens *et al.*, 2015) does not appear to be have been used in digester simulation thus far.

8.2.3 Lagrangian-based CFD modelling

Eulerian-Lagrangian (EL) modelling has been increasingly applied to anaerobic digestion in order to produce complex, efficient models. Lagrangian modelling considers CFD from the perspective of the fluid rather than from the domain (Eulerian). Lagrangian elements are computationally economical, but often large numbers of Lagrangian objects are needed.

The classical application is related to bubbly flows, where bubbles (or bubble clusters) are treated as Lagrangian elements within an Eulerian liquid (Dapelo *et al.*, 2015). Dapelo *et al.* (2015), who developed an EL solution for a lab-scale system (for unconfined gas mixing), and subsequently extended the approach for a full-scale digester (Dapelo & Bridgeman, 2018a, 2018b) studied the effect of non-Newtonian behaviour on sludge flow patterns and mixing quality. They demonstrated better mixing efficiencies with a lower gas mixing input, and the potential for significant energy optimisation by switching biogas injection periodically between different nozzles.

The EL approach to modelling gas-mixed digesters was also applied by Coughtrie *et al.* (2013). The authors simulated an air-lift reactor (ALR) system using a liquid Newtonian fluid and compared both ANSYS Fluent and OpenFOAM CFD results. They compared their results with the single phase approach by Karim *et al.* (2004) and Karim *et al.* (2007), and concluded that the assumptions were not substantially different, except when understanding the effects of the biogas near the draft-tube outlet.

Alternatives to the finite volume method (FVM) used in classical AD CFD studies are the LB (lattice Boltzmann) and SPH (smoothed particle hydrodynamics) methods.

The SPH approach is a promising alternative to simulating engineering fluids, where particles individually simulate the flow of liquid, gas or solid (biomass). Non-linear properties can easily be simulated; substrate definitions can be applied to each individual point-like particle, and bulk fluid properties maintained. One advantage is that a propagating perturbation within a larger liquid domain can be easily tracked (see Figure 8.2). A full SPH approach was used to simulate an egg-shaped digester, incorporating biokinetics, with biokinetic properties carried within each Lagrangian element by Rezavand *et al.* (2019).

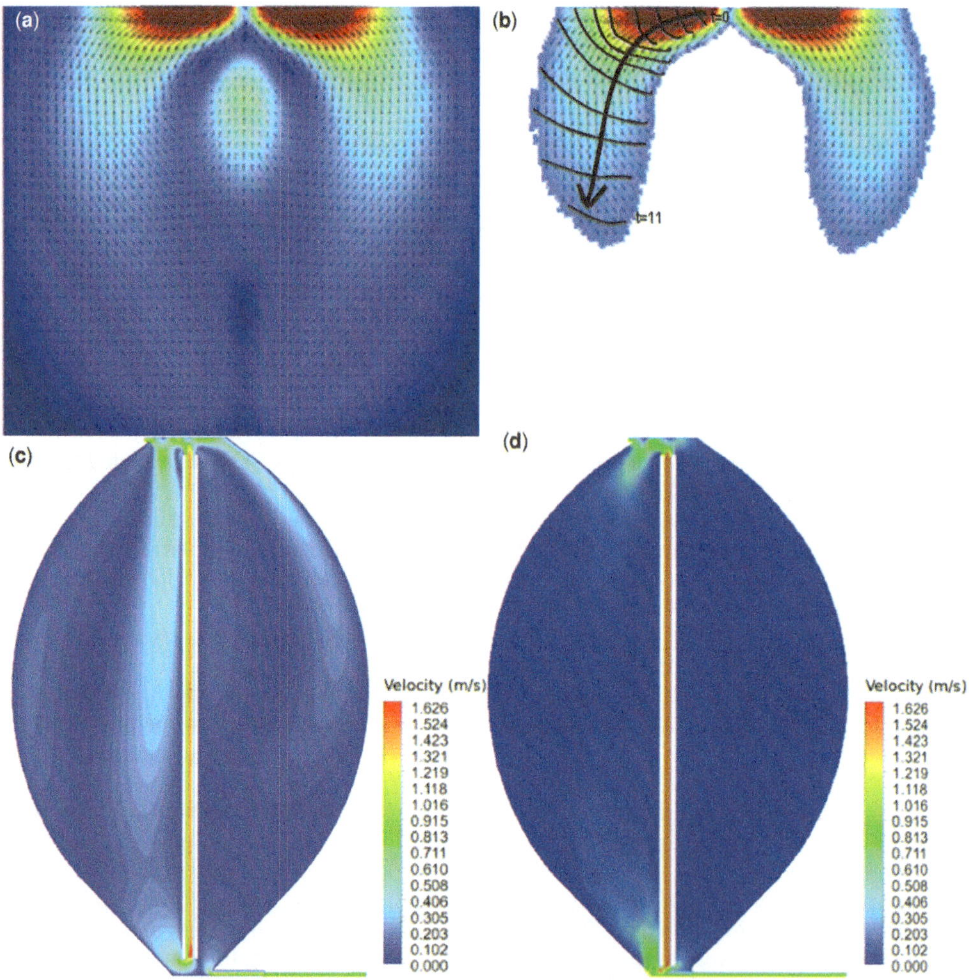

Figure 8.2 A comparison of Eulerian and Lagrangian computational methods for a cylindrical (Dapelo *et al.*, 2015), and an egg-shaped (Rezavand *et al.*, 2019) anaerobic digester: (a) and (c) the Eulerian approach considers the whole domain, while (b) and (d) the Lagrangian approach can focus on propagating, interacting elements.

Lattice-Boltzmann methods solve the Navier-Stokes equations on a discrete lattice, that simulates a series of particle-like propagation and collision processes. Dapelo *et al.* (2019) showed that an LB model for a digester would lead to significant computational efficiencies over CFD-based models, while also modelling the gas-sludge mixing accurately.

8.2.4 Bioreactive modelling

The typical goal of anaerobic digester modelling is to optimise methane gas production, therefore the incorporation of biological reactions has been identified as a major requirement in future CFD digester models (Sadino-Riquelme *et al.*, 2018).

To illustrate the importance of modelling substrate and biomass, Hoffmann *et al.* (2008) gathered hydrodynamic data using CARPT, and used a CFD model to simulate a mechanical mixer (single-phase Newtonian) at different mixing rates to determine a velocity profile. The shear stress distribution near the impeller was then related to the experimental digester microbial community. The results of this analysis found that high shear stresses favoured the development of *Methanosarcina* over *Methanosaeta*. It was hypothesised that the mixing operation mode had a direct impact on both rheology and digester performance, given that these two microorganisms have different susceptibilities to shear because of different cell morphologies.

Karim *et al.* (2005) found that self-generated gas production had a significant impact on overall mixing. Lebranchu *et al.* (2017) found that the mixing regime had a substantial impact on the maximum methane production volume, with methane generation rates linked to effective feed dispersion. It is important to realise that the effect of mixing on digester performance is not only applicable for wastewater treatment plant (WWTP) digestion, but also for other relevant industrial digesters (Jiang *et al.*, 2014).

However, despite the apparent importance of bioreactive modelling in anaerobic digestion, few studies with biological formulae coupled to CFD have been presented. Although different strategies can be applied (Wu, 2012c), the choice of biological model depends on the modelling objectives and its feasibility.

Simple strategies could use a first order hydrolysis-methanogenesis model (Wu *et al.*, 2009), without pH prediction. More complex strategies are likely to use a physicochemical and multiple gas transport model (dependent on local shear rates, pH, etc.). Full implementation in an anaerobic digester should incorporate a complex anaerobic digestion model, such as the standard kinetic model ADM1 (Batstone *et al.*, 2002).

There is scope for all of these options, but the only reference found is the application of a reduced ADM1 (including algebraic equations) in a two-phase Newtonian model provided by Gaden (2013). The author considered solvability issues and implementation more than the practical application of the model, but it demonstrated the capability of current solvers (which have advanced a great deal since then) to solve significant biological CFD problems in AD systems.

It should be noted that there has been extensive single and multi-dimensional modelling of anaerobic biofilm systems (Batstone *et al.*, 2006) which are compatible with CFD, but these are on the micron scale, and not generally applicable to larger-scale CFD simulation.

8.2.5 Conclusions: literature review

Anaerobic digestion appears to be at the forefront of modelling complex, multiphase non-Newtonian systems. However, the CFD community often relies on significant modelling simplifications, such as the use of single-phase techniques for systems which would otherwise be multiphase (e.g., gas injection simulation with momentum source terms). This has largely been driven by the high level of fluid and mechanistic complexity of anaerobic digesters.

With respect to the rheological behaviour of digested sludge (Newtonian vs non-Newtonian), in-reactor concentrations of less than 2% dry solids can be generally modelled by a Newtonian assumption.

When dealing with higher sludge concentrations however, non-Newtonian effects must be included, using PL, BN, or HB rheological models. Although the use of the PL model is the most popular approach, the appropriate rheological model would be the one that correctly predicts shearing forces within the shear-rate ranges experienced by the digester. For instance, HB models are typically better when modelling digesters that are affected by the presence of fluid caverns and overall low velocities.

From the literature, two main research gaps can be identified:

The first involves the need for a satisfactory integrated biokinetic modelling framework to be coupled with conventional CFD hydrodynamic studies; one that predicts methane production accurately.

The second is the consideration of solids as a separate phase, including solids settling via empirical settling equations such as the double Vesilind equation, including compression as required via second-order terms (Bürger & Diehl, 2015) (see also Chapter 6 on sedimentation). While scalar or particulate solid fluxes are readily implemented in most simulation packages, the discontinuities in these equations may make deriving a solution more complex. However, given that one of the main goals of mixing is to disperse the solids gradient, it is surprising that more efforts have not been made in this area.

8.3 PROCESS DESCRIPTION

CFD is most used to describe the extent of mixing and the efficiency of the internal digester hydrodynamics. Unlike dilute, and minimally reactive systems, where geometry is the major determining factor, digesters are high-solids content, reactive systems, and therefore process considerations also play a significant role. The following section outlines the key process considerations, and dynamic and static design elements which need to be considered in developing a CFD model of a digester.

8.3.1 Mixed digester design

Feed concentration needs to be 4–10%, in order to achieve a target in-reactor concentration of 1–5% (Batstone & Jensen, 2011; Metcalf & Eddy *et al.*, 2014). Most primary and activated sludges need pre-thickening by gravity, dissolved air flotation (DAF), or rotary drum thickening to achieve this feed concentration (Metcalf & Eddy *et al.*, 2014). Thickened sludge is fed either continuously or, much more commonly, at intervals (often sequentially across multiple digesters). Sludge may be bottom fed, top fed, or fed through induction piping. The origin of the sludge will depend on the configuration of the plant, and its physical properties will depend on its previous pre-treatments. Table 8.1 summarises key operational factors for digesters.

Table 8.1 Common operational variables in mixed digesters.

Variable	Range	Units
Hydraulic retention time (HRT)[1]	10–30	d
Volume	100–10 000	m^3
Organic loading rate (COD)	1–5	$kg\ COD{\cdot}m^{-3}{\cdot}d^{-1}$
Organic loading rate (VS)[2]	0.5–3	$kg\ VS{\cdot}m^{-3}{\cdot}d^{-1}$
Effluent total solids (TS)[3]	1–5	[-] %
$pH_{average}$	6.5–8.5	[-]
Methane yield	200–600	$L\ CH_4{\cdot}kg^{-1}\ VS$
Biogas composition (CH_4)	50–80	[-] %
Temperature (m-mesophilic; t-thermophilic)	33–39[(m)] 52–65[(t)]	°C

[1]: A safety factor is often introduced to ensure appropriate digestion and process stability (thermophilic digestion allows for lower values). [2]: VS reduction of 50–80% is expected with a 70% sludge biodegradability. [3]: Co-digestion with manure may significantly increase these values by up to 15–20%.

8.3.1.1 Mixing designs

Since the content of an anaerobic digester is a high viscosity fluid, mixing equipment needs to be designed specifically for the application. The standard mixing design manual for anaerobic digesters (which does not consider CFD) is the US EPA digester design guide (US EPA, 1987). A more recent review (which does consider the application of CFD) is Lindmark *et al.* (2014).

Mixing systems are primarily designed to achieve homogeneous conditions within the reactor, avoiding excessive solids and activity gradients, and to prevent scum/foam formation at the slurry surface. Common metrics for mixing performance include: (1) a solids gradient of no more than 10% from the average solids concentration, (2) assessment of the 'active' mixed volume, normally determined experimentally by tracer tests (Verhoff & Echelberger, 1974), or by virtual tracer testing using CFD, and (3) sludge turnover time (the average time for an element of fluid/particle to transit the digester) which is typically in the order of 20–30 minutes. These guidelines are difficult to measure experimentally, but they can be conveniently assessed by CFD-based analysis.

The advantages and disadvantages of different systems have been summarised extensively in the US EPA mixer design guide (US EPA, 1987), but generally are a balance between (1) scalability (gas fed systems are more applicable to larger digesters), (2) maintainability (simple gas injectors and mechanically fed systems are more maintainable), (3) corrosion resistance (gas fed systems are subject to gas corrosion, particularly when seals allow air input), and (4) electrical efficiency (simpler systems are generally less efficient). Therefore, there are two main approaches to effectively mix the contents of an anaerobic digester: either mechanical mixing, or biogas injection.

8.3.1.2 Mechanical mixing

Mechanical agitation is typically achieved by impellers which transmit the energy supplied by rotational motion into fluid momentum. Either top entry mixers or side entry agitators are standard. As an alternative, digesters can be mixed using draft tubes combined with axial impellers (Meister *et al.*, 2018) and these draft tubes may recirculate sludge externally or internally (US EPA, 1987).

The design of impellers for digesters is complicated by the large geometric scale (in both vertical and lateral dimensions), and the high viscosity of the fluid. Mechanical mixers are the preferred mixing strategy when dealing with smaller digesters for this reason. Baffles, while effective to minimise solid body rotation, might have negative impacts including a large dead zone with high viscosity fluids.

Due to the wide variety of impellers available, a general way to quantify their performance is by calculating key relevant dimensionless numbers, as described in Table 8.2. The most useful characterisation

Table 8.2 Relevant dimensionless numbers for mechanical mixing.

Expression	Description
$Re = \frac{\rho N D^2}{\eta} = \frac{\rho U_d D}{\eta}$	The *Reynolds number* relates the inertial and viscous forces and it is used to determine the regime of the flow (laminar, transitional, and turbulent). Several *Re* numbers can be found depending on the definition of the characteristic lengths and velocities, and the apparent viscosity.
$N_p = \frac{P}{\rho N^3 D^5} = \frac{2\pi T}{\rho N^2 D^5}$	The *power number* is the most characteristic dimensionless number for mechanical stirrers. An *Np* vs *Re* plot will usually be a simple way to approximately determine the flow regime, especially if non-Newtonian fluids are involved. In fully turbulent conditions, *Np* will approach a constant value.
$N_Q = \frac{2\pi \int_0^{D/2} u_z r \, dr}{N D^3} = \frac{Q}{N D^3}$	The *flow number* quantifies the movement of the fluid directly induced by the impeller. The flow rate within the integral expresses the axial flow discharged over the impeller swept surface. Other flow rates can be defined by changing the integration limits (e.g., entrained flow rate).

N [s^{-1}] and U_d [ms^{-1}] are the impeller rotational speed and the average velocity inside the draft tube, D and r [m] are the impeller diameter and radius, ρ [kgm^{-3}] is the fluid density, η [Pas] is the apparent dynamic viscosity, P [W] is the total power consumed, T [Nm] is the torque, Q [m^{3}s^{-1}] is the impeller discharged flow rate, u_z [ms^{-1}] is the axial velocity component.

for mechanical stirrers is that provided by the curve N_p vs Re. This relationship has been traditionally used for impeller characterisation based on flow regime classification (Chen, 1981; Paul *et al.*, 2003).

Small digesters may be mixed using standard single shaft axial or radial mixers (often multiple impellers on a single shaft), but these are generally only applicable up to approximately 500 m^3. Larger cylindrical digesters need multiple axial impellers (often wall mounted) positioned at different angles to achieve mixing throughout the fluid column (Vrábel *et al.*, 2000). While this can provide high energy efficiencies while mixing, there is substantial additional mechanical complexity when designing large mechanically mixed digesters. For instance, physical constraints on the shaft are imposed since there is a need to maintain water seals with potential shaft flex due to the high power and moments involved. The power consumption for mechanical mixing is in the order of 5–8 W·m^{-3} (0.1–0.2 kWh·m^{-3}·d^{-1}) (Batstone & Jensen, 2011; US EPA, 1987).

8.3.1.3 Gas mixing

Gas-mixed systems primarily transfer momentum to the fluid by gas buoyancy (as well as dissipation of the gas velocity). These can be unconfined (gas is injected from a lance or a diffuser), or confined (gas is discharged while the fluid passes through a fixed tube), as shown in Figure 8.3.

Gas systems are highly scalable, and confined systems can have a high efficiency, at the cost of increased mechanical and structural complexity. Gas recirculation rates are generally 5–10 m^3(gas)·m^{-3}(digester)·d^{-1} (US EPA, 1987), which is approximately 10 times the native gas production, and has a power consumption of 0.1–0.2 kWh·m^{-3}reactor·d^{-1} (for compressed biogas).

Liquid recirculation is rarely used on its own (due to poor momentum dissipation and fairly low efficiencies), but often combined with agitators or gas mixing through external pipework, or gas-liquid venturi systems (Jensen *et al.*, 2018).

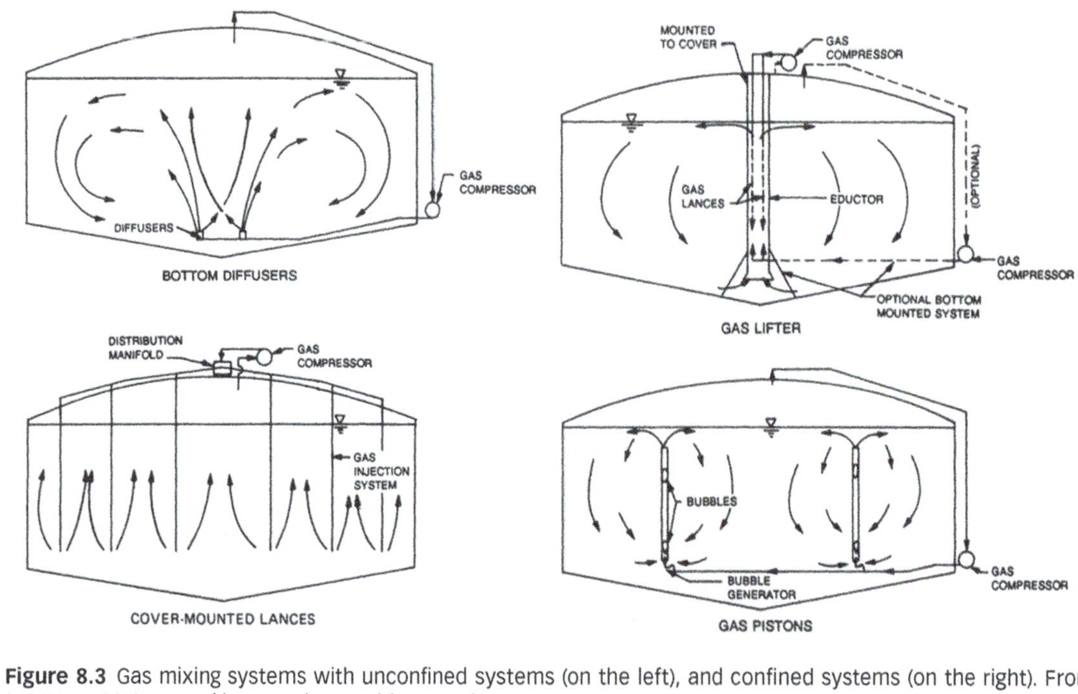

Figure 8.3 Gas mixing systems with unconfined systems (on the left), and confined systems (on the right). From (US EPA, 1987), reused in accordance with US Code 17:105.

8.3.1.4 Gas and liquid collection systems

For biogas collection, the digester roof can either be static or moving (floating roof). A floating roof normally relies on a liquid ring seal, and has the advantage of flexible gas storage capacity and inherent pressure relief (through the liquid seal). However, a disadvantage is the increase in mechanical complexity, and possible diffusive and convective biogas losses through the seal. Liquid collection systems (or overflow systems), are generally an internal or external collection box, which also provides a liquid seal through hydrostatic pressure. Headspace pressure for a floating roof system is regulated by weight on the roof, while fixed roof systems rely on the regulation by mechanical or hydrostatic pressure relief valves. In order to avoid the classification of digesters as pressure vessels, design pressures often target overhead pressures of less than 0.1 bar.

8.3.2 Plug-flow digesters

Plug-flow (PF) digesters are an alternative to the, previously described, 'completely-mixed' reactors. PF digesters can be (1) un-mixed (RCM Digesters Ltd) or mixed (DVO Inc.) (Dvorak, 2014), and (2) rectangular or dual rectangle (in a U-configuration) high-solids digesters, in which material passes from influent to effluent (Batstone & Jensen, 2011; Gould, 2015). These designs are usually internally heated to avoid heat transfer losses. Feed solids concentrations typically range between 7% and 20%, and hence, high loading is achieved. They are often reported to be prone to digester failure and difficult to recover and stabilise (since multiple zones may exist in a reactor). For this reason, and due to the complex hydrodynamics, distributed parameter analysis has focused more on the kinetic analysis of the process rather than CFD (Wu *et al.*, 2009). This is a clear limitation of plug-flow digesters, as active mixing and fluid transport are inherently complex, particularly for a highly viscous fluid.

8.4 CFD CONCEPTS RELEVANT TO AD MODELLING

8.4.1 Shear rate

The concept of shear rate deserves special attention in AD since it determines the viscous response of sludge, and it has been used as a metric to quantify the mixing performance for mechanical stirrers and other gas mixing systems (Bridgeman, 2012; Dapelo *et al.*, 2015; Sindall & Bridgeman, 2013).

In fluid dynamics, the *shear rate* ($\dot{\gamma}$) or *the magnitude of the rate of strain tensor* quantifies the rate at which an element of fluid is being deformed or strained. In three dimensions, its precise definition is given by the second invariant (II) of the shear strain tensor (S) and it contains the normal and shearing strains that an element of fluid can experience:

$$S = \frac{1}{2}\left(\nabla \mathrm{U} + \nabla \mathrm{U}^{\mathrm{T}}\right); S_{ij} = \frac{1}{2}\left(\frac{\partial U_i}{\partial x_j} + \frac{\partial U_j}{\partial x_i}\right)$$
$$\dot{\gamma} = \sqrt{2\mathrm{II}_S} = \sqrt{2S:S} = \sqrt{\frac{1}{2}\sum_{i=1}^{3}\sum_{j=1}^{3}\frac{\partial U_i}{\partial x_j}\left(\frac{\partial U_i}{\partial x_j} + \frac{\partial U_j}{\partial x_i}\right)} \quad (8.1)$$

where S and S_{ij} are the strain rate tensor, U_i is the ith fluid velocity component, x_i is the ith spatial dimension considered, ∇ is the gradient operator, the superscript T is the transpose matrix operation, and the operator ':' is the scalar product of two tensors.

Although it is generally recognised that a wide shear-rate distribution will exist because of the large digester volume (varying in orders of magnitude depending on the distance from the impeller), several correlations for mechanical stirrers are available and they are widely used to calculate the average shear rate in a digester ($\bar{\dot{\gamma}}$), such as the popular Metzner-Otto (MO) correlation:

$$\bar{\dot{\gamma}} = k_s N \quad (8.2)$$

where k_s is a proportionality constant dependent on the type of impeller (higher for radial types of impellers (Paul *et al.*, 2003)).

Note, that although the MO equation is strictly valid only in pure laminar conditions, it can be used for Newtonian and non-Newtonian fluids. Although these types of correlations have been used for a long time, they will fail to provide meaningful values if the flow starts to display turbulence. Since it has been demonstrated that $k_s \propto N_Q$ (Wu *et al.*, 2006), it is recommended that the MO correlation is applied with care, especially when involving non-Newtonian fluids (Fernandes del Pozo *et al.*, 2020). For a more accurate determination of a characteristic shear rate, the user may rely on postprocessing the CFD results instead.

Another approach is to determine the average shear rate (or '*G*' value) in the digester by:

$$\bar{\dot{\gamma}} = \sqrt{\frac{\bar{\epsilon}\rho}{\eta}} = \sqrt{\frac{P}{\eta V}} \tag{8.3}$$

where $\bar{\epsilon}$ [Wkg^{-1}] is the average power dissipated per unit of mass, P[W] is the power supplied from the impeller to the fluid, and V[m^3] is the active volume of the digester.

Although $\bar{\epsilon}$ (or P/V) may provide a crude estimation of the mixing intensity that its being provided to the digester, it does not provide information about how efficiently that energy is being dissipated (applicable to any mixing configuration). Current values may depend on the configuration, but they are found in the range 1–10 W·m^{-3} because of the large volumes involved in full-scale digesters (usually expressed in power per volume).

8.4.2 Non-Newtonian rheology

The problem of describing the rheological characteristics of biological sludges and quantifying their non-Newtonian behaviour is similar to that being performed in more fundamental branches of science such as soft matter physics. The importance of correctly describing the viscous forces while solving the equations of motion applied to AD sludges has been highlighted in recent years by multiple authors, and still constitutes a very active area of research. This dependence can be explicitly shown in the momentum equation in the general conservation equations for mass and momentum (Bird *et al.*, 2002):

$$\begin{gathered}\nabla \cdot \mathbf{U} = 0 \\ \frac{\partial}{\partial t}\rho\mathbf{U} = -\nabla \cdot \rho\mathbf{U}\mathbf{U} - \nabla \mathrm{p} + \nabla \cdot \boldsymbol{\tau} + \rho\mathbf{g}\end{gathered} \tag{8.4}$$

where $\nabla \cdot \boldsymbol{\tau}$ is the divergence of the viscous stress tensor, which includes the effect of viscosity. In order to provide appropriate modelling closure, for isotropic and incompressible fluids this relation can be written as:

$$\boldsymbol{\tau} = 2\eta(\dot{\gamma})S \tag{8.5}$$

Non-Newtonian behaviour refers to the fact that the viscosity of the sludge is no longer constant (Newtonian). It may show dependence with shear rate $\eta(\dot{\gamma})$ (such as shear-thinning), or a time-dependent history (such as thixotropy, where the viscosity reduces depending on the time spent in a shear-intensive environment). This means that the actual viscosity in a digester can follow a highly non-linear relationship with $\dot{\gamma}$.

Rheological characterisation of biological sludges is a specialised field, partly because there is no standard rheological protocol that can be used, and as a result, several inconsistencies have been found among researchers (Ratkovich *et al.*, 2013; Thiène *et al.*, 2019). Consequently, the majority of CFD studies use rheological data from literature that have been obtained for a specific sludge. This is generally regarded as a weak assumption (since they commonly lack information on the effect of solids concentration, e.g.), unless the modeller has no clear alternative.

The following features have often been invoked to describe the rheological properties of AD sludge in a WWTP digester (Baudez *et al.*, 2011).

Shear-thinning (pseudoplastic) power law: Viscosity decreases as the shear rate is increased. This represents the movement fluidity and relaxation of sludge structures. As a rule of thumb, sludge can be considered non-Newtonian above 2% TS. Mathematically, this is translated into the power index having a value $n < 1$, and it is commonly referred to as a power law model.

Yield stress: This refers to the finite shear stress applied to the sludge that needs to be overcome in order to fluidise. The concept of yield stress is most useful and valid when performing the calibration of a rheological model. In a modelling context, a yield stress means that the viscosity limit must be overcome to enable fluid movement. It represents the network structure of sludge at higher concentrations. Use of a yield stress function is referred to as a Bingham plastic model.

Normally a dedicated rheological experiment will be conducted to obtain parameters (yield stress, power law exponent, etc.) for a given sludge concentration. Then, given the response to varying sludge concentration, the parameters will be made concentration dependent. Temperature dependency is usually considered through Arrhenius-type formulae, and generally implies a viscosity adjustment according to the internal temperature of the digester (higher temperatures decreasing the overall viscosity of the sludge).

Additionally, the CFD modelling community often relies on the two following simplifications:

Use of steady-state rheological models: The majority of CFD modelling is run at steady state and does not consider time-dependent stress and elasticity effects (Baudez *et al.*, 2012). Instead, an apparent viscosity is calculated through any of the available generalised rheological models. Although an average digested sludge is a heterogeneous mixture of particles of different sizes and shapes with water, the isotropic assumption has been found to be convenient for modelling purposes. However, this assumption can lead to limitations when the sludge has mm-wide particles and fibrous materials (Liebetrau *et al.*, 2015). Note that most of the rheological characterisations for digested sludge predict shear-thinning behaviour.

Use of clear model fluids: Polymeric fluids such as carboxymethyl cellulose (CMC), XG, and Carbopol have also been used to simulate biological sludges. These fluids are particularly attractive because their concentration can be adjusted to match the rheological behaviour of digested sludge (Dapelo *et al.*, 2015; Eshtiaghi *et al.*, 2012; Fernandes del Pozo *et al.*, 2020; Low *et al.*, 2016). This is particularly useful when validating CFD models.

More detailed information about the rheological behaviour of biological sludges can be found in Baudez *et al.* (2011), Liebetrau *et al.* (2015), Markis *et al.* (2016), Ratkovich *et al.* (2013) and Thiène *et al.* (2019).

8.5 CFD APPROACH

8.5.1 Modelling simplifications

Modelling assumptions are the most important part of the CFD model development, and selecting the appropriate ones is what leads to a truly representative model of the real system.

After selecting the appropriate software, the number of spatial dimensions (2D vs 3D) should be considered. Next, the number of phases (single phase, gas-liquid phase, liquid-liquid, solid-gas-liquid) is determined by the processes within the digester we are interested to model.

Dynamic (time-dependent modelling) vs steady-state modelling is an important distinction. If the user is only interested in the steady-state flow, the time derivative can be set to zero in the conservation equations, allowing for a unique steady-state velocity, pressure (and mass and energy state) solution. However, steady-state solutions can be difficult to converge without a consistent initial estimate, and commonly it is necessary to apply numerical relaxation from a given initial condition by dynamic modelling. An alternative option for hexahedral meshes is a multi-grid approach (Tu *et al.*, 2018), which uses varying mesh sizes to provide consistent initial estimates for large-scale algebraic problems.

Where mass scalars or other processes with large time constants are included, the model will be required to solve stiff partial derivative equations (PDEs), which can make progress towards the solution unstable. One option is to use separate timescales for momentum and other equations (e.g., scalar fluxes). That is, the momentum equations are solved to steady state, and larger time steps are then taken for physical or biological processes with longer time constants. Transient or unsteady simulations do include the time derivative and appropriate time stepping may be needed to capture the main flow features which are governed by the timescales of the process (e.g., the rise of bubbles). Transient simulations can be very time consuming due to restrictions in time stepping, and also computationally intensive due to the amount of data storage CFD simulation requires.

Periodicity is also exploited in CFD by considering the process to be symmetric in one of its dimensions. Thus, in mechanical stirrers, or gas mixing, it can be used to reduce the size of the computational domain.

8.5.2 Eulerian multiphase approach

Multiphase CFD modelling is a challenging task since additional models are needed to describe the interaction between phases. The majority of gas mixing simulations use a Eulerian-Eulerian (E-E) approach due to its maturity in CFD modelling packages, but as noted above, other Euler-Lagrangian approaches have been used.

The E-E approach is particularly recommended for complex multiphase systems, such as bubble columns and fluidised beds, and is usually a starting point for more advanced simulations. In gas mixing in an anaerobic digester, the primary phase is the slurry and the dispersed phase is the biogas. In the E-E approach, the two phases interact with each other, and this is referred to as a two-way coupling process. This interaction can be modelled with the following set of equations (averaged locally in the corresponding volume) which expresses the conservation of mass and momentum for each phase (Andersson *et al.*, 2011):

$$\sum_{k}^{n^{o}phases} \alpha_k = 1$$

$$\frac{\partial(\rho_k\alpha_k)}{\partial t} + \frac{\partial(\rho_k\alpha_k U_{i,k})}{\partial x_i} = -\sum_{l}^{n^{o}phases} \dot{m}_{kl} - \dot{m}_{lk} + S_k \tag{8.6}$$

$$\frac{\partial(\rho_k\alpha_k U_{i,k})}{\partial t} + \frac{\partial(\rho_k\alpha_k U_{i,k}U_{j,k})}{\partial x_j} = -\alpha_k\frac{\partial P}{\partial x_i} + \frac{\partial(\alpha_k\tau_{ij,k})}{\partial x_i} + \alpha_k\rho_k g_i + F_{i,k}$$

where α_k is the volume fraction (in each cell) of the k-phase (liquid and gas), ρ_k is the k-phase bulk fluid density, $\dot{m}_{kl}$ and $\dot{m}_{lk}$ are the mass transport values between the considered phases, S_k is a source term accounting for mass injection (e.g., nozzle), P is the mixture pressure for both phases, and F_k is the total inter-phase force experienced by the k-phase with respect to the other phases.

Appropriate closure of the conservation of momentum equation is obtained by specifying sub-models for the different interphase contributions of F_k. Although many terms can be included, it is recognised that the drag force is the most important (when the slip velocity between phases is approximately constant), and other forces such as lift, virtual force, and turbulence dispersion force might also be important depending on the system and fluid regime (Wei *et al.*, 2019; Wu, 2010b). The most widely used closure for drag forces is that of Schiller and Naumann (1935), which is applicable for generic bubbly flows, assuming spherical bubbles with constant diameter.

One of the most important aspects about multiphase modelling is guaranteeing numerical stability due to the coupling between phases. This may require the user to apply bounding numerical schemes, especially for the gas phase. Drag and boundary equations for Euler-Euler gas-liquid interactions are well established, and generally involve momentum coupling, but are made more complex when inter-phase mass transfer occurs. Additionally, these equations are hard to converge and monitoring integrated quantities is helpful when judging numerical convergence (e.g., constant overall gas fraction; a stable, liquid hydrodynamic pattern has been established, etc.).

8.5.3 Geometry and mesh

The computational domain must include the main physical features that will affect the flow. In mechanical stirrers, it is necessary to describe the shape and spatial dimensions of the digester, and the presence of internal components such as pipes, baffles, and impellers. For gas mixing digesters, the position of the nozzles is also needed. Mesh defeaturing might be considered if full-scale digesters are modelled. Most of the meshing software (especially for automatically generated unstructured meshes) tends to produce highly non-orthogonal cells around complex objects (>70–75°). As a general guideline, the user should refine (and increase the density of) the mesh in local areas where velocity gradients are highest, or when residuals overall exhibit fluctuating behaviour. Insufficient refinement not only leads to poor variable accuracy, but it might also lead to a solution that excludes important flow features (e.g., flow separation).

Boundary conditions are important for any accurate CFD solution and it is important to define the digester liquid surface (either as a fixed frictionless plane, or a gas sink, if biological reactions are included), the flow velocity and contents at inlet/outlet pipes, and accurate boundary conditions for additional momentum sources and phases (e.g., the release of gases at nozzles).

A special boundary condition is related to the rotation of an impeller in mechanically mixed digesters. For steady-state simulations, the multiple reference frame (MRF) approach is a popular and computationally cheap technique (and it is recommended). It assumes the existence of a rotating frame where the momentum equations include an additional rotational source term equal to the impeller speed, and a static frame for the rest of the digester. The sliding mesh (SM) approach is an alternative system which relies on rotating the mesh around the impeller and 'slides' with respect to the rest of the digester mesh. It is important to stress that the quality of the CFD simulation can be affected at the interface of the frames in both approaches and must be solved by appropriate numerical schemes with a fine mesh around the interface.

For multiphase systems, the user may consider simplifying the CFD model to a single-phase simulation. Systems that consist of gas-injection mixing (confined or unconfined) will either require a two-phase approach, or directly model the momentum transfer from the biogas to the fluid. For instance, Karim *et al.* (2007) used the following correlation adapted from Kojima *et al.* (1999) to describe liquid velocity exiting a gas-lift system:

$$U_{L,\mathrm{draft}} = 0.401\left(U_g\left(\frac{D_T}{D_D^i}\right)^2\right)^{0.564}\left(\frac{D_T{}^2 - D_D^{o\,2}}{D_D^{i\,2}}\right)^{-0.182} \tag{8.7}$$

where $U_{L,\mathrm{draft}}$ is the liquid recirculation velocity [m·s^{-1}] through the draft tube, U_g is the superficial gas velocity with respect to tank diameter [m·s^{-1}], D_T is the tank diameter [mm], D^i, is the inner diameter [mm], and D^o is the outer diameter of the draft tube [mm].

Coughtrie *et al.* (2013) found that this boundary condition was a reasonable simplification to be used when compared with more resource intensive Eulerian-Lagrangian two-phase model approaches.

8.5.4 Fluid properties

8.5.4.1 Bulk density

A variable bulk density for the sludge present in the digester might be required when considering high solid concentration sludges. Usually, the correlation of Landry and Roberge (2004) for daily manure has been assumed to be applicable for digested sludges:

$$\rho_{\mathrm{bulk}} = 1000 + 14.6\,\mathrm{TS} - 2.38\,\mathrm{TS}^2 + 0.0367\,\mathrm{TS}^3 \tag{8.8}$$

Note that: (1) density differences might be important depending on the presence of heterogeneous materials, and (2) they are usually important when considering sludges with >5–10% TS.

Experimentally, the determination of the bulk density for any biological fluid can be done simply with a pycnometer.

8.5.4.2 Rheological model selection

The selection of a rheological model will depend on the type of sludge, and the objectives of the CFD model itself. Based on the literature review, in-digester concentrations of up to 2% (feed concentrations up to 3%) can be effectively modelled by a Newtonian assumption, while higher concentrations have been most extensively described by a PL or HB model. Very few models consider variation in rheological model parameters with solids concentrations and temperature, which would be important if varying concentration and temperature fields are to be simulated. Rheological models have been used interchangeably to model a wide variety of sludge substrates in a WWTP such as municipal sewage sludges (primary sludge, secondary activated sludge, digested sludge), animal manure, agricultural crops, silage, and mixtures of all of them. It is also important to mention that pre-treatment of the inlet sludge can significantly alter the rheological characteristics (e.g., gravitational, or mechanical thickening) and that the anaerobic digestion process itself affects the rheology of the sludge (Monteiro, 1997).

If the CFD user has access to a rheometer, it is advisable to perform rheological experiments at different solids concentrations, rather than using data from literature. However, due to the lack of a standard methodology (including variability of sludge samples, different rheometer manufacturers, etc.) and the complexity of the experiments (multiple degrees of freedom, non-standard geometries, etc.), it is only advisable to perform this step if the user has some previous experience of rheological experiments. If this condition is not satisfied, the modelling exercise can suffer from several dangers and pitfalls (over-parameterisation, erroneous calibration, identifiability problems, low-quality data (Barnes *et al.*, 1989; Dochain & Vanrolleghem, 2001; Ratkovich *et al.*, 2013; Thiène *et al.*, 2019)) and the accuracy of the CFD model will suffer due to poor experimental data.

Consequently, alternative available model structures with calibrated parameters have been proposed to model the rheology of biological sludges (Eshtiaghi *et al.*, 2013; Ratkovich *et al.*, 2013; Seyssiecq *et al.*, 2003). Since they do not depend upon the history of deformation, they are termed 'generalised non-Newtonian fluids' and the most common mathematical expressions for each model can be found in Table 8.3.

Although the selection of rheological model is highly dependent on the objectives of the CFD, Figure 8.4 provides the comparison of a few rheological profiles from different sources, each of them reconstructed from a previously calibrated model.

As can be seen in Figure 8.4, there is a great variability in rheological characteristics between different sludges, which is then translated into a wide range of possible viscosities. Additionally, it is important to note that every calibrated rheological model is only strictly valid for a certain shear-rate range, concentration, and for a fixed temperature. This distinction is important when considering rheological differences between mesophilic and thermophilic digesters.

8.5.4.3 Numerical considerations when implementing a rheological model in CFD

When implementing a rheological model in CFD software, numerical considerations are important within the code. In most cases, the CFD model will limit the applicability of the calibrated model between lower and upper limits of shear stress, to deal with the unrealistically high viscosity values that could be produced when the shear rate approaches zero.

For instance, the HB model can be implemented with limiting viscosities in the following way:

$$\begin{cases} \eta = \eta_{\max} & \dot{\gamma} > \dot{\gamma}_{\max} \\ \tau = \tau_0 + K\dot{\gamma}^n & \dot{\gamma}_{\min} \geq \dot{\gamma} \geq \dot{\gamma}_{\max} \\ \eta = \tau_0\dot{\gamma}^{-1} + K\dot{\gamma}^{n-1} & \\ \eta = \eta_{\min} & \dot{\gamma} < \dot{\gamma}_{\min} \end{cases} \tag{8.9}$$

Table 8.3 Description of the main rheological models used for AD CFD.

Model	Mathematical Expression	Comments
Power law (PL, $n<1$)	$\tau = K\dot{\gamma}^{n}$ $\eta = K\dot{\gamma}^{n-1}$	The most commonly used rheological model to describe biological sludges. It is also known as the 'Ostwald-de Waele' model.
Bingham model (BN, $n=1$)	$\begin{cases} \tau = \tau_0 + \mu_{Bn}\dot{\gamma} \\ \eta = \tau_0\dot{\gamma}^{-1} + \mu_{Bn} \end{cases} \tau \geq \tau_0$ $\begin{cases} \dot{\gamma} = 0 \\ \eta = \infty \end{cases} \tau < \tau_0$	The sludge starts to flow with a constant viscosity once the yield stress has been overcome.
Herschel-Bulkley (HB, $n<1$)	$\begin{cases} \tau = \tau_0 + K\dot{\gamma}^{n} \\ \eta = \tau_0\dot{\gamma}^{-1} + K\dot{\gamma}^{n-1} \end{cases} \tau \geq \tau_0$ $\begin{cases} \dot{\gamma} = 0 \\ \eta = \infty \end{cases} \tau < \tau_0$	The combination of a power law and a Bingham model, and the recommended general model to characterise AD sludge in the low to intermediate shear range.
Generalised Herschel-Bulkley (GHB, $n<1$)	$\begin{cases} \tau = \tau_0 + \mu_{\infty}\dot{\gamma} + K\dot{\gamma}^{n} \\ \eta = \tau_0\dot{\gamma}^{-1} + \mu_{\infty} + K\dot{\gamma}^{n-1} \end{cases} \tau \geq \tau_0$ $\begin{cases} \dot{\gamma} = 0 \\ \eta = \infty \end{cases} \tau < \tau_0$	The combination of BN and HB and applicable to a wide range of shear forces.

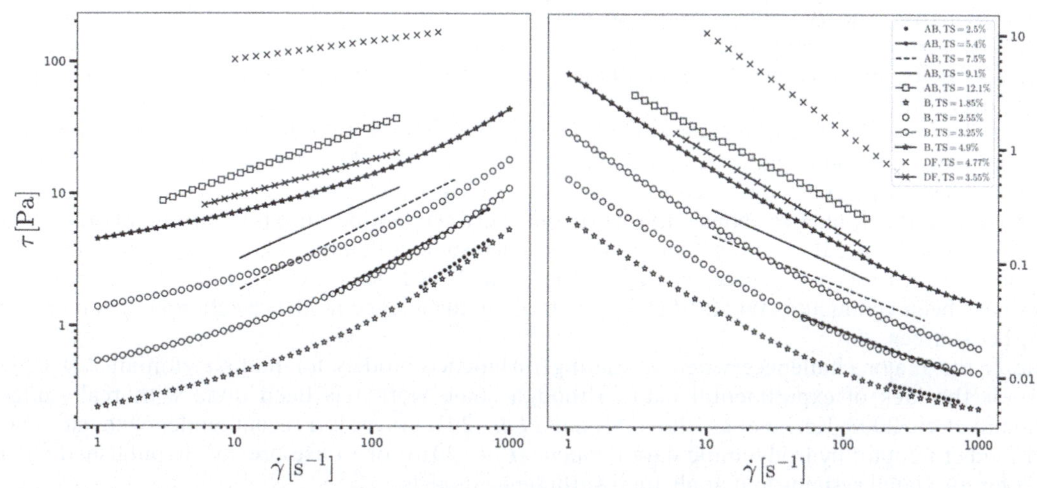

Figure 8.4 Rheological model profiles (τ vs $\dot{\gamma}$ [left] and η vs $\dot{\gamma}$ [right]) of several types of sludges at different concentrations (TS%) encountered in a general anaerobic digestion process. AB: Cattle manure, $T = 35°C$, PL ($K = 0.042$, 0.192, 0.525, 1.052, 5.885; $n = 0.710$, 0.562, 0.533, 0.467, 0.367) (Achkari-begdouri & Goodrich, 1992); B: Digested sludge, $T = 40°C$, GHB ($\tau_0 = 0.088$, 0.201, 0.606, 2.175; $K = 0.148$, 0.331, 0.770, 2.327; $n = 0.308$; $\mu_{\infty} = 0.004$, 0.008, 0.011, 0.022) (Baudez *et al.*, 2013); DF: Activated and digested sludge, $T = 37°C$, PL ($K = 74.92$, 4.92; $n = 0.14$, 0.28) (Dieudé-Fauvel *et al.*, 2014). Axes are shown in logarithmic scale.

The accuracy of the CFD model in the low shear-rate zones will depend on this consideration, especially if a substantial fluid cavern is present in the digester, or if there is a significant low velocity volume inside the digester. It is therefore recommended that the user carefully check the limit details when implementing non-Newtonian parameters in CFD software.

8.5.5 Turbulence modelling

Turbulence modelling is one of the most important key aspects to consider when running a CFD model. Since much of the work in AD CFD has been conducted at steady state, fields are obtained by time-averaging the continuity and mass equations (an approach often cited in literature as Reynolds-averaged Navier-Stokes (RANS)). As a result, to provide closure of the resulting set of equations a model is needed to relate the new terms involving time-averaged velocity fluctuations $(-\rho U_i' U_j')$ with known quantities. Thus, the time-filtered continuity and mass equations (assuming incompressibility) read as:

$$
\begin{aligned}
&\frac{\partial}{\partial x_i}\left(\rho \overline{U_i}\right) = 0 \\
&\frac{\partial}{\partial x_j}\left(\rho \overline{U_i}\,\overline{U_j}\right) = -\frac{\partial}{\partial x_j}\left(\rho \overline{U_i' U_j'}\right) - \frac{\partial \overline{P_i}}{\partial x_i} + \frac{\partial \overline{\tau_{ij}}}{\partial x_j} + \rho g_i
\end{aligned}
\tag{8.10}
$$

where the instantaneous velocity has been decomposed into a mean $(\overline{U_i})$ and a fluctuant (U_i') component prior to the time filtering.

The results of the velocity fluctuations (termed Reynolds stresses) will generate a local momentum exchange due to the convective transport in the flow (performed by a wide variety of different scale eddies), which will cause the surrounding fluid to be mixed faster. Since turbulence is a dissipative process, it will also interact with the dissipation of energy already performed by viscous forces. Since sludge is much more viscous than classical fluids such as water or air, viscous stresses may not be negligible anymore compared to the Reynolds stresses. This constitutes one of the major problems related with turbulent closure in AD CFD, since viscous forces from the sludge will also affect the development of the flow. For instance, the flow will be less chaotic in AD sludge compared to water for a given energy input, so the flow is said to be 'less turbulent'. In a nutshell, the objective of AD CFD turbulence modelling is to use a closed set of equations that will predict the time-averaged velocity and pressure fields of sludges without calculating the complete turbulent flow patterns over time.

Despite its importance, turbulence modelling in non-Newtonian fluids is still an open question and there is no universal turbulence model that can be used. An additional problem is that the flow experiences less turbulence compared to less viscous fluids, which means that the flow regime will likely remain in the 'transition' phase. In this regime, between laminar and fully turbulent flow, even the best (commonly used) turbulence models experience some difficulty in accurately predicting the flow field. This is further complicated by the fact that in an AD CFD context very few detailed validated studies have been conducted (Wu, 2010c), so much of the turbulence model selection criterion comes from other sources.

One of the major challenges when validating turbulence models for non-Newtonian AD CFD studies, is the lack of experimental data. Although some work has been done with real sludge (Hoffmann *et al.*, 2008; Karim *et al.*, 2004; Sindall *et al.*, 2017) there is a recent tendency to use clear model fluids to obtain hydrodynamic data (Dapelo *et al.*, 2015) or to use previously published CFD models for unrelated systems and apply their turbulence models.

The caveats and pitfalls of turbulence models in hydraulics, in the context of activated sludge systems, have been recently reviewed by Karpinska and Bridgeman (2016) and it is noted that turbulence models calibrated from Newtonian fluids have been argued to be generally applicable for non-Newtonian flows based on Reynolds number similarity (Venneker *et al.*, 2010). As a general CFD guideline, the popular Standard $k-\varepsilon$ model should only be used for preliminary results and to generate initial flow fields for more advanced turbulence models. Depending on the availability of computational resources and features to be properly resolved (jet flow, rotating flow, flow separation and reattachment, etc.), other models such as the Realizable $k-\varepsilon$, RNG $k-\varepsilon$, Shear Stress Transport (SST) $k-\omega$, and Reynolds Stress Model (RSM) are recommended (Wang *et al.*, 2019). As a rule of thumb, the user may wish to check the sensitivity of the flow by comparing the results between turbulence models to rule out unphysical simulations.

8.5.6 Monitoring key variables and convergence

The torque or moment of force (T) is a very useful quantity to monitor in mechanical mixing CFD simulations, and it is defined as the sum of the pressure and tangential viscous moments with respect to the axis of rotation of the impeller (integrated over all the cell surfaces of the shaft and the impeller):

$$
\begin{aligned}
T &= \mathbf{r} \times F_{\text{total}} = \int_S \left[\rho\left(P - P_{\text{ref}}\right) \cdot \mathbf{n} + \rho\left(u + u_T\right)\left(\nabla\mathbf{U} + \nabla\mathbf{U}^T\right) \cdot \mathbf{n}\right] \times \mathbf{r}\,\mathrm{d}S \\
P &= 2\pi NT
\end{aligned}
\tag{8.11}
$$

where $\mathbf{r}$ is a vector from the axis to the cell considered, P is the pressure acting normal to the cell, $\mathbf{n}$ is an outward cell vector to the surface of the cell, and u and u_T are the kinematic laminar and turbulent viscosities, respectively. The operands '·' and 'x' are the dot product (projection) and the cross product.

Note that in non-Newtonian fluids, the viscous moment is not negligible compared with the pressure moment and that both terms should be multiplied by the fluid density if they are defined as kinematic variables. Additionally, accurate integration of forces along the shaft and the blades may require a higher number of cells around these objects.

In the CFD field, it is usually required to provide mesh-independent results or at least provide numerical error bounds as a consequence of using a finite number of cells to describe the system (Versteeg & Malalasekera, 2007). The Grid Convergence Index method (GCI) (Celik *et al.*, 2008; Roache, 1998) has become a standard in CFD to report the quality of the mesh with respect to key flow variables. It is based on generalised Richardson extrapolation when, at least, three CFD solutions with different grids are obtained and compared. This method has been used in anaerobic digestion for power consumption, draft flow rate (Craig *et al.*, 2013), and single point velocities (Bridgeman, 2012). This method not only provides a way of assessing the quality of the mesh, but also permits the use a coarser mesh if computational resources are a constraint (Dapelo *et al.*, 2015).

8.5.7 Validation of CFD results

Extensive dedicated validation experiments can be done in the laboratory. The most sophisticated are PIV and CARPT (Karim *et al.*, 2004), which include particle tracing (through imaging).

The standard approach is the use of a clear model fluid which simulates the hydrodynamic properties of sludge, for example XG or CMC, (Dapelo *et al.*, 2015; Eftekharzadeh *et al.*, 2006), or the addition of a dye at a specific time point to trace dispersion. The use of clear fluids with a pH dependent fluorescent dye also constitutes an interesting approach to gather process data (Sajjadi *et al.*, 2016).

While this gives a significant level of independent information at laboratory scale, it is still a challenge to validate full-scale digesters (due to the opaque nature of the contents, and the access to an operating site in steady-state conditions). Generally, the user might want to perform a comparison of CFD results with integrated macroscopic quantities, such as the power consumption.

One common approach at full scale is the use of tracers, which are injected in the inlet and provide a profile of the concentration at the outlet (time-concentration data). Tracers such as common salts can be used in the laboratory (e.g., Hernández-Aguilar *et al.* 2016), but background salt levels are often too high for full-scale systems. Lithium is generally used for full-scale studies (Kamarád *et al.* 2013) but studies are often limited due to cost. The basic method of analysis includes modelling the tracer as a passive dissolved scalar (subject to advection and diffusion). An alternative is the use of Lagrangian tracer particles (Hurtado *et al.*, 2015), either during simulation or in post-processing. The problem with the tracer approach in general is that it is very difficult to interpret model-vs-data differences. Characteristic features (active volume, bypass flows, plug flow vs mixed, oscillatory patterns indicating recycle) can be inferred from the tracer output, but more complex hydraulic behaviour cannot be inferred from a 1D result.

8.6 RESEARCH NEEDS

CFD modelling of elevated-solids digesters uses some of the most sophisticated techniques in wastewater simulation, largely driven by the multiphase nature and non-Newtonian properties of the liquid phase.

As noted in the literature review, the primary needs in this field are more bioreactive modelling, and the ability to model solids concentration gradients in digesters.

In addition, further analysis of the most appropriate rheological model would be beneficial (power laws are still the most widely applied), along with development of a more fundamental rheological model which calculates the viscosity, based on the solids fraction and water content, from first principles.

The consistent acquisition of rheological data; full-scale studies of gas production against mixing efficiency, and a deeper study of turbulent-regime modelling, particularly in areas of high Reynolds number, will all be critical in comparing experimental data with simulation studies in the future.

REFERENCES

Achkari-Begdouri A. and Goodrich P. R. (1992). Rheological properties of Moroccan dairy cattle manure. *Bioresource Technology*, **40**, 149–156, https://doi.org/10.1016/0960-8524(92)90201-8

Andersson B., Andersson R., Hakansson L., Mortensen M., Sudiyo R. and van Wachem B. (2011). Computational Fluid Dynamics for Engineers. Cambridge University Press, Cambridge.

Barnes H. A., Hutton J. F. and Walters K. (1989). An Introduction to Rheology, 1st edn. Elsevier Science Publishers B.V., Amsterdam.

Batstone D. J. and Jensen P. D. (2011). Anaerobic Processes, Treatise on Water Science. Academic Press, Oxford, U.K.

Batstone D. J., Keller J., Angelidaki I., Kalyuzhnyi S. V., Pavlostathis S. G., Rozzi A., Sanders W. T. M., Siegrist H. and Vavilin V. A. (2002). The IWA anaerobic digestion model No 1 (ADM1). *Water Science and Technology*, **45**, 65–73, https://doi.org/10.2166/wst.2002.0292

Batstone D. J., Picioreanu C. and van Loosdrecht M. C. M. (2006). Multidimensional modelling to investigate interspecies hydrogen transfer in anaerobic biofilms. *Water Research*, **40**(16), 3099–3108, https://doi.org/10.1016/j.watres.2006.06.014

Baudez J. C., Markis F., Eshtiaghi N. and Slatter P. (2011). The rheological behaviour of anaerobic digested sludge. *Water Research*, **45**(17), 5675–5680, https://doi.org/10.1016/j.watres.2011.08.035

Baudez J. C., Gupta R. K., Eshtiaghi N. and Slatter P. (2012). The viscoelastic behaviour of raw and anaerobic digested sludge: strong similarities with soft-glassy materials. *Water Research*, **47**(1), 173–180, https://doi.org/10.1016/j.watres.2012.09.048

Baudez J. C., Slatter P. and Eshtiaghi N. (2013). The impact of temperature on the rheological behaviour of anaerobic digested sludge. *Chemical Engineering Journal*, **215–216**, 182–187, https://doi.org/10.1016/j.cej.2012.10.099

Bird R. B., Lightfoot E. N. and Stewart W. E. (2002). Transport Phenomena, 2nd edn. John Wiley and Sons, New York.

Bridgeman J. (2012). Computational fluid dynamics modelling of sewage sludge mixing in an anaerobic digester. *Advances in Engineering Software*, **44**(1), 54–62. CIVIL-COMP, https://doi.org/10.1016/j.advengsoft.2011.05.037

Bürger R. and Diehl S. (2015). On time discretizations for the simulation of the settling-compression process. 9th IWA Symposium on Systems Analysis and Integrated Assessment 1, Goald Coast, Australia, pp. 9–11.

Celik I. B., Ghia U. and Roache P. J. (2008). Procedure for estimation and reporting of uncertainty due to discretization in CFD applications. *Journal of Fluids Engineering-Transactions of the ASME*, **130**, 1–4.

Chen Y. R. (1981). Impeller power consumption in mixing livestock manure slurries. *Transactions of the ASABE*, **24**, 187–192, https://doi.org/10.13031/2013.34222

Conti F., Wiedemann L., Sonnleitner M., Saidi A. and Goldbrunner M. (2019). Monitoring the mixing of an artificial model substrate in a scale-down laboratory digester. *Renewable Energy*, **132**, 351–362, https://doi.org/10.1016/j.renene.2018.08.013

Coughtrie A. R., Borman D. J. and Sleigh P. A. (2013). Effects of turbulence modelling on prediction of flow characteristics in a bench-scale anaerobic gas-lift digester. *Bioresource Technology*, **138**, 297–306, https://doi.org/10.1016/j.biortech.2013.03.162

Craig K. J., Nieuwoudt M. N. and Niemand L. J. (2013). CFD Simulation of anaerobic digester with variable sewage sludge rheology. *Water Research*, **47**(13), 4485–4497, https://doi.org/10.1016/j.watres.2013.05.011

Dapelo D. and Bridgeman J. (2018a). Euler-Lagrange computational fluid dynamics simulation of a full-scale unconfined anaerobic digester for wastewater sludge treatment. *Advances in Engineering Software*, **117**, 153–169, https://doi.org/10.1016/j.advengsoft.2017.08.009

Dapelo D. and Bridgeman J. (2018b). Assessment of mixing quality in full-scale, biogas-mixed anaerobic digestion using CFD. *Bioresource Technology*, **265**, 480–489, https://doi.org/10.1016/j.biortech.2018.06.036

Dapelo D., Alberini F. and Bridgeman J. (2015). Euler-Lagrange CFD modelling of unconfined gas mixing in anaerobic digestion. *Water Research*, **85**, 497–511, https://doi.org/10.1016/j.watres.2015.08.042

Dapelo D., Trunk R., Krause M. J. and Bridgeman J. (2019). Towards lattice-Boltzmann modelling of unconfined gas mixing in anaerobic digestion. *Computers and Fluids*, **180**, 11–21, https://doi.org/10.1016/j.compfluid.2018.12.008

Dieudé-Fauvel E., Héritier P., Chanet M., Girault R., Pastorelli D., Guibelin E. and Baudez J. C. (2014). Modelling the rheological properties of sludge during anaerobic digestion in a batch reactor by using electrical measurements. *Water Research*, **51**, 104–112, https://doi.org/10.1016/j.watres.2013.12.018

Dochain D. and Vanrolleghem P. A. (2001). Dynamic Modelling and Estimation in Wastewater Treatment Processes. IWA Publishing, London, UK.

Dvorak S. W. (2014). Biosolids Digester and Process for Biosolids Production. US Patent US8835155B2.

Eftekharzadeh S., Harrison D., Marx J. J. and Wilson T. E. (2006). Applying rheological techniques to upgrade anaerobic digesters and handle high solids concentrations. *Proceedings of the Water Environment Federation*, **2006**(2), 680–689, https://doi.org/10.2175/193864706783796817

Eshtiaghi N., Yap S. D., Markis F., Baudez J. C. and Slatter P. (2012). Clear model fluids to emulate the rheological properties of thickened digested sludge. *Water Research*, **46**(9), 3014–3022, https://doi.org/10.1016/j.watres.2012.03.003

Eshtiaghi N., Markis F., Yap S. D., Baudez J.-C. and Slatter P. (2013). Rheological characterisation of municipal sludge: a review. *Water Research*, **47**(15), 5493–5510, https://doi.org/10.1016/j.watres.2013.07.001

Fernandes del Pozo D., Liné A., Van Geem K. M., Le Men C. and Nopens I. (2020). Hydrodynamic analysis of an axial impeller in a non-newtonian fluid through particle image velocimetry. *AIChE Journal*, **66**(6), 1–16, https://doi.org/10.1002/aic.16939

Fleming J. G. (2002). Novel simulation of anaerobic digestion using computational fluid dynamics. PhD thesis, North Carolina State University, Raleigh, NC, USA.

Gaden D. L. F. (2013). Modelling anaerobic digesters in three dimensions: integration of biochemistry with computational fluid dynamics. PhD thesis, Manitoba, Canada.

Gould M. C. (2015). Bioenergy and Anaerobic Digestion, Bioenergy. Elsevier Inc, London.

Hernández-Aguilar E., Alvarado-Lassman A., Osorio A. and Méndez-Contreras J. M. (2016). Development of energy efficient mixing strategies in egg-shaped anaerobic reactors through 3D CFD simulation. *Journal of Environmental Science and Health, Part A*, **0**, 1–8.

Hoffmann R. A., Garcia M. L., Veskivar M., Karim K., Al-dahhan M. H. and Angenent L. T. (2008). Effect of shear on performance and microbial ecology of continuously stirred anaerobic digesters treating animal manure. *Biotechnology and Bioengineering*, **100**(1), 38–48, https://doi.org/10.1002/bit.21730

Hurtado F. J., Kaiser A. S. and Zamora B. (2015). Fluid dynamic analysis of a continuous stirred tank reactor for technical optimization of wastewater digestion. *Water Research*, **71**, 282–293, https://doi.org/10.1016/j.watres.2014.11.053

Jayranaiwachira V. and James A. E. (1999). Simulation of non-Newtonian fluids in Anaerobic Digesters. Proceeding of the 13th National Mechanical Engineering Conference, pp. 332–337.

Jensen P. D., Ge H. and Batstone D. J. (2011). Assessing the role of biochemical methane potential tests in determining anaerobic degradability rate and extent. *Water Science and Technology*, **64**, 880–886, https://doi.org/10.2166/wst.2011.662

Jensen M. B., Kofoed M. V. W., Fischer K., Voigt N. V., Agneessens L. M., Batstone D. J. and Ottosen L. D. M. (2018). Venturi-type injection system as a potential H2 mass transfer technology for full-scale in situ biomethanation. *Applied Energy*, **222**, 840–846, https://doi.org/10.1016/j.apenergy.2018.04.034

Jiang J., Wu J., Zhang J., Poncin S. and Li H. Z. (2014). Multiscale hydrodynamic investigation to intensify the biogas production in upflow anaerobic reactors. *Bioresource Technology*, **155**, 1–7, https://doi.org/10.1016/j.biortech.2013.12.079

Kamarád L., Pohn S., Bochmann G. and Harasek M. (2013). Determination of mixing quality in biogas plant digesters using tracer tests and computational fluid dynamics. *Acta Universitatis Agriculturae Et Silviculturae Mendelianae Brunensis*, **LXI**(5), 1269–1278, https://doi.org/10.11118/actaun201361051269

Karim K., Varma R. and Vesvikar M. (2004). Flow pattern visualization of a simulated digester. *Water Research*, **38**, 3659–3670, https://doi.org/10.1016/j.watres.2004.06.009

Karim K., Hoffmann R., Klasson K. T. and Al-dahhan M. H. (2005). Anaerobic digestion of animal waste: effect of mode of mixing. *Water Research*, **39**, 3597–3606, https://doi.org/10.1016/j.watres.2005.06.019

Karim K., Thoma G. and Al-dahhan M. H. (2007). Gas-lift digester configuration effects on mixing effectiveness. *Water Research*, **41**, 3051–3060, https://doi.org/10.1016/j.watres.2007.03.042

Karpinska A. M. and Bridgeman J. (2016). CFD-aided modelling of activated sludge systems – a critical review. *Water Research*, **88**, 861–879, https://doi.org/10.1016/j.watres.2015.11.008

Kojima H., Sawai J., Uchino H. and Ichige T. (1999). Liquid circulation and critical gas velocity in slurry bubble column with short size draft tube. *Chemical Engineering Science*, **54**(21), 5181–5185, https://doi.org/10.1016/S0009-2509(99)00237-7

Landry H. and Roberge M. (2004). Physical and rheological properties of manure products. *Applied Engineering in Agriculture*, **20**(3), 277–288, https://doi.org/10.13031/2013.16061

Langner J. M. (2009). Inverstigation of non-newtonian flow in anaerobic digesters. MSc thesis, University of Manitoba, Manitoba, Canada.

Lebranchu A., Delaunay S., Marchal P., Blanchard F., Pacaud S., Fick M. and Olmos E. (2017). Impact of shear stress and impeller design on the production of biogas in anaerobic digesters. *Bioresource Technology*, **245**, 1139–1147, https://doi.org/10.1016/j.biortech.2017.07.113

Leonzio G. (2018). Study of mixing systems and geometric configurations for anaerobic digesters using CFD analysis. *Renewable Energy*, **123**, 578–589, https://doi.org/10.1016/j.renene.2018.02.071

Liebetrau J., Pfeiffer D. and Thrän D. (2015). Collection of methods for biogas: methods to determine parameters for analysis purposes and parameters that describe processes in the biogas sector. Biomass Energy Use, DBFZ.

Lindmark J., Thorin E., Bel R. and Dahlquist E. (2014). Effects of mixing on the result of anaerobic digestion: review. *Renewable and Sustainable Energy Reviews*, **40**, 1030–1047, https://doi.org/10.1016/j.rser.2014.07.182

Low S. C., Eshtiaghi N., Slatter P., Baudez J. C. and Parthasarathy R. (2016). Mixing characteristics of sludge simulant in a model anaerobic digester. *Bioprocess and Biosystems Engineering*, **39**, 473–483, https://doi.org/10.1007/s00449-015-1530-4

Maier C., Weichselbaum W., Schlerka M. and Harasek M. (2010). Development of agitation systems in biogas plants: investigation of mixing characteristics, improvement of energy efficiency and scale-up using CFD. *Chemical Engineering Transactions*, **21**, 1195–1200.

Mao L., Zhang J., Dai Y. and Tong Y. W. (2019). Effects of mixing time on methane production from anaerobic co-digestion of food waste and chicken manure: experimental studies and CFD analysis. *Bioresource Technology*, **294**, 122177, https://doi.org/10.1016/j.biortech.2019.122177

Markis F., Baudez J. C., Parthasarathy R., Slatter P. and Eshtiaghi N. (2016). Predicting the apparent viscosity and yield stress of mixtures of primary, secondary and anaerobically digested sewage sludge: simulating anaerobic digesters. *Water Research*, **100**, 568–579, https://doi.org/10.1016/j.watres.2016.05.045

Martínez Mendoza A., Montoya Martínez T., Fajardo Montañana V. and López-Jiménez P. A. (2011). Modeling flow inside an anaerobic digester by CFD techniques. *International Journal of Energy and Environment*, **2**(6), 963–974.

Mata-Alvarez J., Dosta J., Macé S. and Astals S. (2011). Codigestion of solid wastes: a review of its uses and perspectives including modeling. *Critical Reviews in Biotechnology*, **31**(2), 99–111, https://doi.org/10.3109/07388551.2010.525496

Meister M., Rezavand M., Ebner C., Pümpel T. and Rauch W. (2018). Mixing non-newtonian flows in anaerobic digesters by impellers and pumped recirculation. *Advances in Engineering Software*, **115**, 194–203, https://doi.org/10.1016/j.advengsoft.2017.09.015

Metcalf & Eddy, Tchobanoglous G., Stensel H. D., Tsuchihashi R. and Burton F. (2014). Wastewater Engineering: Treatment and Resource Recovery, 5th edn. McGraw-Hill, New-York, NY, USA.

Mohammadrezaei R., Zareei S. and Behroozi-Khazaei N. (2018). Optimum mixing rate in biogas reactors: energy balance calculations and computational fluid dynamics simulation. *Energy*, **159**, 54–60, https://doi.org/10.1016/j.energy.2018.06.132

Monteiro P. S. (1997). The influence of the anaerobic digestion process on the sewage sludges rheological behaviour. *Water Science and Technology*, **36**(11), 61–67, https://doi.org/10.2166/wst.1997.0394

Nopens I., Torfs E., Ducoste J., Vanrolleghem P. A. and Gernaey K. V. (2015). Population balance models: a useful complementary modelling framework for future WWTP modelling. *Water Science and Technology*, **71**(2), 159–167, https://doi.org/10.2166/wst.2014.500

Paul E. L., Atiemo-Obeng V. A. and Kresta S. M. (2003). Handbook of Industrial Mixing: Science and Practice. Wiley-Interscience, Hoboken, N.J., New York.

Ratkovich N., Horn W., Helmus F. P., Rosenberger S., Naessens W., Nopens I. and Bentzen T. R. (2013). Activated sludge rheology: a critical review on data collection and modelling. *Water Research*, **47**(2), 463–482, https://doi.org/10.1016/j.watres.2012.11.021

Rezavand M., Winkler D., Sappl J., Seiler L., Meister M. and Rauch W. (2019). A fully lagrangian computational model for the integration of mixing and biochemical reactions in anaerobic digestion. *Computers and Fluids*, **181**, 224–235, https://doi.org/10.1016/j.compfluid.2019.01.024

Roache P. J. (1998). Verification of codes and calculations. *AIAA Journal*, **36**(5), 696–702, https://doi.org/10.2514/2.457

Sadino-Riquelme C., Hayes R. E., Jeison D. and Donoso-Bravo A. (2018). Computational fluid dynamic (CFD) modelling in anaerobic digestion: general application and recent advances. *Critical Reviews in Environmental Science and Technology*, **48**(1), 39–76, https://doi.org/10.1080/10643389.2018.1440853

Sajjadi B., Aziz A., Raman A. and Parthasarathy R. (2016). Fluid dynamic analysis of non-newtonian flow behavior of municipal sludge simulant in anaerobic digesters using submerged, recirculating jets. *Chemical Engineering Journal*, **298**, 259–270, https://doi.org/10.1016/j.cej.2016.03.069

Samstag R. W., Ducoste J. J., Griborio A., Nopens I., Batstone D. J., Wicks J. D., Saunders S., Wicklein E. A., Kenny G. and Laurent J. (2016). CFD For wastewater treatment: an overview. *Water Science and Technology*, **74**(3), 549–563, https://doi.org/10.2166/wst.2016.249

Schiller L. and Naumann L. (1935). A drag coefficient correlation. *Z. Ver. Deutch. Ing.*, **77**(1), 318–320.

Seyssiecq I., Ferrasse J. and Roche N. (2003). State-of-the-art: rheological characterisation of wastewater treatment sludge. **16**, 41–56.

Sindall R. and Bridgeman J. (2013). Velocity gradient as a tool to characterise the link between mixing and biogas production in anaerobic waste digesters. *Water Science and Technology*, **67**(12), 2800–2807, https://doi.org/10.2166/wst.2013.206

Sindall R. C., Dapelo D., Leadbeater T. and Bridgeman J. (2017). Positron emission particle tracking (PEPT): a novel approach to fl ow visualisation in lab-scale anaerobic digesters. *Flow Measurement and Instrumentation*, **54**, 250–264, https://doi.org/10.1016/j.flowmeasinst.2017.02.009

Terashima M., Goel R., Komatsu K., Yasui H., Takahashi H., Li Y. Y. and Noike T. (2009). CFD Simulation of mixing in anaerobic digesters. *Bioresource Technology*, **100**(7), 2228–2233, https://doi.org/10.1016/j.biortech.2008.07.069

Thiène O., Dieudé-Fauvel E. and Baudez J. C. (2019). Impact of mechanical history on sludge rheological properties: role of the organic content. *Water Research*, **157**, 175–180, https://doi.org/10.1016/j.watres.2019.03.040

Tu J., Yeoh G. H. and Liu C. (2018). Computational Fluid Dynamics. Butterworth-Heinemann, Oxford

US EPA. (1987). Anaerobic digester mixing systems. *Journal (Water Pollution Control Federation)*, **59**(3), 162–170.

Venneker B. C. H., Derksen J. J. and Van den Akker H. E. A. (2010). Turbulent flow of shear-thinning liquids in stirred tanks-the effects of reynolds number and flow index. *Chemical Engineering Research and Design*, **88**(7), 827–843, https://doi.org/10.1016/j.cherd.2010.01.002

Verhoff F. H. and Echelberger Jr, W. F. (1974). Mixing in anaerobic digestion. *Biotech*, **XVI**, 757–770.

Versteeg H. K. and Malalasekera W. (2007). An Introduction to Computational Fluid Dynamics: The Finite Volume Method, 2nd edn. Pearson Education Limited, London.

Vesvikar M. S. and Al-dahhan M. (2005). Flow pattern visualization in a mimic anaerobic digester using CFD. *Biotechnology and Bioengineering*, **89**(6), 719–732, https://doi.org/10.1002/bit.20388

Vrábel P., Van Der Lans R. G. J. M., Luyben K. C. A. M., Boon L. and Nienow A. W. (2000). Mixing in large-scale vessels stirred with multiple radial or radial and axial up-pumping impellers: modelling and measurements. *Chemical Engineering Science*, **55**(23), 5881–5896, https://doi.org/10.1016/S0009-2509(00)00175-5

Wang J., Xue Q., Guo T., Mei Z., Long E., Wen Q., Huang W., Luo T. and Huang R. (2018). A review on CFD simulating method for biogas fermentation material fluid. *Renewable and Sustainable Energy Reviews*, **97**, 64–73, https://doi.org/10.1016/j.rser.2018.08.029

Wang P., Reviol T., Ren H. and Böhle M. (2019). Effects of turbulence modeling on the prediction of flow characteristics of mixing non-newtonian fluids in a stirred vessel. *Chemical Engineering Research and Design*, **147**, 259–277, https://doi.org/10.1016/j.cherd.2019.05.001

Wei P., Mudde R. F., Uijttewaal W., Spanjers H., van Lier J. B. and de Kreuk M. (2019). Characterising the two-phase flow and mixing performance in a gas-mixed anaerobic digester: importance for scaled-up applications. *Water Research*, **149**, 86–97, https://doi.org/10.1016/j.watres.2018.10.077

Wiedemann L., Conti F., Saidi A., Sonnleitner M. and Goldbrunner M. (2018). Modeling mixing in anaerobic digesters with computational fluid dynamics validated by experiments. *Chemical Engineering and Technology*, **41**(11), 2101–2110, https://doi.org/10.1002/ceat.201800083

Wu B. (2010a). CFD Simulation of mixing in egg-shaped anaerobic digesters. *Water Research*, **44**(5), 1507–1519, https://doi.org/10.1016/j.watres.2009.10.040

Wu B. (2010b). CFD Simulation of gas and non-newtonian fluid two-phase flow in anaerobic digesters. *Water Research*, **44**(13), 3861–3874, https://doi.org/10.1016/j.watres.2010.04.043

Wu B. (2010c). CFD Investigation of turbulence models for mechanical agitation of non-newtonian fluids in anaerobic digesters. *Water Research*, **45**(5), 2082–2094, https://doi.org/10.1016/j.watres.2010.12.020

Wu B. (2012a). Simulation of mixing for high-solids anaerobic digestion. **109**(8), 2116–2126.

Wu B. (2012b). Large eddy simulation of mechanical mixing in anaerobic digesters. *Biotechnology and Bioengineering*, **109**(3), 804–812, https://doi.org/10.1002/bit.24345

Wu B. (2012c). Integration of mixing, heat transfer, and biochemical reaction kinetics in anaerobic methane fermentation. **109**(11), 2864–2874.

Wu B. (2013). Advances in the use of CFD to characterize, design and optimize bioenergy systems. *Computers and Electronics in Agriculture*, **93**, 195–208, https://doi.org/10.1016/j.compag.2012.05.008

Wu B. (2014). CFD Simulation of gas mixing in anaerobic digesters. *Computers and Electronics in Agriculture*, **109**, 278–286, https://doi.org/10.1016/j.compag.2014.10.007

Wu B. and Chen S. (2008). Simulation of non-newtonian fluid flow in anaerobic digesters. *Biotechnology and Bioengineering*, **99**(3), 700–711, https://doi.org/10.1002/bit.21613

Wu J., Graham L. J. and Mehidi N. N. (2006). Estimation of agitator flow shear rate. *AIChE Journal*, **52**(7).

Wu B., Bibeau E. L. and Gebremedhin K. G. (2009). Three-dimensional numerical simulation model of biogas production for anaerobic digesters. *Canadian Biosystems Engineering*, **51**(1978), 1–7.

Yu L., Ma J. and Chen S. (2011). Numerical simulation of mechanical mixing in high solid anaerobic digester. *Bioresource Technology*, **102**(2), 1012–1018, https://doi.org/10.1016/j.biortech.2010.09.079

Zhu H. and Jing R. (2019). CFD simulation study on mixing experiment of anaerobic digestion tank. E3S Web of Conferences 118(ICAEER 2019), 02047.

doi: 10.2166/9781780409030_0195

Chapter 9

Validation

Min Yang, David Fernandes del Pozo and Ingmar Nopens

BIOMATH, Department of Data Analysis and Mathematical Modelling, Ghent University, Gent, Belgium; Email: Yang.Min@UGent.be; david.fernandesdelpozo@ugent.be/fernandesdelpozodavid@gmail.com; ingmar.nopens@ugent.be

9.1 INTRODUCTION

Computational fluid dynamics (CFD) has become a mature technology in the wastewater treatment sector and continues to gain more ground. CFD modelling is used as a powerful tool to design, investigate, and virtually operate wastewater treatment plants (WWTPs) under various conditions. In view of such decisions, a validation step is included in the Good Modelling Practice guidelines for CFD modelling for wastewater applications (Wicklein *et al.*, 2016) in order to assure credibility in the model and, hence, the decisions based on it. However, there is presently little agreement on standards and procedures for assessing such credibility (note that this is also true for other models used in the water sector). This problem arises largely because of the complexity of fluid mechanics. Besides, it is difficult to have a fixed level of credibility or accuracy that applies to all CFD simulations with different levels of complexity (Wicklein *et al.*, 2016). More specifically, since CFD provides an approximate solution to the governing equations of fluid motion, potential errors and uncertainties may arise in the steps towards final solution such as data collection, definition of geometry, generation of mesh, definition of boundary and initial conditions and physical properties, solution of numerical equations, and even the interpretation of simulation results (Michael & Torsten, 2000). In order to assure the confidence of the CFD model, advanced CFD modelling in wastewater treatment can be validated with techniques such as flow measurements, tracer studies, and component measurements that are used for the investigation of fluid flows and/or species transportation (e.g., dissolved pollutants). It is, however, noteworthy that a validation step is not always needed for a CFD model to be credible or useful.

When a validation is deemed necessary, this calls for the need to establish a clear definition of CFD validation and thus to develop a practical procedure to implement the validation steps accordingly. Many organisations including the US Department of Defense (DoD) (Sanders, 1996) and the Institute of Electrical and Electronics Engineers (Knepell, 1993) have made arduous efforts in this respect. The definition of validation used in this book was taken verbatim from the DoD (AIAA [American Institute of Aeronautics and Astronautics], 1998; Sanders, 1996) as *the process of determining the degree to which a model is an accurate representation of the real world from the perspective of the intended uses of the model.* This definition indicates that validation is a 'deterministic process', which

Editor(s): Julien Laurent, Randal Samstag, Jim Wicks and Ingmar Nopens

means that there is no clearly defined completion point, and assumes that a measure of correctness can be determined. Sufficiency and accuracy are usually determined by practical issues mainly including budgetary constraints and to what depth of complexity the model intends to solve the problem. For water resource recovery facilities (WRRFs), three aspects of complexity in CFD modelling and simulation can be identified:

1. **The complexity of the physics in the treatment processes.** The various treatment processes of WWTPs can exhibit an extraordinarily wide complexity range, including different spatial and temporal fluid dynamics (Ratkovich, 2010), various geometries for different units (Karpinska & Bridgeman, 2016), multiple flow physics (e.g., viscous sub-layer, multiphase) (Yang *et al.*, 2016), and involving multidisciplinarity (e.g., biokinetics, distributed properties) (Littleton *et al.*, 2002; Nopens *et al.*, 2015). Very accurate simulations for complex physics in engineering calculations, while desirable, are not essential since most design changes in practical engineering are incremental. For example, the Standard $k-\varepsilon$ turbulence model simplifies the physics of the near-wall viscous sub-layer, reducing the workload for validation and being computationally much cheaper compared to other approaches such as the large eddy simulation (LES) approach. This model is the most popular and validated turbulence model for various practical engineering systems including WRRFs, where mixing in the middle zone is of more interest rather than the flow physics on the wall (Yang *et al.*, 2011). It is accepted in practical engineering that as long as the trends predicted by the tools are consistent within the design envelope (and an estimate of the error and uncertainty can be made), less-than-perfect accuracy of the simulation is commonly acceptable (AIAA, 1998). Perfect models do, by the way, not exist.
2. **The complexity of physical models.** The increase in model complexity leads to an increase in the number of uncertainty and error sources. Thus, the need for the amount and types of validation information increases accordingly. Like other fields such as chemical engineering (Xu *et al.*, 2013) and haemodynamics (Sun *et al.*, 2020), the wastewater treatment industry is seeing an increasing demand for complex CFD models (Laurent *et al.*, 2014). However, it is noteworthy that the model's predictive power depends on its ability to identify the dominant controlling factors and their influences correctly, not upon its completeness. To reduce the complexity of the model, candid discussions with professionals in sewage treatment and fluid mechanics should be conducted to find critical issues and define proper modelling objectives. This can reduce unnecessary uncertainty and errors, making experimental validation, if even needed, more effective and simulation results more useful and reliable.
3. **The level of prediction difficulty of the various quantities of interest from the simulation.** For instance, validation of the velocity magnitude in membrane bioreactors does not imply that shear force exerted on the membrane has been validated to the same degree of accuracy. The effort required to predict as well as to measure these two quantities can be remarkably different. For example, as reported by Radaei *et al.* (2019), in order to investigate the physics on the membrane filament in a membrane bioreactor configured with the commercial ZW500D MBR (membrane bioreactor) cassette, a total of 32 million grid cells were required. The mesh generation took approximately 16 h to complete. In addition to using video imaging to capture bubble velocity data, they had to conduct the shear measurements using the electrodiffusion method (EDM) (Böhm *et al.*, 2013) to validate the CFD model. On the other hand, to model the mixing in a full-scale MBR wastewater treatment plant, 60 000 grid cells were reported to be sufficient (Wang, 2010). In such studies, all membrane filaments together with the space among them are meshed as a whole block and treated as a porous medium. The shear stress is not of interest and thus the residence time distribution (RTD) experiments turned out to be sufficient to validate these CFD simulations.

9.2 LEVEL CLASSIFICATION OF VALIDATION FOR CFD MODELS

The fundamental strategy of validation is the identification and quantification of error (system error and human error) and uncertainty (potential deficiency due to lack of knowledge) in the computational models (AIAA, 1998). The variables used for validation can either be integral (e.g., resistance, shear) or point values (e.g., mean velocities, turbulence intensity, and power spectra). The validation requirements may be different for each variable (Coleman & Stern, 1997).

So far, the approaches for assessing the credibility of the CFD models in the wastewater sector span a wide variety of complexity levels (Wicklein *et al.*, 2016; Sadino-Riquelme *et al.*, 2018; Satpathy *et al.*, 2020). Following the validation level classification method from Sagaut and Deck (2009), these approaches are labelled and ranked by their complexity to categorise the different levels of validation that exist.

Level 1: Comparison of integral quantities such as mean shear and drag with measurements. The measurements of these integral quantities are very common in aerodynamics (Hedges *et al.*, 2002) and ocean engineering (Patursson *et al.*, 2010). The forces on the surfaces in such cases are essential and easy to measure with a simple instrument such as a loadcell. However, this is rarely used in the wastewater sector except for membrane bioreactors, where the electrodiffusion method is normally used to measure the shear profiles (Chan *et al.*, 2007; Böhm *et al.*, 2013; Radaei *et al.*, 2019).

Level 2: Use of data points of first-order statistics (e.g., mean value of velocity, gas holdup, sludge concentration). Mean velocity is probably the most widely used CFD validation variable, maybe because it is intuitively 'more important' and it is easier to measure and process. Another widely used approach is to conduct a tracer study (see Chapter 10). Some works are found in the literature (Behin & Bahrami, 2012; Brannock *et al.*, 2010; Liu *et al.*, 2010; Qi *et al.*, 2013; Wang, 2010) that use this method, which mostly uses a Lithium-based compound. This method is still very popular to assess mixing quality and is the most extended method in the industry (Jourdan *et al.*, 2019; Karpinska & Bridgeman, 2016).

Level 3: Use second-order statistics for variables such as shear stress, flow heat, and species (normally consists of the mean and the root-mean-square error (RMSE) values). To gather data for time statistics, one needs to enable data sampling for time statistics in the CFD software. Second-order statistics validation examples in literature are nearly non-existent because the transient simulation is time consuming for full-scale WRRFs, and it is the mean flow quantities (not the RMSE values) that are usually of interest from an engineering standpoint. Also, to perform a statistical analysis in good conditions, the signal has to be well sampled over a sufficient duration because the spectral information needs to be averaged on many points to be statistically converged.

Level 4, 5, 6: Use of spectral analysis and high-order statistics (Rowley *et al.*, 2009). When interpreting time-sequence data from a transient solution, it is often useful to look at the data's spectral (frequency) attributes; for instance, computing the spectral distribution of shear stress data recorded at a particular location on a membrane surface, or computing the spectral distribution of turbulent kinetic energy using data for fluctuating velocity that occurred in intermittent systems such as aeration bioreactors or anaerobic digesters. To interpret some of these time-dependent data, the Fast Fourier transform (FFT) algorithm is usually used. In essence, the Fourier transform enables the use of any time-dependent data and to resolve them into an equivalent summation of sine and cosine waves.

Classical spectral analysis (level 4) may be completed with a two-point analysis (level 5) and non-linear coupling analysis (level 6) through a meticulous validation process based on the use of the extensive experimental data sets. These data sets are usually obtained by various sensors (e.g., pressure, heat) and visualisation techniques (e.g., particle image velocimetry, PIV) (Larchevêque *et al.*, 2007).

Level >6: In addition, there is an increasing demand to enhance and broaden the predictive ability of CFD by adding sub-models such as biokinetic modelling (e.g., ASM/ADM1 models (Batstone *et al.*, 2002; Henze *et al.*, 2006)), sludge sedimentation models (e.g., Bürger-Diehl models (Bürger *et al.*,

2011)), population balance modelling (PBM) frameworks (Nopens *et al.*, 2015), discrete element models (Hirche *et al.*, 2020), rheological models (Ratkovich *et al.*, 2013), and mass-transfer models (Prata *et al.*, 2018). The coupling will add a new source of error and uncertainty to the original complexity at all levels of CFD validation. For instance, introducing a bubble size distribution through the PBM model will have an impact on the interphase exchange of momentum (e.g., drag force) between the gas-liquid phases (Lakghomi *et al.*, 2015). On the other hand, although the task of calibration and validation for these sub-models also becomes more difficult when considering non-uniform mixing, the CFD coupled simulations provide more reliable results than those obtained from oversimplified sub-models (Karpinska & Bridgeman, 2016; Lei & Ni, 2014).

In very complex cases, however, it is still possible to use the CFD power to get useful and trustworthy information without much effort. A recommended method is to employ the so-called building-block approach (AIAA, 1998; Marvin, 1995), which divides the complex engineering system of interest into four progressively simpler phases: complete system → subsystem cases → benchmark cases → unit problems (Figure 9.1). Following this order, the error and uncertainty in the measurement decrease, while the completeness of initial and boundary conditions increases. The strategy in this approach is the assessment of how accurately the computational results compare with experimental data at multiple levels of complexity. The level of agreement achieved with the test data is then reviewed in light of the design accuracy requirements to establish the level of code validation for that application

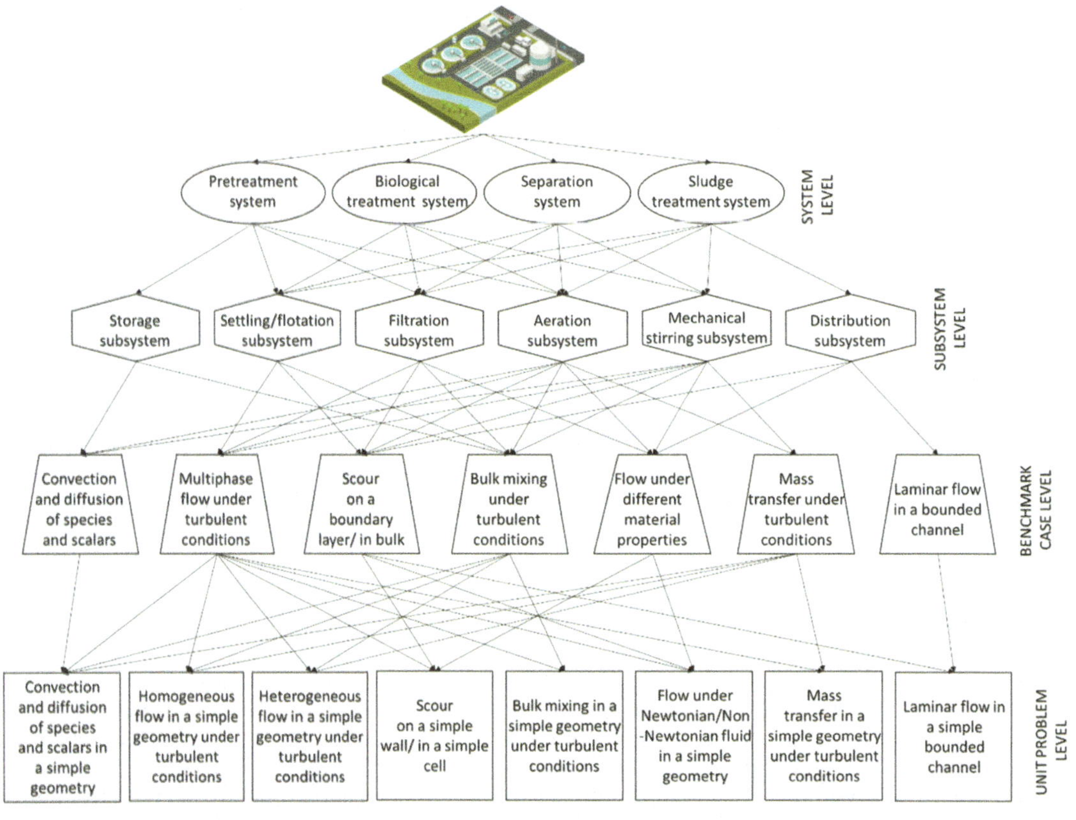

Figure 9.1 Validation options at different system levels of full-scale wastewater treatment plants, inspired by AIAA (1998).

category. Instead of working on WRRFs (complete system) directly, a common practice is to start from a simpler phase. For example, due to the viscoelasticity caused by the high-concentration sludge and the complicated three-phase flow in the anaerobic digester (AD), the validation of CFD in a full-scale AD is comprehensive and challenging (Dapelo *et al.*, 2015; Sindall, 2014). Instead, Fernandes del Pozo (2020) gathered highly detailed velocity measurements for a scaled-down AD, with the transparent Carbopol as the representation of anaerobic sludge. The high-quality data s*et al.*lows the derivation of time-dependent highly detailed velocity measurements as well as derived quantities for a CFD validation study of a high level of complexity. In this benchmark case, the physical models (e.g., rheology) in the code are assessed, and the grid sensitivity studies are conducted to assess the level of refinements necessary to capture key physical effects. Lessons learned from such a benchmark case (lab-scale AD) can then be used to guide the subsystem cases and the complete system activities (full-scale AD).

9.3 MODEL VALIDATION TECHNIQUES

9.3.1 Velocity measurements

Acoustic Doppler current profiler (ADCP), laser Doppler velocimetry (LDV), acoustic Doppler velocimetry (ADV), and particle image velocimetry (PIV) are currently the most widely used and commercially available diagnostic techniques to measure fluid flow velocity in the wastewater sector (Nopens *et al.*, 2020). As point measurement techniques, ADCP, LDV and ADV can offer a good spatial and temporal response in both laboratory and field applications. This makes them ideal for measurements of both time-independent flow statistics, such as moments of velocity (mean, root mean square (RMS), and so on.), and time-dependent flow statistics such as flow spectra and point correlation functions (Rehman *et al.*, 2017; Tropea *et al.*, 2007). Although rakes of these sensors can be built, multi-point measurements are limited mainly due to cost. In comparison to doppler velocimetry, PIV measures many locations simultaneously based on the measurement of the displacement of small tracer particles that are carried by the fluid during a short time interval. The PIV system provides instantaneous 2D or 3D flow fields based on which the mean velocity, turbulent intensity, Reynolds stress, and power spectra can be computed (Foucaut *et al.*, 2004; Ruonan *et al.*, 2016), making it a unique diagnostic tool to measure three-dimensional flow structures and transient phenomenon. However, PIV requires an optically transparent fluid that enables optical access for both the delivery of the light sheet and recording of the images. In addition, since the temporal sampling rate is typically 15 Hz with today's commonly used 30 Hz cross-correlation cameras, the PIV technique is normally used to measure instantaneous velocity fields from which time-independent statistical information can be derived (Jensen, 2004). Cost and processing speed are currently the main limiting factors that influence the temporal sampling rate of PIV. Given the above constraints, the application of PIV in the wastewater sector is limited to laboratory research, and to cases where the sludge can be replaced with a transparent medium (Fernandes del Pozo *et al.*, 2020; Liu *et al.*, 2010; Yang *et al.*, 2017).

9.3.2 Video imaging

Direct observation using a high-speed video camera has been widely used by many researchers to observe flow patterns and behaviour of particle/bubbles in the reactor (Amaral *et al.*, 2018; Lindeboom *et al.*, 2011). Bütehorn *et al.* (2011) used a video camera to investigate the impact of the distance between the aerator header and membranes, the aeration rate, and membrane packing density on fibre movement. The position of the fibre at different conditions was deduced from the recorded images. In a study by Radaei *et al.* (2018), the movement of bubbles throughout a tank, from inception to end at the top of the tank, was tracked by images captured with a high-speed camera and videos recorded with a digital HD video camera. Initial bubble rising velocity for different bubble sizes and frequency was measured by recording bubble flow near the air inlet. Direct observation provides an effective way to study the flow pattern and bubble/particle behaviour. However, the technique also requires optically transparent fluid, which hampers its application in full-scale WRRFs.

9.3.3 Nuclear magnetic resonance imaging (MRI) and computed tomography (CT) scan

The popularity of nuclear MRI and spectroscopy among physical scientists, medical practitioners, engineers and biologists lies in the fact that it is a non-invasive, chemically (and nuclear) specific, motion-sensitive technique that allows the quantitative structural, chemical and dynamical characterisation of many optically opaque systems of scientific interest (Mantle & Sederman, 2003). In the wastewater sector, MRI has mainly focused on water diffusion and distribution in biofilms; transport and fate of heavy metals, trace metals, macromolecular substances and nanoparticles in biofilm research (Herrling *et al.*, 2019; Neu *et al.*, 2010); granular sludge (Ranzinger *et al.*, 2020); and membrane filtration processes in bench-scale MBRs (Bütehorn *et al.*, 2011). X-ray computed tomography (CT) is a non-invasive measurement technique that consists of an X-ray source combined with a detector located at the opposite side of the specimen. A three-dimensional map of the specimen can be created by connecting the 2D cross-sections via image processing. Such a type of 3D model was implemented into a CFD simulation to study the single- and multi-phase flow through submerged hollow-fibre bundles by Bütehorn (2010). De Clercq *et al.* (2005) used a gamma scan to obtain the detailed spatio-temporal solids concentration profiling during batch settling of activated sludge. CT scans are more widely used than MRIs and are typically less expensive. MRIs, however, are thought to be superior regarding the detail of the image. But considering the large size of WRRFs, both nuclear MRI and CT scanning are currently too expensive for CFD validation in WRRFs.

9.3.4 Electrodiffusion method (EDM)

The electrodiffusion method (Sobolík *et al.*, 1998) is presented as a measuring technique suitable for the flow measurement performed in close proximity to a wall, where access would be difficult for other standard anemometric techniques (e.g., PIV, LDV). The shear stress is calculated based on the quasi-steady-state Leveque relationship (Sobolík *et al.*, 1998) as expressed in Equation (9.1):

$$\begin{aligned} I(\tau) &= K\tau^{1/3} \\ K &= 0.862\,nAFC_bD^{2/3}d^{-1/3} \end{aligned} \tag{9.1}$$

where, $I(\tau)$ is the current (A), K is a constant related to the specific testing system, n is the number of exchanged electrons (–), A is the electrode area (m^2), F is the Faraday constant (A s/V), C_b is the concentration of the oxidising ion in the bulk (mole/m^3), D is the diffusion coefficient of the ions (m^2/s), d is the diameter of the probe (m), and τ is the shear stress (Pa). For a full description of the theory and details of the EDM, the reader is referred to Bérubé *et al.* (2006), Chan *et al.* (2007) and Fulton *et al.* (2011).

The EDM is mostly used in the membrane system to retrieve the shear stress data (Böhm *et al.*, 2013; Nikonenko *et al.*, 2010), but it has also been used to investigate the shear rate in the bulk of the fluid. For example, Vlaev *et al.* (2007) used the EDM to investigate the shear rate distribution in the flow fields (stirred vessel of 75 L) of two basic impellers comprising flat and fluid-foil blades. In this study, the shear rate distribution was simulated and validated.

9.3.5 Residence time distribution (tracer study)

RTD is a diagnostic measure of how closely fluid flow approaches either ideal plug flow or perfect mixing. The RTD measurements typically involve the injection of inert tracers to the inlet while high-frequency detectors monitor the tracer concentration along the vessel and/or at the outlet (Bérard *et al.*, 2020). NaCl and KCl solutions are good and affordable tracers, particularly when used with conductivity detectors. A second option is a colourful dye like blue indigo carmine and rhodamine that UV/Vis spectrophotometers and high-speed cameras can detect. Brannock *et al.* (2010) conducted tracer studies on a 2 million litre per day (MLD) hollow fibre (HF) MBR and a 5 MLD flat sheet (FS) MBR by injecting a pulse input of lithium chloride (LiCl) at the MBR inlet (post-screening) with the tracer response being measured in the permeate. This was performed in order to validate CFD models that had been used to simulate the macro-mixing regimes of these two plants. The experimentally determined RTD profiles showed good agreements with the CFD simulated RTDs. It should be borne in mind though that RTD, in contrast to all other methods, is only an indirect measurement of reactor hydrodynamics. Further discussion on the use of physical and virtual tracer tests to study reactor behaviour and/or validate CFD models is presented in Chapter 10.

9.4 CASE STUDIES HIGHLIGHTING THE DIFFERENT LEVELS OF VALIDATION FOR CFD MODELS

9.4.1 Case of a full-scale carrousel ditch

Multiphase modelling of AS (activated sludge) tanks based on gas-liquid 'constant' density simulations has become common practice for the evaluation of both aeration and mixing. It enables faster set-ups for lab-scale validation and it usually involves the use of tap water and air. However, CFD studies on a wide range of types of WRRFs showed that the use of 'constant' density might lead to over-prediction of the mixing degree (Jensen *et al.*, 2006; Samstag *et al.*, 2012). In their case study, Xie *et al.* (2014) introduced a slip velocity model in a density-coupled model, which was used for the prediction of mixing and suspended solids distribution in a full-scale carrousel ditch (26 000 m^3, Pingdingshan, China). The simulation procedure involved a solid-liquid mixture model with the slip velocity sub-model, and the surface rotors and submerged mixers were simulated using moving wall and fan models. A level 2 CFD validation was performed by measuring fluid velocities at 10 test locations (Figure 9.2). Additionally, sludge settling curves for different sludge concentrations were measured to calibrate the density-coupled model. The concept of relative velocity was used to describe the sludge settling. It is defined as the velocity of a sludge phase relative to the velocity of the water phase in the vertical direction. Takács *et al.* (1991) double exponential sedimentation velocity function was applied.

$$v_s = v_2 - v_1 = \min\left(v_0 \times \left(e^{-k_1 X} - e^{-k_2 X}\right),\ V_{s,\max}\right) \tag{9.2}$$

where vs, v_0, v_1, v_2, $V_{s,\max}$, and X are the settling velocity vector, the terminal settling velocity, the velocity of the liquid phase, the velocity of the solid phase, the maximum effective settling velocity, and mixed liquor suspended solids (MLSS) concentration, respectively, and k_1 and k_2 are two fitting constants (see Chapter 6 on sedimentation for further details on sludge settling modelling). To identify the parameters in the equation, the bench-scale experiment data collected from stirred cylinders with different suspended solids concentrations were used. The settling velocities were then determined from the slopes of the straight portions of the lines.

Comparing the predicted average velocities from the mixture model with those from the single-phase model and the measured field data shows that the mixture model matches the experimental data much better than that of the single-phase model (Figure 9.2). Consideration was given to the sludge density, the CFD simulated mixing patterns and solids profiles throughout the ditch. This facilitated identification of the stagnation regions affected by the sludge settling, and resulted in the optimisation of the operation of this unit. This case shows that by coupling the density-coupling sub-model, highly accurate velocity fields can be predicted, and the CFD model can then be used to assess sludge mixing in agitated AS systems.

9.4.2 Case of a commercial ZeeWeed 500D MBR module

This case simulated pulse bubble aeration with high packing density (38 m^2/m^3) of commercial hollow-fibre membrane bundles (0.84 × 0.05 × 2.20 m) (SUEZ Water Technologies & Solutions, Figure 9.3) (Radaei *et al.*, 2018, 2019). Different levels (levels 1, 2, 3) of validation were pursued. First, both

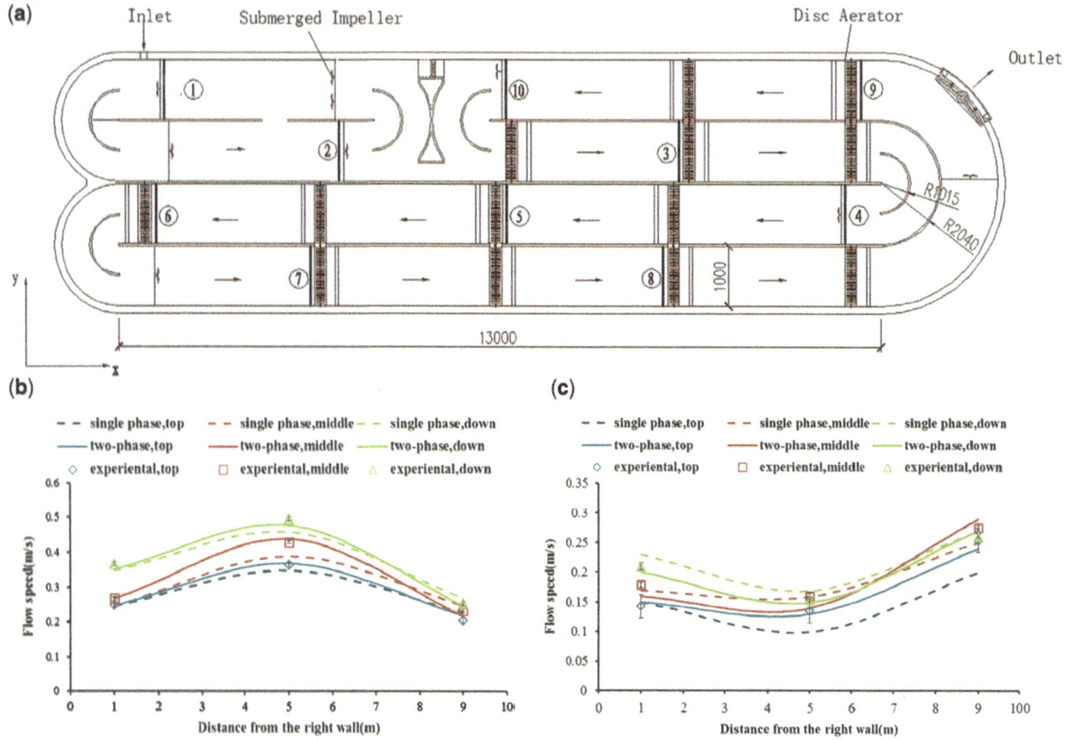

Figure 9.2 The test locations of the full-scale oxidation ditch (a) and the flow speed versus distance from the right wall at test location 3 (b) and test location 6 (c) (Xie *et al.*, 2014).

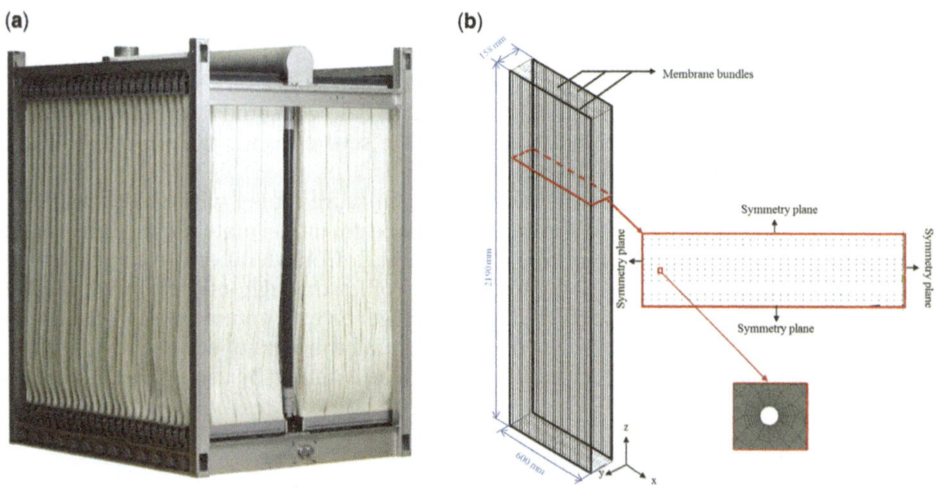

Figure 9.3 ZeeWeed 500D MBR module (a) and computational domain (b) of this module and local mesh near membrane fibre (Radaei *et al.*, 2019).

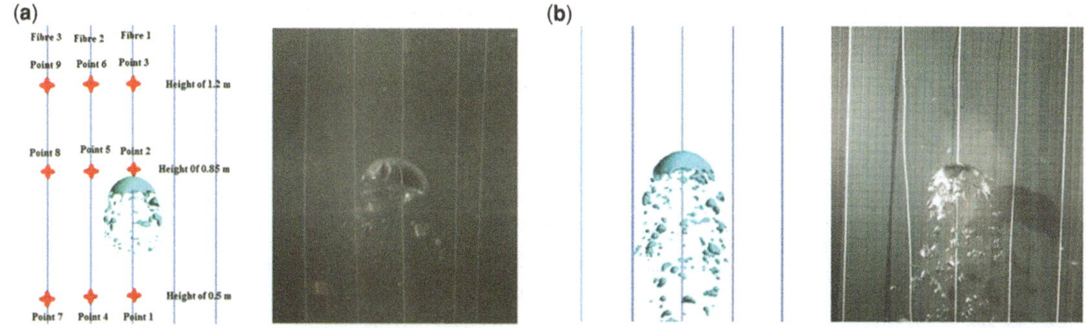

Figure 9.4 Comparison of bubble shape, for an initial air bubble of 115 ml at 0.5 Hz. (a) First bubble released in xanthan gum solution at 0.85 m height. (b) First bubble released in water at 0.85 m height (Radaei *et al.*, 2018).

the CFD simulated bubble shape in xanthan gum solution and water were visually compared to the experimental images (Figure 9.4) (Radaei *et al.*, 2018). The movement of bubbles throughout the tank, from inception to end at the top of the tank, was tracked by images captured with a high-speed camera and videos recorded with a digital HD video camera. The experimental and simulated bubbles formed in xanthan gum solution (Figure 9.4a) and water (Figure 9.4b), have a similar shape and characteristic dimension. Large spherical cap bubbles formed in both liquids with tiny satellite bubbles in the wake of the larger bubble. A larger number of small satellite bubbles were observed in water compared to the xanthan gum solution.

The frame by frame images were then extracted from the recorded videos to calculate bubble velocities by measuring the displacement of the top of the bubble as a function of the frame recorded times (Radaei *et al.*, 2018). The simulated results matched well with the experimental data with a relative difference of 2–5% (Table 9.1). CFD simulations even captured the observed phenomenon that bubble velocity slightly increased from the bottom of the tank to a 0.5 m height, and then the velocity decreased as the bubble moved towards the top of the tank (Table 9.1).

Finally, the CFD model was used to simulate the transient shear stress distribution in a 38 m^2/m^3 module induced by 300 mL pulse bubbles sparged at a constant frequency of 0.5 Hz for a period of 8 s, equivalent to four bubbling periods. The CFD model was validated with the experimental shear profiles (average and RMS shear stress) via the EDM (Radaei *et al.*, 2019) (Figure 9.5a). The simulated data exhibited a similar trend to the experimental results, with an overall difference of 15% (Figure 9.5a). Through the CFD simulation, the authors were able to investigate the impact of packing density and aeration mode on the shear stress. For example, the time-averaged RMS shear stress at point 1 from the simulation time of 6 to 14 s was found to be 1.83 Pa for the 38 m^2/m^3 and was lower than the 2.32 Pa measured for the 1.8 m^2/m^3 membrane module (Figure 9.5b). A 33% increase in the 'Zone of Influence', the flow region where the upward velocity >0.2 m/s, was achieved by moving from a synchronous to an asynchronous form of aeration. This case can be also a benchmark case to guide the subsystem cases (membrane tank) and the complete system activities (full-scale MBR plant).

9.5 DISCUSSION

The validation case studies were chosen to illustrate literature examples where different methodologies/approaches pursuing different levels of validation for CFD models were taken. One take-home message is that what could be thought of as details in the CFD modelling process (i.e., finer mesh or including more physical phenomena such as using better sub-models for density) can often be the reason for insufficient predictive power. Adding certain mechanisms or more geometrical detail can increase

Table 9.1 Experimental and CFD simulated bubble velocity at different heights for a bubble of 115 ml volume at 0.5 Hz frequency in (a) 0.5 g/L xanthan gum solution and (b) water.

	Distance of bubbles from the bottom of tank (m)	CFD predicted (m/s)	Experiment (m/s)	Relative difference (%)
(a)	0	0.5	0.5	–
	0.5	0.58	0.55	−5.45
	0.85	0.52	0.5	−4.0
	1.2	0.49	0.48	−2.08
(b)	0	0.65	0.65	–
	0.5	0.65	0.68	4.4
	0.85	0.61	0.64	4.7
	1.2	0.6	0.62	3.22

predictive power significantly. Another lesson learnt, however, is that data collection is often not straightforward, especially at full-scale. The technique portfolio that can be used at full-scale WRRFs is often limited and, therefore, different from those available at laboratory scale. Obtaining experimental validation data spatially involves the use of several resources. The selection of experimental methods for model validation depends on the required depth of complexity for which the model intends to be validated and is also constrained by budget limitations. The velocity validation performed in the full-scale carrousel ditch case proved the correctness of the bulk-flow field simulations. But the CFD model did not consider the gas phase, which is a factor affecting bulk-flow behaviour and bioreactions (Prata *et al.*, 2018). If the gas phase was included, as it would be in most cases (Lei & Ni, 2014; Wicklein *et al.*, 2016; Yang *et al.*, 2011), the gas holdup, mass transfer rate or even the material properties such as bubble size distribution may also be of interest for validation purposes. A decision on how far the validation needs to go is normally involved in such cases, considering the huge validation workload and model complexity for full-scale studies (Glover *et al.*, 2006). The validation performed in the ZeeWeed benchmark case was very time consuming, and some techniques involve the use of special techniques that can only be used with proper training or guidance. It is often impractical to implement such sophisticated techniques for full-scale WRRFs considering their huge size and the opaque activated sludge that prevents optical observations. However, having a successful validation case available increases the trust in the model.

CFD validation approaches are time and resource demanding, and one can argue whether this effort is always required for every new case. In contrast, it could be more useful to gather cases where validation has been performed successfully and instead list recommendations for model development (choice of appropriate geometry details, mesh, turbulence model, rheological model, etc.) for future users, keeping the community from performing detailed 'redundant' validation experiments for similar systems.

When validation is to be pursued, its required level for each case will depend on the system complexity defined in the building-block approach and the objective. As conceptually illustrated in Figure 9.6, for instance, there are unit problems where simple flow simulations (e.g., laminar flow in a tube) are no longer required to be validated from a practical point of view as the validation can be regarded as common knowledge. In the intermediate complexity region (benchmark cases and subsystem cases), proper guidance is still required to obtain sufficient knowledge to avoid continued massive investments in measurement campaigns. It becomes more cumbersome for highly complex systems with increasing phenomena and scale. A whole train of a WRRF may contain: processes that are significantly impacted by distributed dynamics (e.g., (an)aerobic granular, dissolved air flotation (DAF)) (Nopens *et al.*, 2015), gas-liquid processes (e.g., aeration, ozonation) (Yang *et al.*, 2016), and non-Newtonian processes (e.g., anaerobic digesters, MBRs) (Radaei *et al.*, 2019; Wu, 2010).

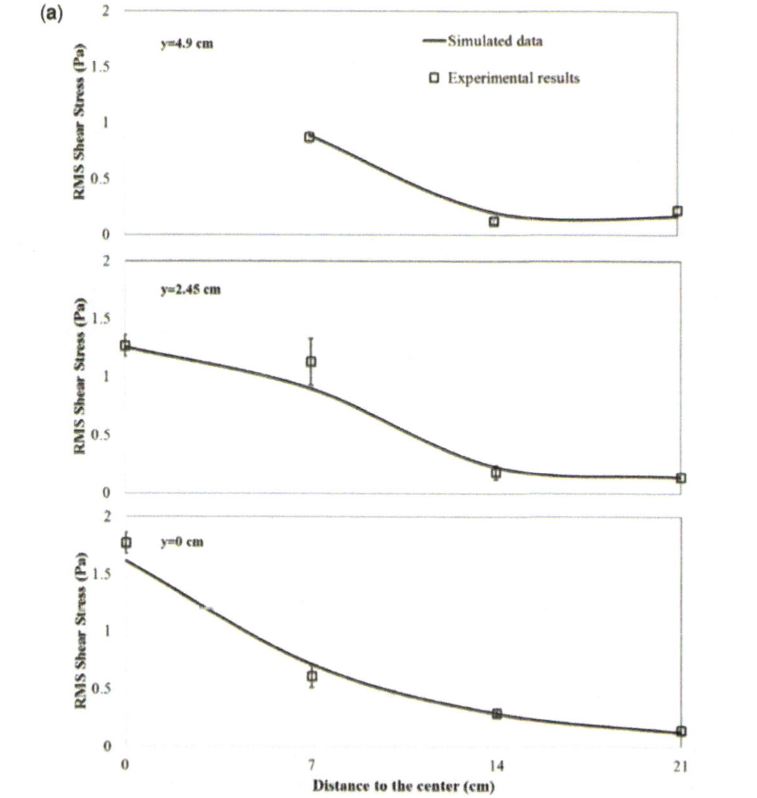

Figure 9.5 Comparison of simulated and experimental RMS shear stress for the membrane module with 38 m²/m³ packing density induced by pulse bubbling at a bubbling frequency of 0.5 Hz (a), and CFD simulated shear stress at point 1 in membrane modules with 38 m²/m (b) (Radaei *et al.*, 2019).

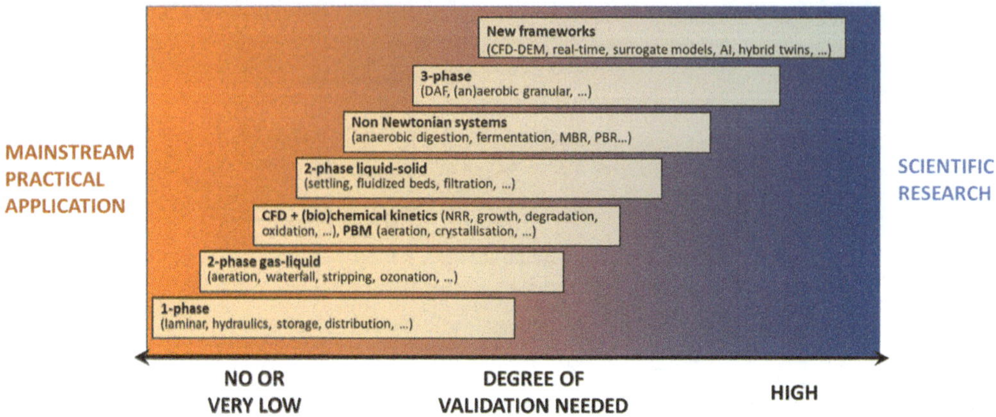

Figure 9.6 Degree of validation needed for different levels of complexity (as presented by Nopens and Audenaert at Watermatex 2019). AI: artificial intelligence; DEM: discrete element method; PBR: photo-bioreactor; NRR: nutrient removal and recovery.

It has to be pointed out, though, that validation is still a requirement to test hypotheses with respect to newly developed CFD models or extensions from a scientific point of view. This is much less the case in a full-scale setting where there are typically practical constraints, and the client decides on how much uncertainty is acceptable. Missing validation should in other words not refrain one from using models, providing the modeller understands the potential limitations.

In the light of this discussion, one should also not lose the link with the modelling objective. Depending on the goal, validation needs might be different. Dissolved air flotation (DAF) is a good example where different levels of validation can be implemented depending on the goals. If one is interested in the design of the nozzle and outlet structures and position, and how they affect flow, a mixture multiphase model with mono bubble size simplification suffices and a tracer test could be used for validation (Shawwa & Smith, 1998). However, if one is interested in the details of microbubble coalescence and capturing of solids by the microbubble, a more sophisticated two-phase model or even three-phase model would be needed, along with a DPM (discrete phase model) sub-model to trace the particles (Satpathy *et al.*, 2020) and a PBM sub-model to account for bubble coalescence (Lakghomi *et al.*, 2015; Yang *et al.*, 2021). Accordingly, comparable experimental tools, for instance, ADV and PIV or even a bubble spectrometer, may be needed for model validation (Fanaie *et al.*, 2019). It is noteworthy that, apart from its value for model improvement, validation drastically increases the practitioner's trust in the model. High trust leads to increased weight of decisions people take based on model outcomes, markedly increasing its value.

9.6 CONCLUSION AND RECOMMENDATIONS

In view of the increasing usage of CFD, the demand for predictive power through validation is expected to arise continuously and will still be important in the near future. It is proposed to the modelling community that this demand should be put in perspective and not force people to eternally validate systems for which knowledge has already been built up and proven. A validation repository consisting of validated CFD cases could be developed that scrutinises the need for explicit validation of future CFD models in practical engineering. The need for validation will in this way dynamically shift towards more complex modelling cases. For simple cases (unit problems, benchmark cases), models could be used without extensive validation. Populating such a repository with cases for intermediate and high complexity CFD problems can aid the practical usage of CFD models and put their power into practice.

REFERENCES

AIAA. (1998). Guide: Guide for the Verification and Validation of Computational Fluid Dynamics Simulations (AIAA G-077-1998 (2002)), American Institute of Aeronautics and Astronautics. American Institute of Aeronautics and Astronautics, Inc., Washington, DC.

Amaral A., Bellandi G., Rehman U., Neves R., Amerlinck Y. and Nopens I. (2018). Towards improved accuracy in modeling aeration efficiency through understanding bubble size distribution dynamics. *Water Research*, **131**, 346–355, https://doi.org/10.1016/j.watres.2017.10.062

Batstone D. J., Keller J., Angelidaki I., Kalyuzhnyi S. V., Pavlostathis S. G., Rozzi A., Sanders W. T. M., Siegrist H. and Vavilin V. A. (2002). The IWA anaerobic digestion model No 1 (ADM1). *Water Science and Technology*, **45**, 65–73, https://doi.org/10.2166/wst.2002.0292

Behin J. and Bahrami S. (2012). Modeling an industrial dissolved air flotation tank used for separating oil from wastewater. *Chemical Engineering and Processing: Process Intensification*, **59**, 1–8, https://doi.org/10.1016/j.cep.2012.05.004

Bérard A., Blais B. and Patience G. S. (2020). Experimental methods in chemical engineering: residence time distribution–RTD. *Canadian Journal of Chemical Engineering*, **98**(4), 848–867, https://doi.org/10.1002/cjce.23711

Bérubé P. R., Afonso G., Taghipour F. and Chan C. C. V. (2006). Quantifying the shear at the surface of submerged hollow fiber membranes. *Journal of Membrane Science*, **279**(1–2), 495–505, https://doi.org/10.1016/j.memsci.2005.12.043

Böhm L., Drews A. and Kraume M. (2013). Bubble induced shear stress in flat sheet membrane systems-serial examination of single bubble experiments with the electrodiffusion method. *Journal of Membrane Science*, **437**, 131–140, https://doi.org/10.1016/j.memsci.2013.02.036

Brannock M., Wang Y. and Leslie G. (2010). Mixing characterisation of full-scale membrane bioreactors: CFD modelling with experimental validation. *Water Research*, **44**(10), 3181–3191, https://doi.org/10.1016/j.watres.2010.02.029

Bürger R., Diehl S. and Nopens I. (2011). A consistent modelling methodology for secondary settling tanks in wastewater treatment. *Water Research*, **45**(6), 2247–2260, https://doi.org/10.1016/j.watres.2011.01.020

Bütehorn S. (2010). Experimental and Numerical Investigation of the Hydrodynamics of Microfiltration Processes Using a Multi-Scale Approach (Experimentelle und numerische Untersuchung der Hydrodynamik von Mikrofiltrationsprozessen unter Verwendung eines Multi-Skalen-Ansatzes. MSc thesis, Technischen Hochschule, Aachen, Germany.

Bütehorn S., Volmering D., Vossenkaul K., Wintgens T., Wessling M. and Melin T. (2011). CFD Simulation of single- and multi-phase flows through submerged membrane units with irregular fiber arrangement. *Journal of Membrane Science*, **384**(1–2), 184–197, https://doi.org/10.1016/j.memsci.2011.09.022

Chan C. C. V., Bérubé P. R. and Hall E. R. (2007). Shear profiles inside gas sparged submerged hollow fiber membrane modules. *Journal of Membrane Science*, **297**(1–2), 104–120, https://doi.org/10.1016/j.memsci.2007.03.032

Coleman H. W. and Stern F. (1997). Uncertainties and CFD code validation. *Journal of Fluids Engineering*, **119**(4), 795–803, https://doi.org/10.1115/1.2819500

Dapelo D., Alberini F. and Bridgeman J. (2015). Euler-Lagrange CFD modelling of unconfined gas mixing in anaerobic digestion. *Water Research*, **85**, 497–511, https://doi.org/10.1016/j.watres.2015.08.042

De Clercq J., Jacobs F., Kinnear D. J., Nopens I., Dierckx R. A., Defrancq J. and Vanrolleghem P. A. (2005). Detailed spatio-temporal solids concentration profiling during batch settling of activated sludge using a radiotracer. *Water Research*, **39**(10), 2125–2135, https://doi.org/10.1016/j.watres.2005.03.023

Fanaie V. R., Khiadani M. and Ayres T. (2019). Effects of internal geometry on hydrodynamics of dissolved air flotation (DAF) tank: an experimental study using particle image velocimetry (PIV). *Colloids and Surfaces A: Physicochemical and Engineering Aspects*, **575**, 382–390, https://doi.org/10.1016/j.colsurfa.2019.05.027

Fernandes del Pozo D., Liné A., Van Geem K. M., Le Men C. and Nopens I. (2020). Hydrodynamic analysis of an axial impeller in a non-Newtonian fluid through particle image velocimetry. *AIChE Journal*, **66**(6), 1–16, e16939, https://doi.org/10.1002/aic.16939

Foucaut J. M., Carlier J. and Stanislas M. (2004). PIV Optimization for the study of turbulent flow using spectral analysis. *Measurement Science and Technology*, **15**(6), 1046–1058, https://doi.org/10.1088/0957-0233/15/6/003

Fulton B. G., Redwood J., Tourais M. and Bérubé P. R. (2011). Distribution of surface shear forces and bubble characteristics in full-scale gas sparged submerged hollow fiber membrane modules. *Desalination*, **281**(1), 128–141, https://doi.org/10.1016/j.desal.2011.07.050

Glover G. C., Printemps C., Essemiani K. and Meinhold J. (2006). Modelling of wastewater treatment plants–how far shall we go with sophisticated modelling tools? *Water Science and Technology: A Journal of the International Association on Water Pollution Research*, **53**(3), 79, https://doi.org/10.2166/wst.2006.078

Hedges L. S., Travin A. K. and Spalart P. R. (2002). Detached-Eddy simulations over a simplified landing gear. *Journal of Fluids Engineering, Transactions of the ASME*, **124**(2), 413–423, https://doi.org/10.1115/1.1471532

Henze M., Gujer W., Mino T. and van Loosdrecht M. (2006). Activated Sludge Models ASM1, ASM2, ASM2d and ASM3. IWA Publishing, London, UK.

Herrling M. P., Lackner S., Nirschl H., Horn H. and Guthausen G. (2019). Recent NMR/MRI Studies of Biofilm Structures and Dynamics, 1st ed. Annual Reports on NMR Spectroscopy. Elsevier Ltd, Amsterdam, Netherlands.

Hirche D., Chew J. W. and Hinrichsen O. (2020). CFD-DEM study of geometry changes in an AnFMBR towards particle momentum. *Chemical Engineering Journal*, **379**(1), 122336, https://doi.org/10.1016/j.cej.2019.122336

Jensen K. D. (2004). Flow measurements. *Journal of the Brazilian Society of Mechanical Sciences and Engineering*, **26**(4), 400–419, https://doi.org/10.1590/S1678-58782004000400006

Jensen M. D., Ingildsen P., Rasmussen M. R. and Laursen J. (2006). Computational fluid dynamics modelling of hydraulics and sedimentation in process reactors during aeration tank settling. *Water Science and Technology*, **53**(12), 257–264, https://doi.org/10.2166/wst.2006.428

Jourdan N., Neveux T., Potier O., Kanniche M., Wicks J., Nopens I., Rehman U. and Le Moullec Y. (2019). Compartmental modelling in chemical engineering: a critical review. *Chemical Engineering Science*, **210**, 115196, https://doi.org/10.1016/j.ces.2019.115196

Karpinska A. M. and Bridgeman J. (2016). CFD-aided modelling of activated sludge systems – a critical review. *Water Research*, **88**, 861–879, https://doi.org/10.1016/j.watres.2015.11.008

Knepell P. L. and Arangno D. C. (1993). Simulation Validation: A Confidence Assessment Methodology. Los Alamitos, CA: Wiley-IEEE Computer Society Press.

Lakghomi B., Lawryshyn Y. and Hofmann R. (2015). A model of particle removal in a dissolved air flotation tank: importance of stratified flow and bubble size. *Water Research*, **68**, 262–272, https://doi.org/10.1016/j.watres.2014.09.053

Larchevêque L., Sagaut P. and Labbé O. (2007). Large-eddy simulation of a subsonic cavity flow including asymmetric three-dimensional effects. *Journal of Fluid Mechanics*, **577**, 105–126, https://doi.org/10.1017/S0022112006004502

Laurent J., Samstag R. W., Ducoste J. M., Griborio A., Nopens I., Batstone D. J., Wicks J. D., Saunders S. and Potier O. (2014). A protocol for the use of computational fluid dynamics as a supportive tool for wastewater treatment plant modelling. *Water Science and Technology*, **70**(10), 1575–1584, https://doi.org/10.2166/wst.2014.425

Lei L. and Ni J. (2014). Three-dimensional three-phase model for simulation of hydrodynamics, oxygen mass transfer, carbon oxidation, nitrification and denitrification in an oxidation ditch. *Water Research*, **53**(4), 200–214, https://doi.org/10.1016/j.watres.2014.01.021

Lindeboom R., Smith G., Jeison D., Temmink H. and van Lier J. B. (2011). Application of high speed imaging as a novel tool to study particle dynamics in tubular membrane systems. *Journal of Membrane Science*, **368**(1–2), 95–99, https://doi.org/10.1016/j.memsci.2010.11.029

Littleton H. X., Daigger G. T., Strom P. F. and Jin R. (2002). Simulation of bio p removal in CFD environment analysis of macro-environment variations in simultaneous biological nutrient removal systems. *Proceedings of the Water Environment Federation*, **2002**(14), 754–770, https://doi.org/10.2175/193864702784248386

Liu N. Y., Zhang Q. D., Chin G.-L. L., Ong E.-H. H., Lou J., Kang C.-W. W., Liu W. J. and Jordan E. (2010). Experimental investigation of hydrodynamic behavior in a real membrane bio-reactor unit. *Journal of Membrane Science*, **353**(1–2), 122–134, https://doi.org/10.1016/j.memsci.2010.02.042

Mantle M. D. and Sederman A. J. (2003). Dynamic MRI in chemical process and reaction engineering. *Progress in Nuclear Magnetic Resonance Spectroscopy*, **43**(1–2), 3–60, https://doi.org/10.1016/S0079-6565(03)00005-0

Marvin J. G. (1995). Perspective on computational fluid dynamics validation. *AIAA Journal*, **33**(10), 1778–1787, https://doi.org/10.2514/3.12727

Michael C. and Torsten W. (2000). Ercoftac best practice guidelines for CFD. European Research Community on Flow, Turbulence and Combustion.

Neu T. R., Manz B., Volke F., Dynes J. J., Hitchcock A. P. and Lawrence J. R. (2010). Advanced imaging techniques for assessment of structure, composition and function in biofilm systems. *FEMS Microbiology Ecology*, **72**(1), 1–21, https://doi.org/10.1111/j.1574-6941.2010.00837.x

Nikonenko V. V., Pismenskaya N. D., Belova E. I., Sistat P., Huguet P., Pourcelly G. and Larchet C. (2010). Intensive current transfer in membrane systems: modelling, mechanisms and application in electrodialysis. *Advances in Colloid and Interface Science*, **160**(1–2), 101–123, https://doi.org/10.1016/j.cis.2010.08.001

Nopens I., Torfs E., Ducoste J., Vanrolleghem P. A. and Gernaey K. V. (2015). Population balance models: a useful complementary modelling framework for future WWTP modelling. *Water Science and Technology*, **71**(2), 159–167, https://doi.org/10.2166/wst.2014.500

Nopens I., Sudrawska D., Audenaert W., Fernandes del Pozo D. and Rehman U. (2020). Water and wastewater CFD and validation: are we losing the balance? *Water Science and Technology*, **81**(8), 1636–1645, https://doi.org/10.2166/wst.2020.181

Patursson Ø., Swift M. R., Tsukrov I., Simonsen K., Baldwin K., Fredriksson D. W. and Celikkol B. (2010). Development of a porous media model with application to flow through and around a net panel. *Ocean Engineering*, **37**(2–3), 314–324, https://doi.org/10.1016/j.oceaneng.2009.10.001

Prata A. A., Santos J. M., Timchenko V., Stuetz R. M., Prata Jr, A. A., Santos J. M., Timchenko V. and Stuetz R. M. (2018). A critical review on liquid-gas mass transfer models for estimating gaseous emissions from passive liquid surfaces in wastewater treatment plants. *Water Research*, **130**, 388–406, https://doi.org/10.1016/j.watres.2017.12.001

Qi W.-K. K., Guo Y.-L. L., Xue M. and Li Y.-Y. Y. (2013). Hydraulic analysis of an upflow sand filter: tracer experiments, mathematical model and CFD computation. *Chemical Engineering Science*, **104**, 460–472, https://doi.org/10.1016/j.ces.2013.09.035

Radaei E., Liu X., Tng K. H., Wang Y., Trujillo F. J. and Leslie G. (2018). Insights on pulsed bubble control of membrane fouling: effect of bubble size and frequency. *Journal of Membrane Science*, **554**(15), 59–70, https://doi.org/10.1016/j.memsci.2018.02.058

Radaei E., Liu X., Tng K. H., Merendino G., Trujillo F. J., Bérubé P. R. and Leslie G. (2019). Numerical and experimental investigation of pulse bubble aeration with high packing density hollow-fibre MBRs. *Water Research*, **160**, 60–69, https://doi.org/10.1016/j.watres.2019.05.056

Ranzinger F., Matern M., Layer M., Guthausen G., Wagner M., Derlon N. and Horn H. (2020). Transport and retention of artificial and real wastewater particles inside a bed of settled aerobic granular sludge assessed applying magnetic resonance imaging. *Water Research X*, **7**, 100050, https://doi.org/10.1016/j.wroa.2020.100050

Ratkovich N. (2010). Understanding hydrodynamics in Membrane Bioreactor systems for wastewater treatment: two-phase empirical and numerical modelling and experimental validation. PhD, Ghent University, Ghent, Belgium.

Ratkovich N., Horn W., Helmus F. P., Rosenberger S., Naessens W., Nopens I. and Bentzen T. R. (2013). Activated sludge rheology: a critical review on data collection and modelling. *Water Research*, **47**(2), 463–482, https://doi.org/10.1016/j.watres.2012.11.021

Rehman U., Audenaert W., Amerlinck Y., Maere T., Arnaldos M. and Nopens I. (2017). How well-mixed is well mixed? Hydrodynamic-biokinetic model integration in an aerated tank of a full-scale water resource recovery facility. *Water Science and Technology*, **76**(8), 1950–1965, https://doi.org/10.2166/wst.2017.330

Rowley C. W., Mezi I., Bagheri S., Schlatter P. and Henningson D. S. (2009). Spectral analysis of nonlinear flows. *Journal of Fluid Mechanics*, **641**, 115–127, https://doi.org/10.1017/S0022112009992059

Ruonan B., Liekai C., Xingkui W. and Danxun L. (2016). Comparison of ADV and PIV measurements in open channel flows. *Procedia Engineering*, **154**, 995–1001, https://doi.org/10.1016/j.proeng.2016.07.588

Sadino-Riquelme C., Hayes R. E., Jeison D. and Donoso-Bravo A. (2018). Computational fluid dynamic (CFD) modelling in anaerobic digestion: general application and recent advances. *Critical Reviews in Environmental Science and Technology*, **48**(1), 39–76, https://doi.org/10.1080/10643389.2018.1440853

Sagaut P. and Deck S. (2009). Large eddy simulation for aerodynamics: status and perspectives. *Philosophical Transactions of the Royal Society A: Mathematical, Physical and Engineering Sciences*, **367**(1899), 2849–2860, https://doi.org/10.1098/rsta.2008.0269

Samstag R. W., Wicklein E. A., Reardon R. D., Leetch R. J., Parks R. M. and Groff C. D. (2012). Field and CFD analysis of jet aeration and mixing. *Proceedings of the Water Environment Federation*, **2012**(12), 4113–4139, https://doi.org/10.2175/193864712811708301

Sanders P. (1996). DoD Modeling and Simulation (M&S) Verification, Validation, and Accreditation (VV&A). Office of the under Secretary of Defense (Acquisition and Technology). Washington DC.

Satpathy K., Rehman U., Cools B., Verdickt L., Peleman G. and Nopens I. (2020). CFD-based process optimization of a dissolved air flotation system for drinking water production. *Water Science and Technology*, **81**(8), 1668–1681, https://doi.org/10.2166/wst.2020.028

Shawwa A. R. and Smith D. W. (1998). Hydrodynamic characterization in dissolved air flotation (DAF) contact zone. *Water Science and Technology*, **38**(6 pt 5), 245–252, https://doi.org/10.2166/wst.1998.0258

Sindall R. C. (2014). Increasing the efficiency of anaerobic waste digesters by optimising flow patterns to enhance biogas production. PhD thesis, University of Birmingham, Birmingham, UK.

Sobolík V., Tihon J., Wein O. and Wichterle K. (1998). Calibration of electrodiffusion friction probes using a voltage-step transient. *Journal of Applied Electrochemistry*, **28**(3), 329–335, https://doi.org/10.1023/A:1003224018571

Sun L., Gao H., Pan S. and Wang J. X. (2020). Surrogate modeling for fluid flows based on physics-constrained deep learning without simulation data. *Computer Methods in Applied Mechanics and Engineering*, **361**, 1–43.

Takács I., Patry G. G. and Nolasco D. (1991). A dynamic model of the clarification-thickening process. *Water Research*, **25**(10), 1263–1271, https://doi.org/10.1016/0043-1354(91)90066-Y

Tropea C., Yarin A. L. and Foss J. F. (2007). Handbook of Experimental Fluid Mechanics. Springer Berlin, Heidelberg.

Vlaev S., Georgiev D., Nikov I. and Elqotbi M. (2007). The CFD approach for shear analysis of mixing reactor: verification and examples of use. *Journal of Engineering Science and Technology*, **2**(2), 177–187.

Wang Y. (2010). Evaluation of membrane bioreactor mixing performance via computational fluid dynamics modelling. PhD thesis, University of New South Wales, NSW, Australia.

Wicklein E., Batstone D. J., Ducoste J., Laurent J., Griborio A., Wicks J., Saunders S., Samstag R., Potier O. and Nopens I. (2016). Good modelling practice in applying computational fluid dynamics for WWTP modelling. *Water Science and Technology*, **73**(5), 969–981, https://doi.org/10.2166/wst.2015.565

Wu B. (2010). CFD Simulation of gas and non-newtonian fluid two-phase flow in anaerobic digesters. *Water Research*, **44**(13), 3861–3874, https://doi.org/10.1016/j.watres.2010.04.043

Xie H., Yang J., Hu Y., Zhang H., Yang Y., Zhang K., Zhu X., Li Y. and Yang C. (2014). Simulation of flow field and sludge settling in a full-scale oxidation ditch by using a two-phase flow CFD model. *Chemical Engineering Science*, **109**, 296–305, https://doi.org/10.1016/j.ces.2014.02.002

Xu Y., Liu M. and Tang C. (2013). Three-dimensional CFD–VOF–DPM simulations of effects of low-holdup particles on single-nozzle bubbling behavior in gas–liquid–solid systems. *Chemical Engineering Journal*, **222**, 292–306, https://doi.org/10.1016/j.cej.2013.02.065

Yang Y., Yang J., Zuo J., Li Y., He S., Yang X. and Zhanga K. (2011). Study on two operating conditions of a full-scale oxidation ditch for optimization of energy consumption and effluent quality by using CFD model. *Water Research*, **45**(11), 3439–3452, https://doi.org/10.1016/j.watres.2011.04.007

Yang M., Yu D., Liu M., Zheng L., Zheng X., Wei Y., Wang F. and Fan Y. (2016). Optimization of MBR hydrodynamics for cake layer fouling control through CFD simulation and RSM design. *Bioresource Technology*, **227**, 102–111, https://doi.org/10.1016/j.biortech.2016.12.027

Yang M., Liu M., Yu D., Zheng J., Wu Z., Zhao S., Chang J. and Wei Y. (2017). Numerical simulation of scaling-up for AEC-MBRs regarding membrane module configurations and cyclic aeration modes. *Bioresource Technology*, **245**(Part A), 933–943, https://doi.org/10.1016/j.biortech.2017.09.027

Yang M., del Pozo D. F., Torfs E., Rehman U., Yu D. and Nopens I. (2021). Numerical simulation on the effects of bubble size and internal structure on flow behavior in a DAF tank: a comparative study of CFD and CFD-PBM approach. *Chemical Engineering Journal Advances*, **7**, 100131, https://doi.org/10.1016/j.ceja.2021.100131

doi: 10.2166/9781780409030_211

Chapter 10

How other simulation methods and digital/experimental tracer experiments can be useful for CFD with reactions

Olivier Potier[1], Julien Laurent[2], Rainier Hreiz[1] and Damien Batstone[3]

[1]*Laboratoire Réactions et Génie des Procédés, Université de Lorraine, CNRS, LRGP, F-54000 Nancy, France. Email: Olivier.Potier@univ-lorraine.fr*
[2]*ICube, ENGEES, CNRS, Université de Strasbourg, Strasbourg, France. Email: julien.laurent@engees.unistra.fr*
[3]*Australian Centre for Water and Environmental Biotechnology, University of Queensland, Brisbane, Australia. Email: d.batstone@uq.edu.au*

10.1 INTRODUCTION

For accurate computational fluid dynamics (CFD) simulations involving (bio-)chemical reactions, it is important to correctly simulate the transport of chemical/biological compounds, which act as reactants and products. The different concentrations are represented in CFD by scalars. This involves the general mass continuity equation as described in Chapter 2 in the intensive form (concentration) for the property being simulated (Equation (10).(1), see also Equation (10).(20)):

$$\frac{\partial}{\partial t}(C) = -U_j \frac{\partial C}{\partial r_j} + \frac{\partial}{\partial r_j}\left(D\frac{\partial C}{\partial r_j}\right) + S \tag{10.1}$$

The source term S represents reaction kinetics (positive for production, negative for consumption). The net output represents the sum of the outputs minus the inputs through the boundaries of the cell with the surrounding cells; input or output depending on the directions of flows through the cell boundary. Advective flow is explicitly contained in the second term. However, diffusion does not include just molecular diffusion, it can also include eddy diffusion (due to turbulence). The latter usually dominates for scalars. To estimate this, a classic approach is deducing a turbulent diffusion coefficient (D_t) from the turbulent kinematic viscosity (ν_t) calculated by the CFD code. It is based on the turbulent Schmidt number (Sc_t):

$$Sc_t = \frac{\nu_t}{D_t} \tag{10.2}$$

The Sc_t value is chosen before the simulation. Then, D_t (and the diffusive mass flux) can be deduced from ν_t everywhere in the simulation domain. However, the precise turbulent Schmidt number value can significantly vary depending on the situation (Gualtieri *et al.*, 2017). Thus, the weakness of the method is the determination of the convenient Sc_t value, which then acts as an adjustable coefficient. Indeed, a little change in Sc_t value induces significant change in the diffusive mass flux and therefore modifies simulation outcomes (Hreiz *et al.*, 2019). This makes it an important contestable parameter (i.e., one that can justifiably change based on observations). Moreover, the phenomenon is very sensitive and CFD-configuration dependent: for a given CFD simulation project, with all the configuration simulation parameters remaining constant, changing only the turbulence model could induce a significant change in the diffusive flow, and then weaken the validity of the simulation (Potier *et al.*, 2019). In this situation, the turbulent Schmidt number must be adjusted so as to better fit the experimental data. The best method is to conduct tracing on the real plant (residence time distribution), and to compare it with virtual tracing made with the CFD code, and to progressively adjust Sc_t until the two curves match perfectly (Le Moullec *et al.*, 2008b). It is also important to identify where changing a parameter such as Sc_t is reasonable vs underlying models or geometry to avoid inappropriate adjustment of parameters. Finally, conducting tracer tests on real plants is not always possible, especially when dealing with the design of new processes. Best knowledge of Sc_t value in wastewater treatment applications is therefore still required.

It is therefore necessary to fully understand the theory of tracer tests and residence time distribution (RTD) in order to:

(1) Consider this issue appropriately in CFD simulations involving reactions (see also Chapters 4, 7 and 8);
(2) Interpret correctly experimental/virtual tracer tests and use these results for model validation (see also Chapters 7 and 9);
(3) Use the results of CFD simulations and experimental/virtual tracer tests to derive simpler models useful for process simulations where computational time can be an issue (see Chapter 1 and Laurent *et al.*, 2014).

As explained in Chapters 1 and 2, there are three main kinds of modelling approaches used in reaction engineering science and then wastewater treatment modelling:

(1) CFD with reaction kinetics, the focus of this report;
(2) *Systemic modelling*, which was developed in the twentieth century (Levenspiel, 1972, 2013; Villermaux, 1993), and is uniformly used and still improved today for reactive design and analysis (Hangos & Cameron, 2001; Jourdan *et al.*, 2019);
(3) *Compartmental modelling* developed during the last 20 years.

The most suitable approach depends on the objectives of the study and the constraints of each method. CFD with reactions (scalar transformation) provides the highest mechanistic detail, while the systemic approach gives an overview of the hydrodynamics approximating it using networks of mixed tanks or 1D models with dispersion. The compartmental modelling lies in the middle; being often based on discretization of the velocity fields obtained by CFD.

From a hydrodynamic detail point of view, CFD provides the most detail. However, it requires high computational resources in comparison with compartmental simulation, and even very high in comparison with systemic simulations. The compartmental approach is suitable for very large and complex processes and can deal with multi-scale problems with quite good precision (Jourdan *et al.*, 2019). The systemic modelling has two other characteristics: firstly, (chemical) reaction engineering science is based on this approach and, therefore, the systemic approach is still useful to globally design new types of processes and study new trends; secondly, residence time distribution (RTD), which is the chemical engineering method to handle tracing data, is also based on the systemic approach.

In this chapter, we successively focus on the three modelling approaches for reactor simulations:

- Systemic modelling (especially ideal reactors and an introduction to more complex reactors): this enables understanding of the mass balance for advective, and diffusive processes;

- CFD with reaction kinetics (a rapid outlook on scalar transport issues),
- Compartmental modelling.

Afterwards, the basics of residence time distribution and tracing technique are presented. Finally, the concepts of scalar transport and virtual tracer experiments in CFD, emphasising the importance of turbulent Schmidt number calibration, are discussed.

10.2 SYSTEMIC MODELLING

At a low level of complexity, the systemic approach uses the concept of ideal reactors. They are simplified concepts and generally do not correspond to real reactors. In simple cases however, they are useful tools to rapidly simulate or design a process. There are two kinds of ideal flow reactor models:

- The continuous stirred-tank reactor (CSTR) which is a fully mixed flow reactor. A CSTR may vary in volume or pressure over time but it is assumed (within the volume) to have a spatially uniform concentration, pressure, and energy density (temperature). The batch reactor is a specific case of the CSTR with no flow across such a reactor, it is not described in this chapter.
- The plug-flow reactor (PFR), which is often a very long flow reactor in which the mass flux gradually progresses; there are strong concentration gradients inside. A PFR is generally fixed in volume but varies in concentration axially, not radially. An 'ideal' PFR has no axial diffusion, and real systems approaching an 'ideal' PFR are rare, particularly in hydraulic systems where there is significant turbulent diffusion.

In real plants, simple reactors can be modelled by a single ideal reactor. For more complex hydrodynamics, the systemic approach enables the use of interconnected multiple units of virtual reactors (in parallel, in series, in parallel and series, recirculation, etc.).

10.2.1 Ideal reactor models

This section aims at introducing the modelling basics of the two kinds of ideal flow reactors: CSTR and PFR. As explained previously, the batch reactor is also an ideal one, but is not a continuous flow reactor: thus, it is not described in this chapter.

10.2.1.1 Continuous stirred-tank reactor (CSTR)

In a CSTR, it is assumed that the reactants mix instantly. The concentration of a given reactant (or product) is uniform across the reactor volume. Moreover, it is equal to the outlet concentration (Figure 10.1).

Let us write the corresponding material balance in a CSTR (Equation 10.3):

$$\text{Input} + \text{Reaction} = \text{Output} + \text{Accumulation} \tag{10.3}$$

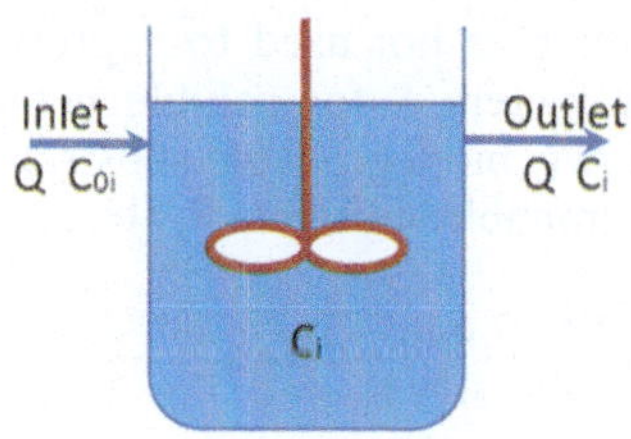

Figure 10.1 Material balance over a CSTR.

For continuous units in wastewater treatment, volume is commonly fixed (though some units have variable volume). Following this hypothesis, input and output instantaneous flowrates are identical ($Q_{in} = Q_{out} = Q$), and volume is fixed ($dV/dt = 0$), the net mass balance is then:

$$\frac{dC_i}{dt} = \frac{Q}{V}(C_{0i} + C_i) + r_i \tag{10.4}$$

where:

- Q is the volumetric flow rate
- V is the volume of the CSTR
- i is the element i, which is a reactant and/or a product (i.e., substrate, biomass) and even a non-reactive compound
- C_{0i} is the concentration of element i at the inlet
- C_i is the concentration of element i in the reactor and at the outlet
- t is time
- r_i is the rate of the reaction (conversion of i)

As the volume is fixed and input and output flowrates are constant, the equations can be rewritten as a function of τ (Equation 10.5), which is the theoretical residence time, also called the hydraulic retention time (HRT) in the water treatment community and space-time in the chemical engineering community.

$$\tau = \frac{V}{Q} \tag{10.5}$$

$$\frac{dC_i}{dt} = \frac{1}{\tau}(C_{0i} - C_i) + r_i \tag{10.6}$$

As an example using Activated Sludge Model #1 (ASM1) notation (Henze *et al.*, 1987), the rate of consumption of soluble substrate S_S for aerobic heterotrophic bacterial growth is:

$$r_{Ss} = -\frac{1}{Y_H}\hat{\mu}_H\left(\frac{S_S}{K_S + S_S}\right)\left(\frac{S_O}{K_{O,H} + S_O}\right)X_{B,H} \tag{10.7}$$

where:

- Y_H is the heterotrophic growth yield
- $\hat{\mu}_H$ is the heterotrophic maximal growth rate
- S_S is readily (soluble) dissolved substrate concentration
- S_O is the dissolved oxygen concentration
- K_S and $K_{O,H}$ are the half saturation constants for substrate and oxygen, respectively
- $X_{B,H}$ is the heterotrophic biomass concentration

Remark: with ASM1, the character C is not used to express the concentration parameter. The concentrations are written with characters S for soluble compounds and readily biodegradable substrate, or X for non-soluble and slowly biodegradable substrate.

The mass balance of readily biodegradable substrate S_S thus gives:

$$\frac{dS_S}{dt} = \frac{1}{\tau}(S_{S0} - S_S) - \frac{1}{Y_H}\hat{\mu}_H\left(\frac{S_S}{K_S + S_S}\right)\left(\frac{S_O}{K_{O,H} + S_O}\right)X_{B,H} \tag{10.8}$$

This kind of equation can be easily solved with numerical integration methods such as Runge-Kutta.

10.2.1.2 Plug-flow reactor (PFR)

The plug-flow reactor (PFR) is also an ideal reactor in chemical engineering terminology. It corresponds to a theoretical concept and does not exist in practice. Plug-flow reactors are generally defined as those where advection along the length of the reactor dominates, and dispersion can be excluded as a transport mechanism. In chemical engineering systems, particularly long tubular reactors with narrow diameters can be reasonably described by this model. In wastewater systems however, there is significant axial dispersion that makes an 'ideal' plug-flow reactor not an appropriate model.

Remark: the term plug-flow reactor is very often used with misuse of language in the wastewater community. In full-scale WWTPs, even for a very long reactor, the hydrodynamic behaviour is very far from being an ideal PFR. Using the term PFR could lead to misunderstanding as it should be called PFR with dispersion (PFR-AD, see Section 10.2.2).

An ideal PFR with constant reactor cross-section is defined by the general single dimensional form of the general conservation equation with no diffusion:

$$\frac{\partial C_i}{\partial t} = -U_x \frac{\partial C_i}{\partial x} + r_i \tag{10.9}$$

where:

- $\partial C_i/\partial t$ is the accumulation term
- x is the length (on the reactor axis 0-x)
- U_x is the average velocity of the fluid along axis 0-x from the inlet to the outlet of the PFR

At steady-state, since $U_x = L/\tau$ (L = length of reactor, τ = HRT), this results in:

$$\frac{dC_i}{dx} = \frac{\tau r_i}{L} \tag{10.10}$$

10.2.2 Non-ideal reactor models

CSTR and ideal PFR are the two models describing ideal flow reactors. They describe the extremes between complete axial dispersion, and no axial dispersion. In chemical engineering, when the flow is non-ideal but not too complex, and somewhere between the CSTR and the PFR, the hydrodynamics can be represented using two main models:

- Plug-flow reactor with axial dispersion: if the dispersion tends to zero, the plug-flow reactor with axial dispersion tends to the ideal plug-flow reactor studied in the previous section. The plug-flow reactor with axial dispersion is characterized by the Peclet number (*Pe*), where a high Peclet number indicates that advection dominates over diffusion.
- CSTRs in series (also called tanks in series (TIS), or cascade of CSTRs): in this case, the number of CSTRs (often represented by the character *J*) defines the hydrodynamics profile. *J* is then the most important factor to characterize a cascade of CSTRs. When the number *J* tends to infinity, the hydrodynamic behaviour tends towards that of an ideal PFR while a *J* of unity corresponds to a 'single' CSTR.

It is worth noting that the CSTRs in series or TIS model can perform as a PFR with axial dispersion, being an empirical representation of the same system. These models can be derived from residence time distribution (RTD) analysis.

More complex hydrodynamics of real reactors that cannot be successfully modelled by CSTRs in series or PFR with dispersion can be described using more complex virtual networks of CSTRs, PFRs and PFRs with dispersion, organized in parallel, series, recirculation, and so on. (see Section 10.2.4 or Laurent *et al.* 2015). This is also a systemic approach, but less used in the wastewater community.

10.2.2.1 Definition of the dispersion coefficient

The axial dispersion coefficient, D_i, quantifies the global diffusion over the reactor, which is the result of several physical phenomena occurring at different scales:

(1) Molecular diffusion: this mechanism of mass transfer occurs in the presence of a concentration gradient and is generally defined by the first Fick's law which can be written in its simplified form as (here in 1D along axis z):

$$\phi_i = -D_{m,i}\frac{\partial C_i}{\partial z} \tag{10.11}$$

where ϕ_i is the diffusive mass (or molar) flux of species I per unit surface (kg m^{-2} s^{-1} or mol m^{-2} s^{-1}), and $D_{m,i}$ is its molecular diffusion coefficient in the given medium (m^2 s^{-1}).

(2) Turbulent diffusion or eddy diffusion: this is related to turbulent motions in the case of a turbulent flow. It occurs at different scales depending on the size of the eddies. The turbulent diffusion coefficient is written $D_{t,i}$ (m^2 s^{-1}). In the case of turbulent flows, it is generally much higher than molecular diffusion.

(3) Kinematic diffusion (due to the non-uniformity of the flow): this is due to fluid layers flowing at different velocities.

While molecular diffusion is linked to the respective properties of the chemical species under consideration and the carrier fluid, turbulent diffusion and kinematic diffusion primarily depend on the flow characteristics within the reactor. Moreover, depending on the flow Reynolds number, $D_{m,i}$ is generally much smaller (even negligible) than the axial dispersion coefficient D_i (which is also expressed in m^2 s^{-1}).

10.2.2.2 Plug-flow reactor with axial dispersion

The plug-flow reactor with axial dispersion (PFR-AD) is a more sophisticated model than the simple PFR. It combines an advective flow (like the PFR model) with a diffusive effect that is characterized by the axial dispersion coefficient D_i. The PFR-AD enables a wide range of hydrodynamic behaviours to be described, mainly for reactors having a high length-to-width ratio.

The PFR-AD is described similarly to the ideal PFR (Equation 10.9) but considering the diffusion term:

$$\frac{\partial C_i}{\partial t} = -U_x\frac{\partial C_i}{\partial x} + D_i\frac{\partial^2 C_i}{\partial x^2} + r_i \tag{10.12}$$

In chemical engineering literature (e.g., Levenspiel, 1998), the Peclet number, Pe, is often used to characterize the PFR with axial dispersion. It is a dimensionless number that represents the ratio between the mass transport by advection and dispersion (a high Peclet number indicates the system is convection/advection dominated). It is also sometimes called Bodenstein number.

$$Pe = \frac{uL}{D_i} \tag{10.13}$$

The dispersion number, d, is also used. It corresponds to the inverse of Pe, and is defined by:

$$d = \frac{D_i}{uL} \tag{10.14}$$

Since Equation (10.12) is a second order PDE (partial differential equation) in space, and first order in time, the initial condition throughout the reactor (time boundary) and two spatial boundary

Table 10.1 Characterization of dispersion according to *d* and *Pe* values.

Pe	*d*	Characteristics
$Pe = \infty$	$d = 0$	Ideal plug flow: no dispersionNot for real reactors
$Pe \rightarrow \infty$ (in practice Pe > 100)	$d \rightarrow 0$ (in practice d < 0.01)	Tends to ideal plug flow
$Pe > 100$	$d < 0.01$	Low dispersion
$4 < Pe < 100$	$0.01 < d < 0.25$	Moderate dispersion
$Pe < 4$	$d > 0.25$	High dispersion
$Pe \rightarrow 0$	$d \rightarrow \infty$	Tends to perfect mixing (CSTR) if closed-closed boundary conditions

conditions are required. The latter depend on boundary conditions at the inlet and outlet. Is it open or closed to the diffusion? If it is an open-open PFR-AD, the species will cross the outlet and inlet limits by both diffusion and advection. In contrast, if it is a closed-closed PFR-AD, the species will only cross the outlet and inlet limits by advection. The consequences depend on the *Pe*. If the *Pe* is higher than 100, there is no difference between the two kinds of reactors. But below 100, the difference is significant. For a *Pe* close to zero, a closed-closed PFR-AD is similar to a single CSTR, but the difference for open-open PFR-AD is very important, since the species would instantaneously diffuse outside the reactor.

The closed-closed PFR-Ads are suitable for simulation of long and complete reactors, considering advective flow across the inlet. The open-open PFR-ADs are suitable for simulation of a part of long reactors far from the inlet and outlet walls (then not considering them). It can also be used for river simulation.

In practice, we can define the following ranges of *d* and *Pe* to characterize the axial dispersion within reactors (Table 10.1):

10.2.2.3 Tanks-in-series

The CSTRs in series, also called the Tanks-in-Series (TIS), approach also represents the behaviour between a single CSTR and an ideal PFR. It enables approximation and modelling of the global dispersion in some reactors by representing them as a series of CSTRs. It should be underlined that the TIS model does not enable representation of all kinds of reactors. If significant lateral flows occur in reactors for which the width is not too different from the length, the TIS model is no longer suitable.

A tank-in-series approach was originally used physically in chemical engineering (to replace a single CSTR) where reactor kinetics control, and use of a plug-flow system was not appropriate or economical. This has been directly transposed to process modelling and optimization of this real system. However, progressively, it has also gained use to characterize a single and long reactor in which global behaviour approaches that of several CSTRs in series. The main parameter characterizing the TIS model is *J*, the number of CSTRs of the series (Figure 10.2).

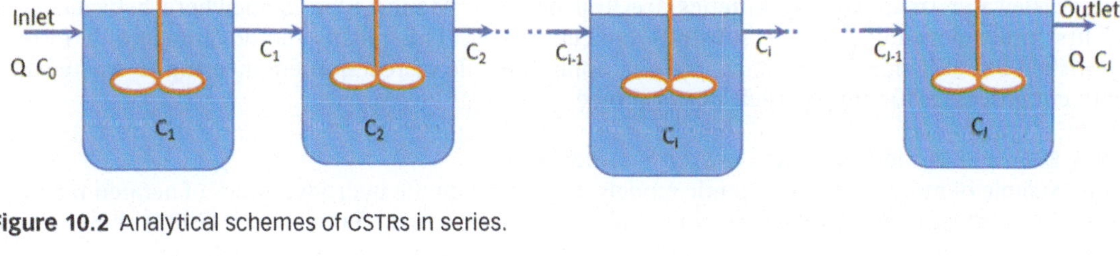

Figure 10.2 Analytical schemes of CSTRs in series.

From the mass balance performed on each CSTR, we obtain the concentration in each at steady state:

$$QC_{i-1} + \frac{V_T}{J} r_i = QC_i \tag{10.15}$$

where:

- V_T is the total volume of the cascade
- J is the number of CSTRs in series

Introducing the HRT of the whole cascade ($\tau = V_T/Q$), Equation (10.15) yields:

$$C_i = C_{i-1} + \frac{\tau}{J} r_i \tag{10.16}$$

The concentration at the outlet of the cascade, C_J, is related to the concentration at the inlet, C_0, by the general equation:

$$C_J = C_0 + \frac{\tau}{J} \sum_{i=1}^{J} r_i \tag{10.17}$$

and for the specific case of first order kinetics:

$$C_J = C_0 \left(1 + k\frac{\tau}{J}\right)^{-J} \tag{10.18}$$

where τ is the space-time (or HRT) of the whole cascade ($\tau = V_T/Q$).

Note that this is valid when the system is symmetrical (i.e., each CSTR in series has the same volume). Individual models are used when an asymmetrical system is simulated.

10.2.3 Comparison of reactors

In chemical engineering literature, when micro-mixing conditions can be assumed, and except where reaction rates are zeroth order ($r_i \neq I$), reaction conversion efficiencies are dependent on reactor type and hydraulic configuration. Optimizing reactors considering reaction kinetic effects, is one of the objectives of the science called 'reaction engineering' or 'chemical reaction engineering'. For kinetics of order higher than zero the reactor with the lowest conversion efficiency is a single CSTR, and the best is theoretically the 'ideal' PFR (but it does not exist in real processes). In this rationale, 'best reactor' means the transformation yield is better at identical reactor volumes. For a similar reason, the volume of the PFR will be smaller than that of the CSTR to reach the same conversion extent. In between behaviours of the CSTR and PFR, there are the TIS and the PFR-AD. The reasoning is the same for both: the highest J (or the highest Peclet number) achieves the maximum conversion extent. This of course needs to consider also competing and side reactions, which change what might otherwise be optimal from a conversion extent perspective.

In wastewater treatment the kinetics are first order or Monod type (somewhere between zero and first order depending on the substrate concentration). This means the advantage of a PFR is diminished but still significant. Then, we can apply the rule of reaction engineering to wastewater treatment processes to improve their performance.

10.2.4 Example of more complex systemic models

As an example of more complex systemic models, we can study the hydrodynamics of aerated reactors and their hydrodynamic change with time and even along the day. Potier *et al.* (2005) showed that

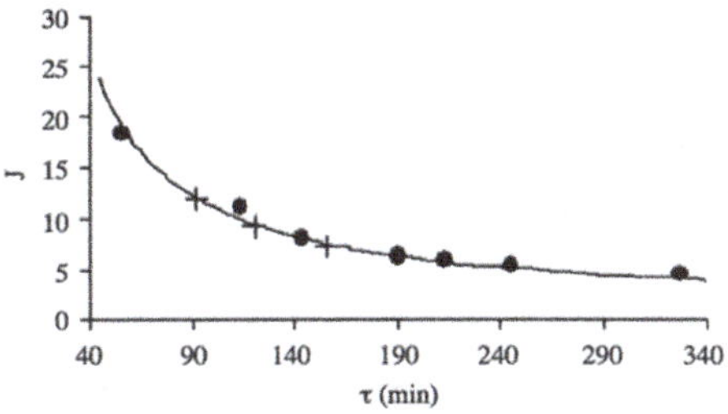

$$Pe = \frac{L^2}{\tau . D}$$

with $Pe = 2J$ (which is precise for high Pe)

Figure 10.3 Potier *et al.* (2005) model. Change of activated sludge reactor hydrodynamics with the liquid flowrate (Q) and the hydraulic residence time ($\tau = V/Q$). Here V is the volume of the reactor; L is its length; J is the number of TIS; Pe is the Peclet number; and D is the axial dispersion coefficient.

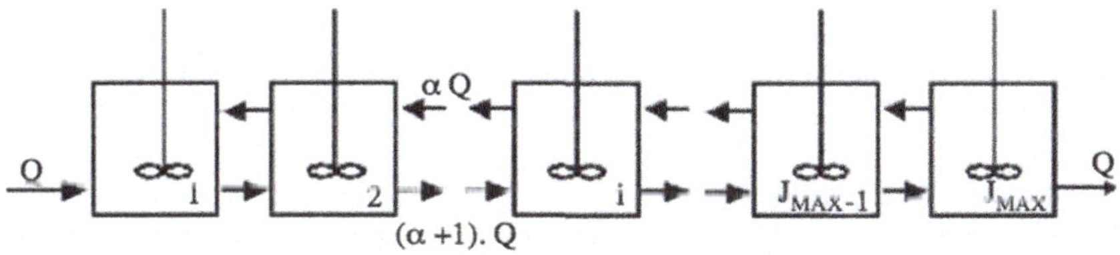

Figure 10.4 Potier *et al.* (2005) model. Schematic representation and relation between the apparent number of TIS (J_{app}) and the back-mixing coefficient α.

rectangular aerated reactors can be well modelled as PFR-AD and also as tanks in series. However, the model significantly changes with the daily change of wastewater flow rate, that is, τ: the HRT (Figure 10.3). Then, the aerated rectangular reactor is always equivalent to a TIS, but the number of tanks in series (J) changes with the liquid flow rate: the higher the liquid flow rate, the higher the J value.

Simulating reactors of this type at constant flow rate with the systemic approach is easy using the tanks-in-series model. It directly gives the programme structure; each tank representing a calculation step of the mass balance equations integration. But simulation of a whole day or a longer period of time leads to problems, since hydrodynamics change and hence the TIS model configuration also changes.

To address this, Potier *et al.* (2005) proposed a dynamic TIS model with back-mixing (Figure 10.4, Equation 10.19) which is able to simulate variations of the hydrodynamics (J_{app} represents the apparent number of CSTRs, i.e., the one that represents the hydrodynamics at a given time) with the flowrate by incorporating a maximal fixed number of CSTRs (J_{max}, the maximum value reached by the hydrodynamic model) and a variable backflow rate (α). This model is valid for $J_{app}>2.5$.

$$\alpha = \frac{1}{2}\left[(J_{max} - 1) - \left(1 + J_{max}^2\left(1 - \frac{2}{J_{app}}\right)\right)\right] \tag{10.19}$$

10.3 CFD WITH REACTIONS

The use of CFD to simulate flows and their properties, such as velocity profiles, in WWTP reactors became common in the last two decades, enabling the investigation of WWTP unit process performance from a fluid mechanics point of view. However, the CFD simulation of the whole process of contaminants removal is still challenging, and capacity in both industry and academia to do so is limited. It requires reactions kinetics and material transport to be added within the CFD conventional simulation. For example, the insertion of biokinetic models into CFD simulations of WWTP processes as well as their validation provided significant and reliable insights into complex contaminant removal performance in these processes (Climent *et al.*, 2018; Glover *et al.*, 2006; Gresch *et al.*, 2009; Le Moullec *et al.*, 2010a, 2010b; Le Moullec *et al.*, 2011; Pereira *et al.*, 2012; Rehman, 2016; Rehman *et al.*, 2017; Sobremisana *et al.*, 2012).

10.3.1 Protocol

(1) First, the procedure is the same as conventional CFD simulation and is described in Chapter 2 using scalar operators, as described also at the beginning of this chapter,
(2) Second, the kinetics rate equations must be implemented in the CFD code as source/sink terms. For biological processes, a similar formulation to ASM models is used (some examples are presented in Chapter 4). Disinfection simulation also relies on this type of approach (Chapter 7). Note that sometimes CFD codes offer the possibility to use pre-implemented kinetic models, but they are often very simple, such as zeroth or first order kinetics, and not suitable for simulation of biological processes.
(3) Finally, the mass transport must be implemented and calibrated. Concentrations of species (reactants and products) are represented by scalars in CFD simulation. They are modelled by the scalar transport and reaction mass balance equations.

10.3.2 Scalar transport and reactions

The mass balance equation for a given species (reactant or product whose concentration is C) can be defined as described in Equation (10.1), but splitting the diffusion term into two contributions:

$$\frac{\partial}{\partial t}(\rho C)+\nabla\cdot\left(\rho\vec{U}C\right)=\nabla\cdot\left(\left(\rho D_m+\frac{\mu_t}{Sc_t}\right)\nabla C\right)+\rho r \tag{10.20}$$

There are transport-reaction scalar equations for all compounds (reactants and products). The four terms of this equation are, from the left to the right: the accumulation, the convective flow, the diffusive flow, and the net production by reactions (negative if compound consumption). r is the reaction rate. The diffusive flow combines both molecular (D_m) and turbulent (μ_t/Sc_t) diffusion. Note that most of the time the molecular diffusion is negligible compared to the turbulent diffusion.

Remark: the same equation is used to simulate the transport of a passive tracer. A tracer moves in a reactor in the same manner as the reactant/product, but there is no reaction with a passive tracer (i.e., the fourth term of the equation is null). Then, using virtual tracing is a useful method to analyse passive tracer (scalar) transport within a vessel with CFD simulation.

The turbulent diffusion must be calculated in every computational cell from the value of the turbulent viscosity (μ_t) as explained in the chapter introduction (Section 10.1). This calculation is based on the turbulent Schmidt number as described by Equation (10.21). In the case of a turbulent flow, the turbulent transport of momentum and concentration is stipulated in terms of turbulent (or eddy) viscosity and turbulent (or eddy) diffusivity. The turbulent Schmidt number, Sc_t, is defined as the ratio of turbulent kinematic viscosity (or momentum diffusivity) to diffusivity (Donzis *et al.*, 2014). The turbulent viscosity is computed by application of the RANS (Reynolds-averaged Navier Stokes) approach for turbulence modelling.

$$Sc_t=\frac{\nu_t}{D_t} \tag{10.21}$$

However, the turbulent Schmidt number (Sc_t) value is not obvious and not always easy to determine. It can significantly vary depending on convective flowrates, level of turbulence, fluid properties, and so on. (Gualtieri *et al.*, 2017). The bottleneck of the method is the determination of Sc_t, which is an adjustable coefficient. Indeed, artefacts in the simulation results can be introduced by an incorrect Sc_t value: prediction of diffusive mass flow is highly sensitive to this value (Hreiz *et al.*, 2019). Moreover, CFD model set-up itself will affect scalar transport simulation: the choice of turbulence model determines the calculation of the eddy viscosity and thus eddy diffusivity, inducing a significant change in the diffusive flow: this is likely to weaken the validity of the simulation (Potier *et al.*, 2019). To address this, the turbulent Schmidt number must be adjusted. It is then recommended to perform a tracer test on the actual plant (residence time distribution), and to compare it with virtual tracing from the CFD code. One can finally adjust the Sc_t value until the two curves match (Le Moullec *et al.*, 2008b). Fortunately, Sc_t is not compound specific, and a passive tracer can be used to infer its value for non-passive scalars. Best knowledge of Sc_t in wastewater treatment plants is still needed, as explained and discussed later in Section 10.7.

10.4 COMPARTMENTAL MODELLING

10.4.1 General description

As explained previously, the properties of compartmental modelling/simulation lie between the systemic modelling/simulation and the CFD with reactions. Compartmental modelling is often obtained from discretization of CFD data. Its benefits are summarized in Figure 10.5 (Jourdan *et al.*, 2019).

10.4.2 Definition

Here is a unified definition proposed in the review article of Jourdan *et al.* (2019):

> 'Compartmental modelling approaches are often mistaken for systemic approaches due to their reactor network structure. But the systemic reactor network construction is solely based on global information whereas the compartment network definition is based both on local and system scales information.
>
> A compartmental model is a representation of a complex system based on the division of this system in functional compartments. It is based on a multi-scale description of the system fluid

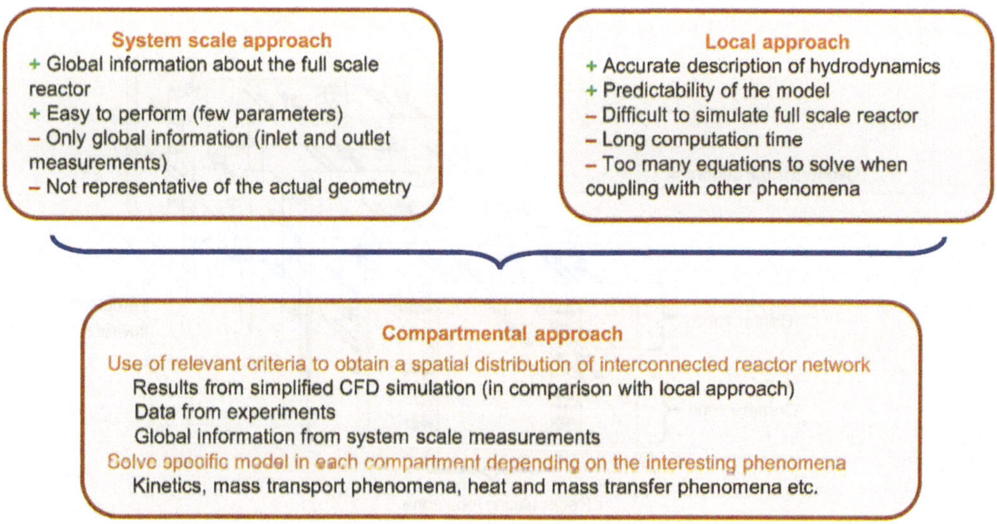

Figure 10.5 Benefits of compartmental approach (Jourdan *et al.*, 2019).

dynamics behaviour defining the system as a network of compartments (often CSTR, plug flow, reactor networks...). This compartment network is representative of the system geometry and spatial distributions of occurring phenomena: fluid dynamics, chemical reactions, biological reactions, transport phenomena, transfer phenomena, shear-induced physical processes (e.g., agglomeration, breakage, and coalescence), and so on. The compartment boundaries definition is based on the homogeneity of selected state variables set as construction criteria. The choice of these criteria depends on studied phenomena: homogeneity of reagent concentration to study (bio) chemical reactions, homogeneity for turbulence intensity to study mixing quality, the same flow rate direction to study transport phenomena, etc.

Compartments aggregate local information about fluid dynamics from CFD simulations and local measurements. They also use data from system scale calculation and system scale experimental characterizations. Convection and turbulent exchange fluxes between compartments are calculated for the complete network structures. Each compartment can be implemented with a complementary model according to their specificity within the studied system (e.g., ASM1 for water treatment, gas-liquid transfer, and algal development model). For each compartment, a dedicated set of sub-models representing physical phenomena taking place in this compartment can be selected. This set of sub-models can be different from one compartment to the other.'

10.4.3 Examples

A compartmental model was developed by Le Moullec *et al.* (2010b) to simulate a long activated sludge pilot-plant reactor. The whole reactor was modelled as a network of spatially distributed CSTRs

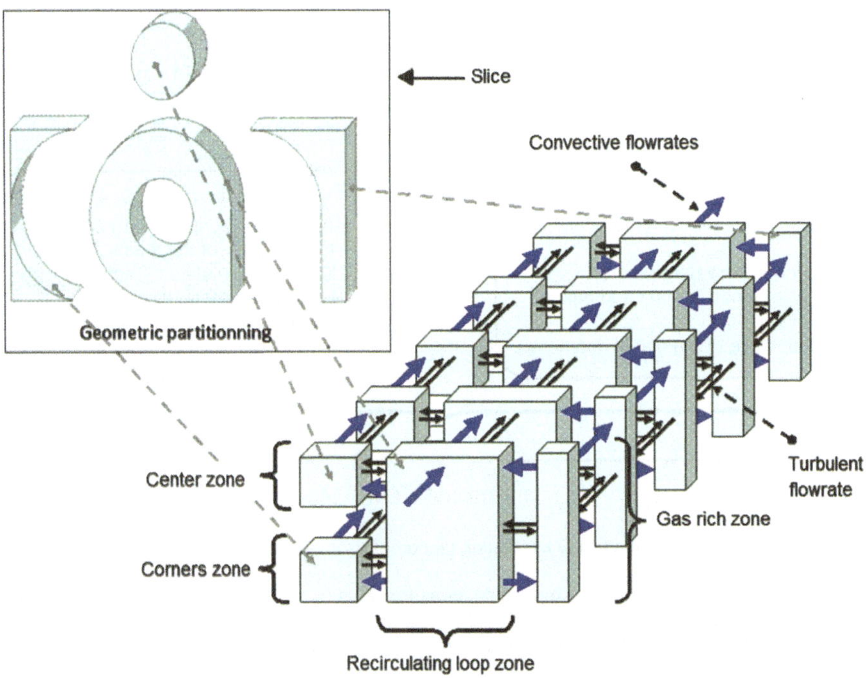

Figure 10.6 Structure of the compartmental model (Source: Le Moullec *et al.*, 2011).

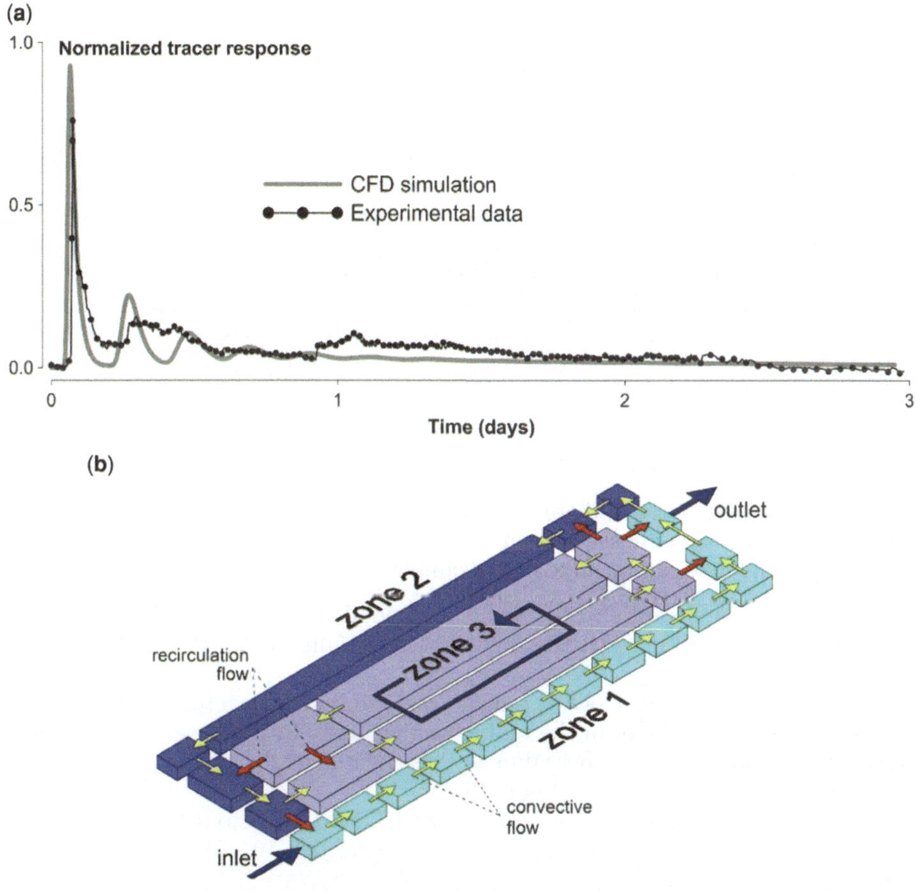

Figure 10.7 (a) Comparison of RTD curves obtained in compartmental modelling versus TIS and tracer experiment. (b) Compartmental model layout of the studied maturation pond (Source: Alvarado *et al.*, 2012).

called compartments (Figure 10.6). The number, size, and spatial distribution of these CSTRs as well as the exchange rates between them were deduced from selected parameters such as gas fraction, velocity and turbulence fields.

The work of Le Moullec *et al.* (2010a) demonstrated the possibility to accurately simulate reactors, not only with a detailed CFD-biokinetic model, but also with a simpler compartmental model.

Another interesting example is the work of Alvarado *et al.* (2012) that showed how compartmental modelling can be used to simulate large systems – in this study, a full-scale waste stabilization pond. Figure 10.7b shows the compartmental model layout of the pond and Figure 10.7a the comparison between simulated and experimental RTDs.

10.5 FUNDAMENTALS OF TRACING EXPERIMENTS AND RTD

10.5.1 Tracing and RTD methods

Tracing and RTD analysis are important tools to build or validate all the models discussed in this chapter. The concept is that a given mass property (scalar) or parcel of fluid (very small volume, theoretically an infinitesimal one) will spend a given residence time in the reactor (detectable as the delay from when the material enters the reactor, to when it exits the reactor). In real reactors fluid parcels generally have different residence times. For all reactors, except ideal plug-flow reactors, different mass parcels will have different residence times, due to dispersion. The range of retention times is represented by a residence time distribution (RTD).

The area under the RTD curve will always equal the cumulative mass added in the relevant period (from the point at which the sample is injected, without a theoretical endpoint), but the height:width ratio and shape determines the overall degree of mixing within the vessel, a criterion which is largely used for classifying reactors. Indeed, the mixing efficiency governs the dispersion of reactants (and other species) across the reactor and hence, their local concentrations.

This residence time distribution can have a great influence on the overall reaction conversion efficiency and so on the local conversion efficiencies. Coupling hydrodynamics and kinetics is therefore mandatory to understand the performance of the process under study.

The two 'ideal' configurations are:

- Completely mixed (CSTR) (see Section 10.6.1.1 for its RTD derivation)
- Ideal plug-flow (PFR with no axial dispersion) (see Section 10.6.1.2 for its RTD derivation)

Between these two configurations, there are several potential causes for the deviation from these ideal conditions which is almost always present.

The non-ideality of the flow can be linked, for example, to:

(a) Short circuiting or preferential paths: they correspond to a portion of the inflow that crosses the reactor faster than the aggregate volume. This results in incomplete dilution, and lower reaction extent.
(b) Dead zones: these have low velocity gradients and minimal interaction with the main volume, and hence do not participate in the main reactive zone. In the case of an activated sludge process, sludge can accumulate over time in these insufficiently aerated zones which may lead to undesirable local anaerobic conditions.
(c) Plug-flow (with axial) dispersion: the peak of the RTD is shifted to the right (compared with a stirred tank), but reactive volume is maintained (see Section 10.2.2).

(a) and (b), above, are detrimental. (c) is beneficial from a reactive point of view (a plug-flow reactor has better conversion efficiency for the same kinetic model), but may be detrimental from a controllability point of view and/or result in unexpected outcomes.

Factors leading to non-ideal flows (Figure 10.8) are numerous, especially including:

- Density differences: when water entering the reactor is significantly colder or warmer than the water already present in the system and in the case of poor mixing, density gradients may cause a portion of the water to go directly to the reactor outlet. This process may be significant in poorly mixed reactors.
- Wind driven recirculation: in shallow reactors (e.g., waste stabilization ponds), wind-induced shear can lead to rapid transport of a portion of the incoming water to the outlet.
- Inappropriate mixing: without a sufficient supply of energy, dead zones can appear.
- Inadequate design: the position of the inlets and outlets as well as geometry (ratios length:width:height) of the reactor are all factors that can lead to dead zones, short-circuiting and inappropriate mixing.

Real reactors are therefore characterized by variable residence time distributions and a non-uniformity of species concentrations across the vessel. Several modelling approaches could be used to describe flow in real reactors, and tracer tests are often one the only ways to investigate the physical system. They can serve to build the model itself (e.g., systemic approach) or to validate

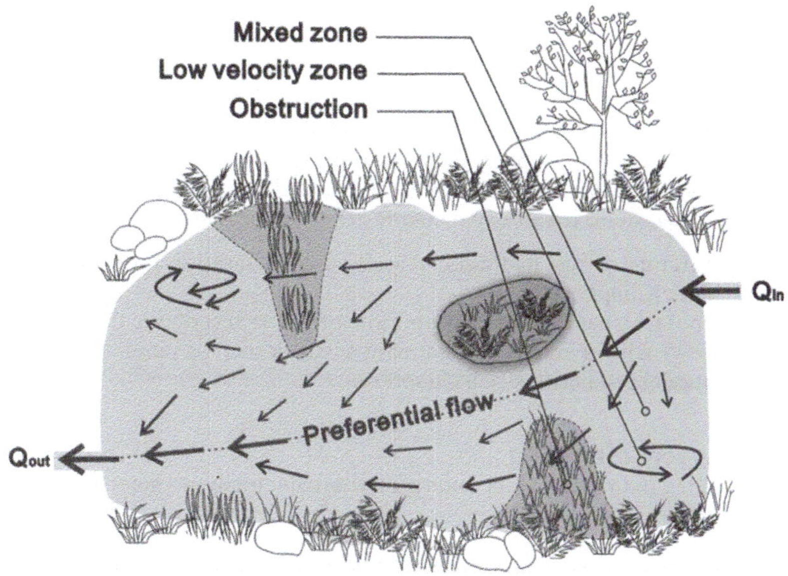

Figure 10.8 Non-idealities of real reactors: example of a constructed wetland (Source: Su *et al.*, 2009).

it (e.g., CFD) and both for compartmental modelling. Note that they can also serve to adjust with precision the dispersion transport process of CFD with reactions (turbulent Schmidt number calibration).

10.5.2 Tracing experiments

In order to evaluate the reactor hydrodynamics, the use of tracers is one of the simplest and most effective methods. It enables evaluation of the global hydrodynamic behaviour without looking directly at the local flow characteristics (velocity field, etc.) within the vessel (black-box approach).

A tracer experiment consists of the injection of a detectable molecule generally by pulse injection, that is, a given quantity is injected in a very short period (other methods are sometimes used, e.g., step-injection). This substance, called a tracer, is generally non-reactive. It may be detectable by a physical property (such as electrical conductivity, radioactivity, etc.) or measurable by sampling at the output (chemical analysis). This enables determination of tracer concentration at the outlet as a function of time. This leads to two concentration signals: one at the inlet which is known and imposed and a second at the outlet which must be measured (Figure 10.9).

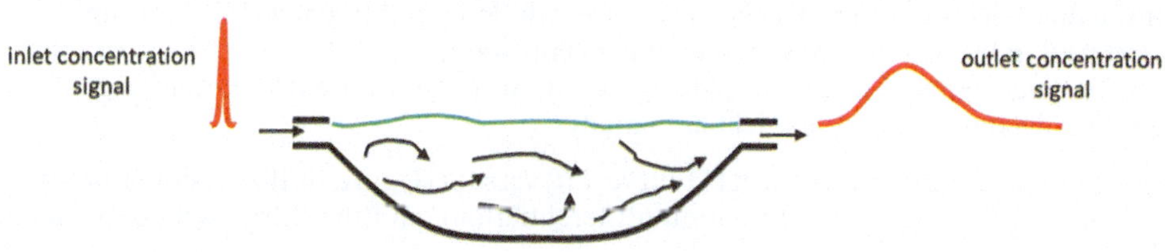

Figure 10.9 Tracer concentration signals at the inlet and the outlet.

The global hydrodynamic behaviour is deduced by comparison of inlet and outlet signals. If there is no loss of tracer (nor reaction, nor sorption), the shape of the outlet signal is only due to the hydrodynamics within the vessel and the shape of the inlet signal.

The signals can be evaluated by a statistical approach and normalized by inlet mass to develop a residence time distribution (RTD) (Danckwerts, 1953). This method applies to any portion of space traversed by flowing elements that stay there for varying periods of time: industrial devices, river basins, organs of a living being, highway the day before a weekend, and so on.

In the field of water treatment, tracer experiments are used for many purposes such as:

- Detecting short-circuits within hydraulic vessels such as aeration tanks, digesters (Chapters 4 and 8) and clarifiers (Chapter 6);
- Evaluating real contact time in chlorination units or UV reactors (see Chapter 7);
- Characterizing flows within extensive treatment systems such as ponds (see Chapter 5);
- Validating CFD model predictions (see Chapter 9).

10.5.3 Choice of the tracer compound

In order to obtain suitable data that can be interpreted satisfactorily, a tracer must meet a number of characteristics:

- No flow perturbation: its density and viscosity must be equivalent to that of the fluid of the reactor;
- Conservative: non-reactive and does not fractionate to the solid phase by sorption or precipitation;
- Low background concentration to avoid interference with the signal;
- High analytical sensitivity;
- Inexpensive and does not perturb the process or cause environmental issues (which may include changes in colour/odour);
- Can be analysed (by sampling followed by laboratory analysis or on-line probe measurment);
- Possibility of injection within a short time interval (pulse) or step injection;
- Low molecular diffusion.

Common types of tracers in the field of water treatment include salts; radioactive tracers or dyes, such as: fluorescein, rhodamine; and metals such as sodium chloride, lithium chloride, and so on.

10.5.4 Conducting the tracing experiment

A tracer experiment consists of injecting the chosen tracer at the reactor inlet and measuring its outlet concentration over time, either by collecting samples for analysis in the laboratory thereafter or by acquiring data from sensors (conductivity meter, fluorometer, etc.).

The injection method determines the type of response obtained at the output. Three injection modes can be considered depending on the configuration:

- Injection of a given quantity of tracer in a very short period, that is, a pulse (Dirac): the duration of the pulse must be lower than a hundredth part of the space-time (HRT);
- Continuous injection of the tracer during the whole operation: step injection until the outlet concentration becomes equal to the inlet concentration;
- Complex inlet signal (e.g., due to natural variation in salinity) but the outlet signal treatment requires a deconvolution or model-based analysis.

The first method leads directly to the RTD curve: $E(t)$ versus t (Figure 10.10) via simple normalization, the second method by simple transformation (differentiation), and the third method by model-based reformulation.

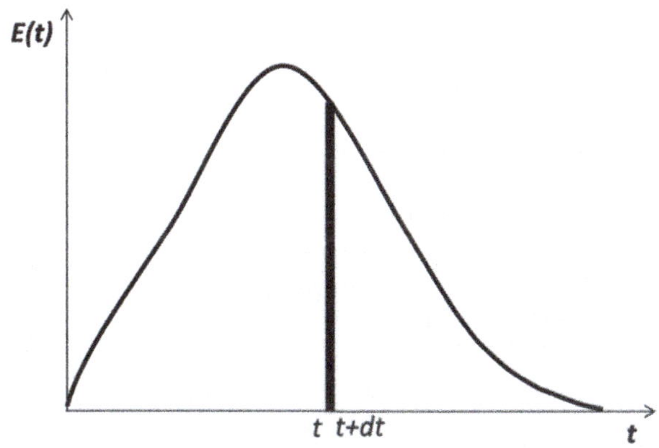

Figure 10.10 Response to a conservative tracer injection.

Pulse injection is ideal, since it minimizes the quantity of tracer used, and requires minimal processing to generate an RTD. Continuous injection is used mainly for specific circumstances (e.g., control system design). The complex inlet signal method is only used where a background inert tracer with natural variation is to be used (generally used for very large hydraulic volumes). We develop the mathematical treatment of the pulse method in the following paragraphs.

10.5.5 Determination of the residence time distribution (RTD)

The description of reactor hydrodynamic characteristics enables the determination of the duration for which a given volume of fluid will stay in the reactor and its average 'age'. The average age (mean residence time) can then be related to reactor efficiency based on reaction kinetics considered.

To standardize the analysis of response curves, their normalization is required: the measured concentration is divided by the area under the curve which represents the whole quantity of tracer that has been injected (note that the recovery is the area under the curve divided by the total tracer added and generally should be ~100%). The standard curve determined from normalization is known as the residence time distribution (RTD). The RTD is a statistical distribution of residence times of fluid particles exiting the reactor.

In the case of a pulse injection, the RTD, denoted $E(t)$, is defined through Equation (10.22) (which is valid when the inlet flowrate is constant):

$$E(t) = \frac{C(t)}{\int_0^{\infty} C(t)\,dt} \tag{10.22}$$

We have therefore:

$$\int_0^{\infty} E(t)\,dt = 1 \tag{10.23}$$

The analysis of this curve enables the determination of several key parameters (Table 10.2).

Table 10.2 Main parameters determined from tracer response curves.

Parameter	Definition	Expression
τ	Space-time, also called theoretical residence time, or hydraulic residence time (HRT)	V/Q
t	Time when the tracer is detected at the outlet	Measured
Mean residence time $\bar{t}$	Average time for a particle of fluid to reach the outlet	$\bar{t} = \dfrac{\int_0^\infty tC(t)dt}{\int_0^\infty C(t)dt}$ or $\bar{t} = \int_0^\infty tE(t)dt$
Variance σ^2	Determines the dispersion coefficient, that is, the broadness of the distribution	$\sigma^2 = \dfrac{\int_0^\infty t^2C(t)dt}{\int_0^\infty C(t)dt} - \bar{t}^2$ or $\sigma^2 = \int_0^\infty (t-\bar{t})^2 E(t)dt$
Residence time distribution (RTD) $E(t)$	See below	See below
Feed flow rate	Easy to calculate from RTD curve if the mass of tracer injected is known	$Q = \dfrac{m}{\int_0^\infty C(t)dt}$ where m is the total amount of tracer that was injected (moles or kg)
Reactor volume accessible to the fluid V_A	Reactor volume effectively involved in mixing	$V_A = Q\bar{t}$
Dead volume V_m	Reactor volume not involved in mixing	$V_m = V_R - V_A$ where V_R=reactor geometrical volume (control volume)

10.6 RESIDENCE TIME DISTRIBUTION OF CLASSIC SYSTEMIC MODELS

10.6.1 Ideal reactors

10.6.1.1 CSTR

Consider injecting tracer as a pulse at the inlet of a CSTR fed with a constant flow rate of clear water. In a CSTR, it is assumed that the tracer mixes instantly. Let us write the corresponding material balance (see also Figure 10.1).

$$\text{Accumulation} = \text{Inlet} - \text{Outlet} + \text{Conversion} \tag{10.24}$$

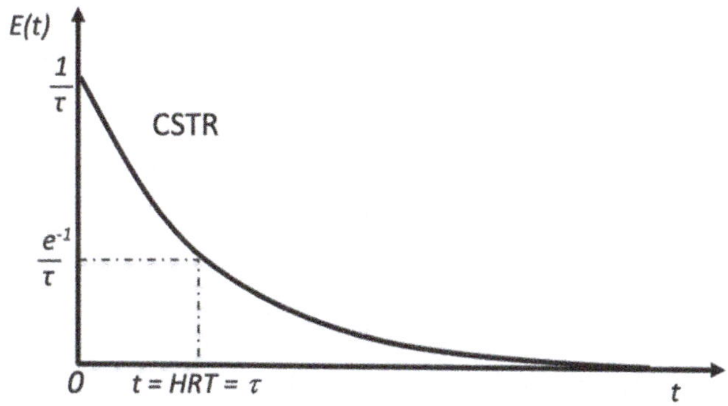

Figure 10.11 Response of a CSTR to a tracer pulse injection.

The tracer is conservative thus the conversion term is 0. One can write:

$$V\frac{dC}{dt} = QC_0 - QC \tag{10.25}$$

For a pulse injection, it may be assumed during measurements after injection that $C_0 \approx 0$. Rewriting the balance gives:

$$\frac{dC}{dt} = -\frac{Q}{V}C \tag{10.26}$$

The integral between the limits $C = C_0$ to $C = C$ and $t = 0$ to $t = t$ is:

$$\int_{C_0}^{C}\frac{dC}{C} = -\frac{Q}{V}\int_{0}^{t}dt \tag{10.27}$$

The resulting expression after integration is:

$$C = C_0 e^{-(Q/V)t} \tag{10.28}$$

We define the following parameters:

- $\tau = V/Q$ = Space-time (theoretical residence time or HRT)
- $\theta = t/\tau$ = normalized residence time (dimensionless)

This allows Equation (10.28) to be rewritten as follows:

$$C = C_0 e^{-(t/\tau)} = C_0 e^{-\theta} \tag{10.29}$$

The corresponding theoretical curve can be plotted (solid lines in Figure 10.11):

10.6.1.2 Plug-flow reactor

The mass balance of the conservative tracer (no reaction) on an elementary volume defined on the cross section of a plug-flow reactor gives the following relation (Equations 10.30 and 10.31):

$$\text{Accumulation in the elementary volume} = \text{inlet of elementary volume} - \text{outlet of elementary volume} \tag{10.30}$$

which gives:

$$\frac{\partial C_i}{\partial t} = -\frac{\partial (uC_i)}{\partial z} \tag{10.31}$$

The response of an ideal plug-flow reactor to a conservative tracer injection can be represented as follows (Figure 10.12):

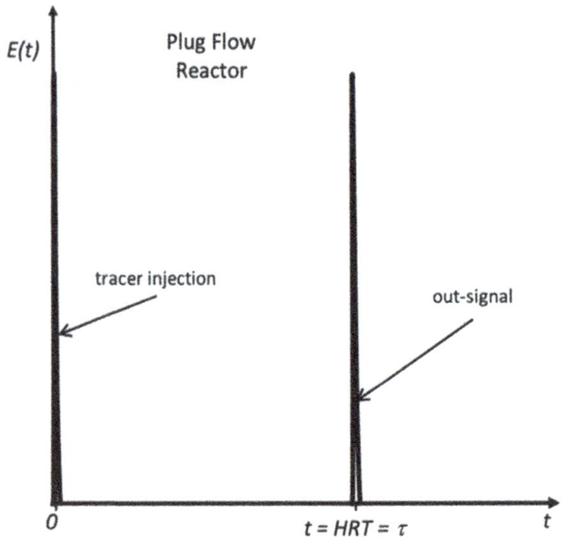

Figure 10.12 PFR response to a conservative tracer pulse injection.

10.6.2 Non-ideal reactors: taking into account dispersion

10.6.2.1 Plug-flow reactor with axial dispersion

The establishment of the mass balance is deduced from Equation (10.32):

$$\frac{\partial C_i}{\partial t} = D_i \frac{\partial^2 C_i}{\partial z^2} - \frac{\partial (uC_i)}{\partial z} \tag{10.32}$$

where D_i is the axial dispersion coefficient ($m^2.s^{-1}$).

The RTD curve for a plug-flow reactor with axial dispersion open to diffusion is given in Equation (10.33), $E(t)$ versus t (Figure 10.13):

$$E(t) = \frac{1}{2}\sqrt{\frac{Pe}{\pi \tau t}} \exp\left[-\frac{Pe(\tau - t)^2}{4\tau t}\right] \tag{10.33}$$

We can also define the mean and variance as follows:

$$\bar{t} = \tau \tag{10.34}$$

$$Pe = 2\frac{\bar{t}^2}{\sigma^2} = \frac{uL}{D} \tag{10.35}$$

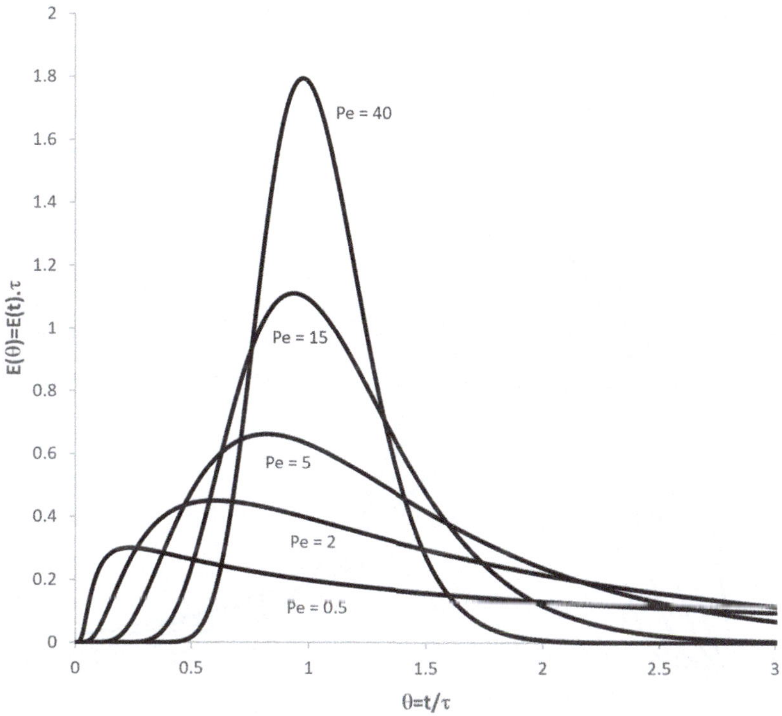

Figure 10.13 Response to a pulse for different *Pe* values; plug-flow reactor with axial dispersion open to diffusion.

10.6.2.2 Tank-in-series

As seen previously, it is possible to approximate and model a long reactor (e.g.: long activated sludge reactor) by representing it as CSTRs in series (Figure 10.2). This model is well known as a cascade of CSTRs, also called the tank-in-series (TIS) model. From the mass balance performed on each CSTR, we finally obtain the RTDs for different J values (Equation 10.36, Figure 10.14):

$$E(t) = \left(\frac{J}{\tau}\right)^{J} \frac{t^{J-1}}{(J-1)!} e^{-(Jt/\tau)} \tag{10.36}$$

10.7 TRACING EXPERIMENTS AND VIRTUAL TRACER TESTS FOR CALIBRATION OF CFD SIMULATION

10.7.1 Turbulent Schmidt number: pitfalls and recommendations

The turbulent Schmidt number (Sc_t) is of fundamental importance, but its value is not obvious to determine. In the wastewater treatment field, users often use the default value provided by commercial software packages (often the recommended value is 0.7 for gas-liquid flows), but the Sc_t value can significantly vary depending on the situation (Gualtieri *et al.*, 2017). As stated earlier in this chapter, Sc_t value as well as CFD model set-up may change significantly the prediction of diffusive mass fluxes, weakening the validity of the simulation (Hreiz *et al.*, 2019; Potier *et al.*, 2019). Calibration of the turbulent Schmidt number has to be performed, if possible, by comparison between experimental and virtual tracer tests datasets.

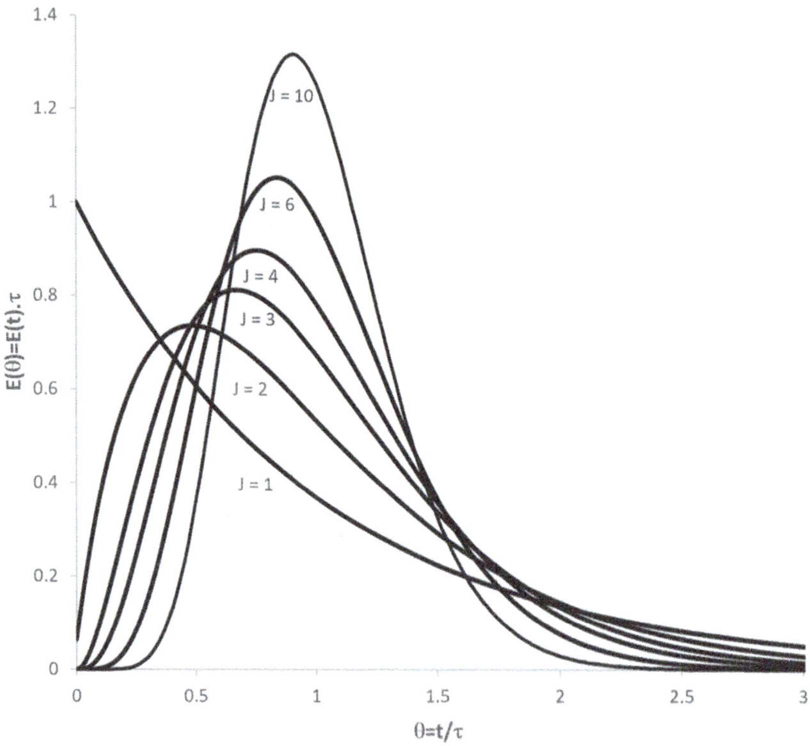

Figure 10.14 Response to a pulse for different CSTRs in series.

10.7.2 Virtual tracer tests in CFD

There are two ways to perform a virtual tracer test in CFD (Le Moullec *et al.*, 2008b). In the case of a steady-state flow, the flow velocity field must first be calculated. Then, two options are available to perform a virtual tracing experiment:

- Define a tracer (passive scalar) with the same properties as the flowing fluid, inject it at the reactor inlet and follow the concentration of this pseudo compound at the outlet. This method requires solving the transport of the tracer in transient conditions, which can be computationally expensive.
- Define solid particles having the same density as the flow fluid and a very small diameter, inject a sufficiently large quantity (to yield statistically significant results) at the reactor's inlet and observe the time it takes for each particle to exit the system. The quantity of particles to be injected depends on the required accuracy.

Afterwards, virtual tracer test results must be subjected to the same methods of analysis as the experimental tracings: analysis of the curve (concentration vs time), setting up of the RTD curve ($E(t)$ vs time), calculation of mean residence time, variance, dispersion coefficient, and so on.

10.7.2.1 Scalar transport

The scalar (tracer) is injected by modifying the tracer input concentration for a time step (by a modification of the boundary condition). In turbulent flow, the transport equation is given by:

$$\frac{\partial}{\partial t}(\rho C) + \nabla \cdot \left(\rho \vec{U} C\right) = \nabla \cdot \left(\left(\rho D_m + \frac{\mu_t}{Sc_t}\right)\nabla C\right) \tag{10.37}$$

where D_m is the molecular diffusion coefficient (m^2/s) and Sc_t is the turbulent Schmidt number.

10.7.2.2 Lagrangian approach

In this approach, the residence time distribution can be deduced from the analysis of the trajectories of a large number of particles in the reactor. The trajectory of a particle is predicted by integrating the forces balance applied to the particle in a Lagrangian reference frame (Ansys, 2011). The particles are chosen to have the same density as the continuous phase so that the gravitational forces have no effect and only the drag force is to be considered. The turbulent dispersion of particles is taken into account by a stochastic approach which considers that the particle interacts with a succession of eddies generated by turbulence, the fluctuating velocities within these eddies are considered to be isotropic and obeying a Gaussian distribution.

10.7.2.3 Choice between the two methods

The particle tracking method (i.e., the Lagrangian approach) has the advantage of not being affected by numerical diffusion (i.e., numerical errors leading to a diffusion-like effect). More information can be extracted from it since the complete trajectory of each particle is obtained. The accuracy of the simulation can be adjusted by choosing a larger or smaller number of particles.

Solving the transport equation of a tracer (passive scalar) enables easier assessment and implementation of chemical or biological reaction models. If one wants to couple a biokinetic model, the transport equation of a tracer is relevant because during the simulation, the chemical and biological concentrations will be represented by scalars, too. Moreover, the chemical and biological reactions will be modelled by source/sink terms implemented in similar transport equations (Equation 10.37).

Then, to check or adjust the turbulent Schmidt number value for CFD with reactions, tracing by solving the transport equation for a passive scalar is the only option, since it uses the same transport equation as the chemical species (reactants and products).

10.7.3 Example of dispersion modelling

This example comes from several papers of Le Moullec *et al.* (2008a, 2008b, 2010a, 2010b, 2011), with RTD data from multiple tracer experiments by Potier *et al.* (2005). Describing experimental and CFD-simulated RTDs, this example is representative of what could be seen by the application of the protocol introduced above.

As for all long activated sludge reactors, the studied reactor can be described by the plug flow with axial dispersion model (PFR-AD). Therefore, the dispersion coefficient is a good estimation of the overall mixing within the reactor. The simulated dispersion coefficient can be considered as the sum of four contributions at different scales: molecular diffusion, spatial dispersion, turbulent dispersion and numerical diffusion (Le Moullec *et al.*, 2008b):

$$D_{\text{total}} = D_{\text{molecular}} + D_{\text{convection}} + D_{\text{turbulent}} + D_{\text{numerical}} \tag{10.38}$$

- Molecular diffusion (D_m): this is the result of Brownian motion. This diffusion term cannot be avoided but it is very often negligible in turbulent flows.
- Spatial or convective dispersion (D_s): spatial dispersion lumps macroscopic inhomogeneities of the flow field. These inhomogeneities may result from, among others, edge effects, recirculation loops or preferential paths and dead zones.
- Turbulent dispersion (D_t): this is the result of the combined motion of all vortices generated by turbulence.

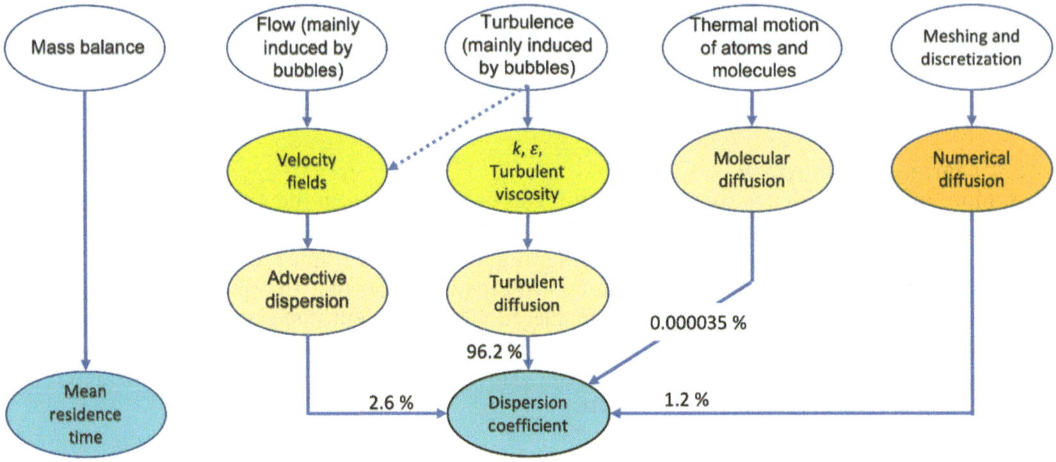

Figure 10.15 Relationship between global quantities of a residence time distribution and local phenomena (inspired by Le Moullec, 2008).

- Numerical diffusion: in CFD, the numerical resolution and discretization of transport equations can result in approximation errors due to for example, spatial, and temporal discretization schemes, mesh quality (spatial discretization) and the time step (temporal discretization).

In their simulation, Le Moullec *et al.* (2008b) obtained these different values:

- The total simulated dispersion coefficient: $D_{total} = 4.26 \times 10^{-4}$ m^2 s^{-1}
- Molecular diffusion: $D_{molecular} = 1.48 \times 10^{-10}$ m^2 s^{-1} which represents 0.000035% of the total simulated dispersion coefficient
- Numerical diffusion: $D_{numerical} = 5.00 \times 10^{-6}$ m^2 s^{-1} which represents 1.2% of the total simulated dispersion coefficient
- Convective dispersion: $D_{convection} = 1.14 \times 10^{-5}$ m^2 s^{-1} which represents 2.7% of the total simulated dispersion coefficient
- Turbulent diffusion: $D_{turbulent} = 4.10 \times 10^{-4}$ m^2 s^{-1} which represents 96.2% of the total simulated dispersion coefficient

The main dispersion source is thus turbulence, which confirms that a precise estimation of Sc_t is crucial.

Figure 10.15 schematically displays the relationships between the physical phenomena and the different sources of dispersion that can be considered in a CFD simulation of gas-liquid reactor. The white areas represent the causes and the blue the consequences.

10.7.4 Comparison of experimental and simulated RTD

To illustrate the comparison between CFD simulated and experimental RTD, here is another example of a study of a full-scale aerated activated sludge (3300 m^3, 4 m depth, 8 m width and 102 m length). Three different flowrates were studied. We can observe from Figure 10.16 that the simulation perfectly matches the experiments. The comparison between virtual and experimental tracing is a useful and precise tool.

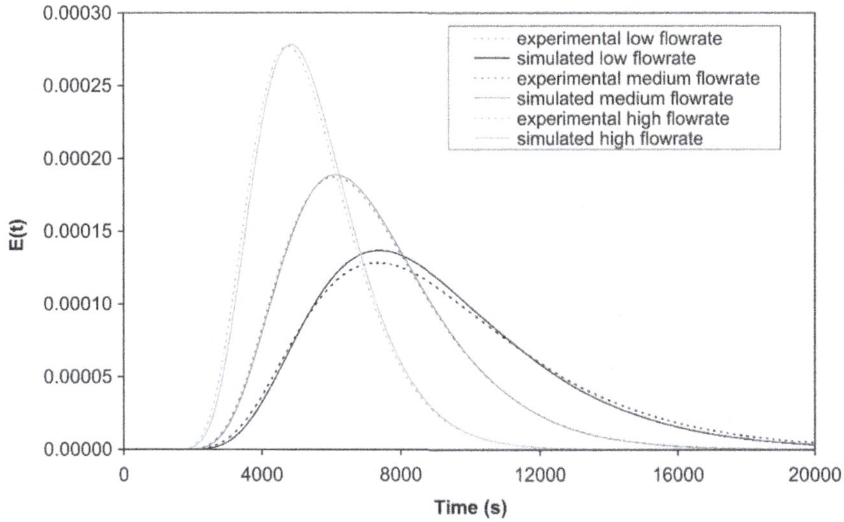

Figure 10.16 Comparison between experimental and CFD-simulated RTD on a full-scale aerated activated sludge reactor (3300 m^3, 4 × 8 × 102 m) for three different liquid flowrates (Source: Le Moullec *et al.*, 2008b).

10.8 CONCLUSION

CFD model validation by tracer testing is a relatively inexpensive and well-documented procedure and can be used for validation of both solids (Bender & Crosby, 1984) and solutes transport modelling for many wastewater treatment applications. There are two main types of method to simulate RTD with CFD: Lagrangian particle tracking (Stropky *et al.*, 2007; Thýn *et al.*, 1998) and solving the transport equation for a passive tracer with a scalar representing the tracer concentration (Le Moullec *et al.*, 2008b; Talvy *et al.*, 2007; Zhang *et al.*, 2007). The second method is sensitive to numerical diffusion, so care is required when it is used. However, if the subsequent goal is to implement biokinetics, the best numerical RTD approach is to use the scalar method, because simulations that include biokinetics also utilize the transport equation, and it is important to create similar conditions for both CFD-RTD and the subsequent integrated CFD biokinetics model.

When performing virtual tracer tests using a CFD simulation, one must define the value of the turbulent Schmidt number. Its value is dependent on many factors, including the choice of the turbulence model (Donzis *et al.*, 2014; Gualtieri *et al.*, 2017). This value can have a great impact on the prediction of solute transport since turbulent diffusivity is often the major component of overall dispersion in reactors (Hreiz *et al.*, 2019; Le Moullec *et al.*, 2008b; Potier *et al.*, 2019). The turbulent Schmidt number must therefore be calibrated, ideally by performing tracing experiments on the real plant.

One last pitfall is well known in the (chemical) reaction engineering science: similar RTDs could be produced for multiple internal vessel geometries. Therefore, these tests should be used in conjunction with other methods when possible (see also Chapter 9 for discussion about validation of CFD models). It is important to note that the hydrodynamics depends not only on the shape and size of the reactors and physical properties of the liquid, but also on the energy engaged in the process and the flow rates (Potier *et al.*, 2005). Hence, reactor hydrodynamics strongly vary over the course of the day, since liquid flow rate fluctuates significantly.

REFERENCES

Alvarado A., Vedantam S., Goethals P. and Nopens I. (2012). A compartmental model to describe hydraulics in a full-scale waste stabilization pond. *Water Research*, **46**(2), 521–530, https://doi.org/10.1016/j.watres.2011.11.038

Ansys I. (2011). ANSYS FLUENT Theory Guide. Southpointe, 275 Technology Drive, Canonsburg, PA 15317, USA.

Bender J. and Crosby R. M. (1984). Hydraulic characteristics of activated sludge secondary clarifiers. U.S. Environmental Protection Agency.

Climent J., Basiero L., Martínez-Cuenca R., Berlanga J. G., Julián-López B. and Chiva S. (2018). Biological reactor retrofitting using CFD-ASM modelling. *Chemical Engineering Journal*, **348**, 1–14, https://doi.org/10.1016/j.cej.2018.04.058

Danckwerts P. V. (1953). Continuous flow systems: distribution of residence times. *Chemical Engineering Science*, **2**(1), 1–13, https://doi.org/10.1016/0009-2509(53)80001-1

Donzis D. A., Aditya K., Sreenivasan K. R. and Yeung P. K. (2014). The turbulent Schmidt number. *Journal of Fluids Engineering*, **136**(6), https://doi.org/10.1115/1.4026619

Glover G. C., Printemps C., Essemiani K. and Meinhold J. (2006). Modelling of wastewater treatment plants – how far shall we go with sophisticated modelling tools? *Water Science and Technology*, **53**(3), 79–89, https://doi.org/10.2166/wst.2006.078

Gresch M., Brügger R., Meyer A. and Gujer W. (2009). Compartmental models for continuous flow reactors derived from CFD simulations. *Environmental Science and Technology*, **43**(7), 2381–2387, https://doi.org/10.1021/es801651j

Gualtieri C., Angeloudis A., Bombardelli F., Jha S. and Stoesser T. (2017). On the values for the turbulent Schmidt number in environmental flows. *Fluids*, **2**(2), 17, https://doi.org/10.3390/fluids2020017

Hangos K. M. and Cameron I. T. (2001). Process Modelling and Model Analysis, 1st edn. Academic Press, San Diego, CA, USA.

Henze M., Grady Jr, C. P., Gujer W., Marais G. V. R. and Matsuo T. (1987). Activated Sludge Model No. 1. IAWQ Scientific and Technical Report No. 1, London, UK.

Hreiz R., Potier O., Wicks J. and Commenge J. M. (2019). CFD Investigation of the effects of bubble aerator layouts on hydrodynamics of an activated sludge channel reactor. *Environmental Technology (United Kingdom)*, **40**(20), 2657–2670.

Jourdan N., Neveux T., Potier O., Kanniche M., Wicks J., Nopens I., Rehman U. and Le Moullec Y. (2019). Compartmental modelling in chemical engineering: a critical review. *Chemical Engineering Science*, **210**, 115196, https://doi.org/10.1016/j.ces.2019.115196

Laurent J., Samstag R. W., Ducoste J. M., Griborio A., Nopens I., Batstone D. J., Wicks J. D., Saunders S. and Potier O. (2014). A protocol for the use of computational fluid dynamics as a supportive tool for wastewater treatment plant modelling. *Water Science and Technology*, **70**(10), 1575–1584, https://doi.org/10.2166/wst.2014.425

Laurent J., Bois P., Nuel M. and Wanko A. (2015). Systemic models of full-scale surface flow treatment wetlands: determination by application of fluorescent tracers. *Chemical Engineering Journal*, **264**, 389–398, https://doi.org/10.1016/j.cej.2014.11.073

Le Moullec Y. (2008). Comparaison des approches systémique, mécanique des fluides numérique et compartimentale pour la modélisation des réacteurs: application à un réacteur canal à boues activées. PhD thesis, Institut National Polytechnique de Lorraine, Nancy Université.

Le Moullec Y., Potier O., Gentric C. and Leclerc J. P. (2008a). A general correlation to predict axial dispersion coefficients in aerated channel reactors. *Water Research*, **42**(6–7), 1767–1777, https://doi.org/10.1016/j.watres.2007.10.041

Le Moullec Y., Potier O., Gentric C. and Pierre Leclerc J. (2008b). Flow field and residence time distribution simulation of a cross-flow gas–liquid wastewater treatment reactor using CFD. *Chemical Engineering Science*, **63**(9), 2436–2449, https://doi.org/10.1016/j.ces.2008.01.029

Le Moullec Y., Gentric C., Potier O. and Leclerc J. P. (2010a). CFD Simulation of the hydrodynamics and reactions in an activated sludge channel reactor of wastewater treatment. *Chemical Engineering Science*, **65**(1), 492–498, https://doi.org/10.1016/j.ces.2009.03.021

Le Moullec Y., Gentric C., Potier O. and Leclerc J. P. (2010b). Comparison of systemic, compartmental and CFD modelling approaches: application to the simulation of a biological reactor of wastewater treatment. *Chemical Engineering Science*, **65**(1), 343–350, https://doi.org/10.1016/j.ces.2009.06.035

Le Moullec Y., Potier O., Gentric C. and Leclerc J. P. (2011). Activated sludge pilot plant: comparison between experimental and predicted concentration profiles using three different modelling approaches. *Water Research*, **45**(10), 3085–3097, https://doi.org/10.1016/j.watres.2011.03.019

Levenspiel O. (1972). Chemical Reaction Engineering, 1st edn. John Wiley & Sons, Hoboken, NJ, USA.

Levenspiel O. (1998). Chemical Reaction Engineering, 3rd edn. John Wiley & Sons, Hoboken, NJ, USA.

Levenspiel O. (2013). The Chemical Reactor Omnibook. Oregon State University, Corvallis, Oregon.

Pereira J. P., Karpinska A. M., Gomes P. J., Martins A. A., Dias M. M., Lopes J. C. B. and Santos R. J. (2012). Activated sludge models coupled to CFD simulations. In: Single and Two-Phase Flows in Chemical and Biomedical Engineering, R. Dias, R. Lima, A. Martins and T. M. Mata (eds.), Bentham Science Publishers, Dubaï, UAE, pp. 153–173.

Potier O., Leclerc J.-P. and Pons M.-N. (2005). Influence of geometrical and operational parameters on the axial dispersion in an aerated channel reactor. *Water Research*, **39**(18), 4454–4462, https://doi.org/10.1016/j.watres.2005.08.024

Potier O., Hreiz R. and Laurent J. (2019). Turbulent Schmidt number: a necessary tool but also a trap for CFD simulation of tracing experiments or reactions, (CFD with reactions, Scalar transport of reactants and products, Calibration of Scalar transport for virtual tracing). TRACER 8 Conference, Da Nang, Vietnam.

Rehman U. (2016). Next generation bioreactor models for wastewater treatment systems by means of detailed combined modelling of mixing and biokinetics. PhD thesis, Ghent University. Faculty of Bioscience Engineering, Ghent, Belgium.

Rehman U., Audenaert W., Amerlinck Y., Maere T., Arnaldos M. and Nopens I. (2017). How well-mixed is well mixed? Hydrodynamic-biokinetic model integration in an aerated tank of a full-scale water resource recovery facility. *Water Science and Technology*, **76**(8), 1950–1965, https://doi.org/10.2166/wst.2017.330

Sobremisana A. P., Ducoste J. J. and de los Reyes F. L. (2012). Combining CFD, FLOC dynamics, and biological reaction kinetics to model carbon and nitrogen removal in an activated sludge system. *Proceedings of the Water Environment Federation*, **2011**(13), 3272–3282, https://doi.org/10.2175/193864711802721569

Stropky D., Pougatch K., Nowak P., Salcudean M., Pagoria P., Gartshore I. and Yuan J. (2007). RTD (residence time distribution) predictions in large mechanically aerated lagoons. *Water Science and Technology*, **55**(11), 29–36, https://doi.org/10.2166/wst.2007.346

Su T.-M., Yang S.-C., Shih S.-S. and Lee H.-Y. (2009). Optimal design for hydraulic efficiency performance of free-water-surface constructed wetlands. *Ecological Engineering*, **35**(8), 1200–1207, https://doi.org/10.1016/j.ecoleng.2009.03.024

Talvy S., Cockx A. and Liné A. (2007). Modeling hydrodynamics of gas–liquid airlift reactor. *AIChE Journal*, **53**(2), 335–353, https://doi.org/10.1002/aic.11078

Thýn J., Ha J. J., Střasák P. and Žitný R. (1998). RTD Prediction, modelling and measurement of gas flow in reactor. *Nukleonika*, **43**(1), 95–114.

Villermaux J. (1993). Génie de la Réaction Chimique. Tec & Doc, Paris, France.

Zhang J., Huck P. M., Anderson W. B. and Stubley G. D. (2007). A computational fluid dynamics based integrated disinfection design approach for improvement of full-scale ozone contactor performance. *Ozone: Science & Engineering*, **29**(6), 451–460, https://doi.org/10.1080/01919510701613420

Index

T

U

V

Y

IWA Publishing's authorised EU representative for General Product Safety Regulations is Diane D'Arras, 15 rue Duret, 75116 Paris, France, e-mail: safety@iwap.co.uk.

Printed and bound by CPI Group (UK) Ltd, Croydon, CR0 4YY
16/06/2026
02140350-0001